NITROGEN IN SOILS OF CHINA

Developments in Plant and Soil Sciences
Volume 74

Nitrogen in Soils of China

Edited by:

Zhu Zhao-liang

and

Wen Qi-xiao
Institute of Soil Science, Academia Sinica, PO Box 821, Nanjing,
People's Republic of China

and

J. R. Freney
CSIRO, Division of Plant Industry, GPO Box 1600, Canberra, ACT 2601 Australia

SPRINGER SCIENCE+BUSINESS MEDIA, B.V.

A C.I.P. Catalogue record for this book is available from the Library of Congress

ISBN 978-0-7923-4372-1 ISBN 978-94-011-5636-3 (eBook)
DOI 10.1007/978-94-011-5636-3

Printed on acid-free paper

Typeset by EXPO Holdings, Malaysia

Contents

Foreword

The study of soil nitrogen has long been an active field, but it was generally pivoted on agricultural and forestry production, and animal husbandry. With the rapid increase in the use of fertilizer nitrogen, more attention has been paid to the relationship between nitrogen management and environmental quality and human health. In addition, the study of soil nitrogen has become more comprehensive with the development of related sciences. The quantitative study of the processes in nitrogen cycling and their interrelationships has been an important part of this project and has attracted the attention of scientists all over the world.

Nitrogen is one of the most important nutrients for plant growth and the application of fertilizer nitrogen is playing an important role in agricultural production. The annual consumption of fertilizer nitrogen in the world has reached 70 million tons, and China has an annual consumption of more than 15 million tons and is the largest fertilizer nitrogen consumer in the world. However, the efficiency of nitrogen fertilizer is low and losses are large. It is estimated that nitrogen losses from agriculture in our country can be as high as 40–60% of the nitrogen applied. Some of the lost nitrogen enters the atmosphere and contributes to the greenhouse effect and some enters water bodies to pollute the water. Consequently, it is important for scientists all over the world to improve the efficiency of use of fertilizer nitrogen, to promote the biological fixation of nitrogen and to increase the nitrogen supplying potential of soils.

Soil nitrogen research in our country began in the 1930s and it is only in the last three decades that it has developed rapidly. Great progress has been made in research on slow-release forms of ammonium bicarbonate, the deep placement of ammonium bicarbonate supergranules, nitrogen loss from fertilizer nitrogen in paddy soils, the nitrogen supplying capacities of paddy soils, recommendations for optimum nitrogen application rates, nitrification, denitrification and nitrogen loss in agroecosystems and biological fixation of nitrogen. The results have benefited agricultural production in China.

Some monographs and proceedings on soil nitrogen have been published abroad. In China, the proceedings of a workshop on soil nitrogen were published in 1986, but no account of the systematic study of the problems and research achievements in soil nitrogen research has been published. The publication of this book meets this need well. Such a book, based on the progress in the study of soil nitrogen in one country, is rarely seen abroad.

This book was written by soil scientists in the Institute of Soil Science, Academia Sinica and it was one of the goals of the Institute's Director during his term. We believe that the publication of this book will further promote the study of soil nitrogen in China. Suggestions and criticism from all colleagues at home and abroad are welcome.

Zhao Qi-guo
Institute of Soil Science
Academia Sinica
P.O. Box No. 821, Nanjing
People's Republic of China

Abbreviations used in Tables

AA	=	aqua ammonia
AAS	=	anthropogenic alluvial soil
ABC	=	ammonium bicarbonate
AC	=	ammonium chloride
AN	=	ammonium nitrate
AS	=	ammonium sulfate
ATC	=	4-amino-1,2,4-triazole hydrochloride
Bn	=	banded
Bs	=	applied as basal fertilizer
Bn-Bs	=	banded as basal
CAN	=	calcium ammonium nitrate
CFA	=	calcareous fluvo-aquic soil
DCD	=	dicyandiamide
DI	=	drained incorporation
DI-Del-Ref	=	reflooded 20 hr after incorporation
DI-Ref	=	reflooded immediately after incorporation
DP-Ir-3	=	deep placed at 3 leaf-stage and irrigated
DP-S	=	deep placed at seedling stage
DP-SG	=	ABC supergranules deep placed
DP-TR	=	deep placed at transplanting
DP	=	deep placed at 6 cm
GTU	=	guanylthiourea
HQ	=	hydroquinone
Inc	=	incorporation
Inc-F	=	incorporated in the presence of floodwater
Inc-Tr	=	incorporated at transplanting
Ir-TR	=	topdressed at revival after irrigation
IT	=	initial tillering
NCPS	=	non-calcareous paddy soil
PI	=	panicle initiation
PPD	=	phenyl phosphorodiamidate
PU-TR	=	traditionally applied prilled urea
Ra	=	apparent recovery
Rt	=	recovery estimated by tracer technique
SB	=	surface broadcast
SB-FW	=	surface broadcast into floodwater
SB-Ir	=	surface broadcast and irrigated

SB-Ir-3	=	surface broadcast at 3 leaf-stage and irrigated
SB-Ir-E	=	surface broadcast at elongation stage and irrigated
SB-Ir-R	=	surface broadcast at revival in spring and irrigated
SB-Pd	=	ABC power surface broadcast
SB-PI	=	surface broadcast at panicle initiation
SB-Ti	=	surface broadcast at tillering
SB-TR	=	surface broadcast at transplanting
SBDr-Ref	=	surface broadcast onto drained soil surface and reflooded
SBn-S	=	surface-banded at seedling stage
SP	=	superphosphate
SS	=	applied at sowing
TD-E	=	top-dressed at elongation
TD-H	=	top-dressed at heading
TD-PI	=	top-dressed at panicle initiation
TD-S	=	top-dressed at sowing
TD-Ti	=	top-dressed at tillering
TD-3	=	top dressed at 3 leaf-stage
TE-Ir	=	top-dressed at elongation stage and irrigated
TR-Ir	=	top-dressed at revival in spring and irrigated
TU	=	thiourea
U	=	urea
UP	=	urea powder
UPA	=	urea phosphoric acid
UP	=	urea powder
USG-1	=	urea supergranules (1.09 g granule^{-1})
USG-2	=	urea supergranules (2.17 g granule^{-1})
USG-DP	=	deep placed urea supergranules
WCU	=	wax-coated urea

List of Contributors

All of the contributors are from the Institute of Soil Science, Chinese Academy of Science, P.O. Box No 821, Nanjing, Peoples' Republic of China.

1
Forms and amount of nitrogen in soil

WEN QI-XIAO

1.1. Introduction

It has long been recognized that nitrogen (N) plays an important role in soil fertility. Even in well-fertilized fields about 50%, and in some cases more than 70%, of the N absorbed by cereal crops is derived from the soil N pool (Zhu 1988). Obviously, knowledge of the forms, distribution and amount of N in soil is essential for understanding the N supply characteristics of a soil and for the efficient management of fertilizer N. In this chapter, a brief account of the forms, distribution and amounts of N in the soils of China is presented.

1.2. Amount and distribution of nitrogen in soil

The bulk of the N in soil is bound to organic matter and very little mineral N is present at any one time. As the content of organic matter is governed by the relative intensity of biological accumulation and decomposition, factors controlling these reactions, such as input of plant and animal remains, environmental conditions and soil texture, significantly affect the amounts of organic matter and N in soil.

1.2.1. Nitrogen content of the surface layer of soils under natural vegetation

Table 1.1 presents the amounts of organic matter and N in the surface layer of soils under natural vegetation in China. The data show that in the temperate zone, across a transect from northeast to northwest China, the amounts of organic matter and N decrease in the following order; black soil > chernozem > castanozem > desert soil. This sequence can be related to the decrease in rainfall and increase in evaporation along the transect. The change in precipitation/evaporation leads to a reduction in plant biomass production and an intensification of organic matter decomposition compared with accumulation.

Along a north–south transect the amounts of organic matter and N decrease significantly in the order black soil > dark-brown soil > albic soil > brown soil > yellow-brown soil. This is presumably due to a more rapid increase in biological decomposition than plant biomass production with the increase in temperature. However, proceeding further south from the yellow-brown soil region, with additional increases in temperature and rainfall, the amounts of soil organic matter and

Zhu Zhao-liang et al. *(eds.): Nitrogen in Soils of China, 1–30.*

Table 1.1. Organic matter and total N (mean ± standard deviation) in the surface layer (0–20 cm) of virgin soils in China.[1]

Soil type	Organic matter (g/kg)	Total N (g/kg)	C/N
Black soil	107 ± 42	5.03 ± 2.00	12.4 ± 1.2
Chernozem	62.7 ± 27.8	3.13 ± 1.40	11.7 ± 1.6
Castanozem	24.2 ± 9.1	1.42 ± 0.51	9.9 ± 1.5
Sierozem	12.3 ± 6.4	0.74 ± 0.37	9.7 ± 2.2
Desert soil	5.7 ± 3.8	0.44 ± 0.27	7.7 ± 1.8
Dark brown soil	92.0 ± 42.9	3.74 ± 1.89	14.6 ± 2.7
Albic soil	73.9 ± 45.3	3.41 ± 1.93	12.5 ± 2.4
Brown soil	35.4 ± 20.7	1.69 ± 0.91	12.2 ± 2.6
Yellow-brown soil	26.7 ± 17.1	1.47 ± 0.99	10.6 ± 2.3
Red soils			
undisturbed	43.9 ± 17.8	1.73 ± 0.76	15.8 ± 4.2
eroded	15.6 ± 6.6	0.71 ± 0.30	12.2 ± 4.0
Latosols and lateritic red earths			
undisturbed	40.4 ± 14.1	1.67 ± 0.61	14.7 ± 3.8
eroded	17.3 ± 0.39	0.80 ± 0.27	13.4 ± 4.4
Yellow soil	66.9 ± 34.3	2.58 ± 1.22	16.2 ± 4.7
Alpine and subalpine meadow soils	82.9 ± 37.6	4.06 ± 1.91	12.2 ± 3.3
Alpine and subalpine steppe soils	24.9 ± 13.7	1.66 ± 0.98	8.8 ± 1.3
Phospho-calcic soil	94.4 ± 58.6	6.39 ± 4.20	8.8 ± 1.2

[1] Wen and Lin (1983); Zhu (1988).

N increased from yellow-brown soil to red soil and latosol. This is presumably due to a greater increase in plant production relative to the increased decomposition of organic matter.

In the high altitude region of southwest China where yellow soils are dominant, the temperature is low and rainfall is high; these conditions are favourable for the accumulation of soil organic matter. Thus the yellow soils contain more soil organic matter and organic N than the red soils and latosols at lower altitude. In addition the yellow soils contain almost as much organic matter and organic N as the albic soils.

In the very high mountainous regions the soils are frozen for much of the year. Thus the amounts of organic matter and N in the alpine soils are high, regardless of plant biomass production, because of the very low rate of decomposition.

1.2.2. Nitrogen in the plow layer of cultivated soils

The amounts of organic matter and N in cultivated soils are governed not only by the natural factors mentioned above, but also by human activities such as cultivation and fertilization. The effect of the natural factors is shown by the similarity in amounts of organic matter and N in the cultivated and virgin soils. Table 1.2 shows that the cultivated soils in the black soil region of northeast China contain the most organic matter and N, those from the Loess Plateau and the Huang-Huai-Hai Plain the least, while the soils from the remaining regions are intermediate.

Table 1.2. Organic matter and total N (mean ± standard deviation) in the plow layer of cultivated soils in different regions of China.[1]

Region	Land use	Organic matter (g/kg)	Total N (g/kg)	C/N
Black soil region in northeastern China	Upland	56.7 ± 25.5	2.63 ± 1.04	12.4 ± 1.9
	Paddy	49.6 ± 15.0	2.58 ± 0.77	11.2 ± 0.8
Qinghai and Tibet	Upland	27.7 ± 16.7	1.44 ± 0.64	11.0 ± 3.7
	Paddy	24.6 ± 10.1	1.43 ± 0.59	10.0 ± 1.6
South China and South Yunnan	Upland	26.8 ± 12.0	1.39 ± 0.77	11.9 ± 3.3
	Rubber	24.3 ± 8.9	1.13 ± 0.43	12.7 ± 2.2
	Paddy	28.5 ± 12.4	1.50 ± 0.67	11.1 ± 2.0
Yunnan-Guizhou Plateau and Sichuan	Upland	19.3 ± 12.8	1.09 ± 0.57	9.7 ± 2.2
	Paddy	27.3 ± 24.1	1.49 ± 1.12	10.5 ± 2.6
Inner Mongolia and XinJiang	Upland	18.3 ± 9.1	1.10 ± 0.53	9.7 ± 1.6
Middle and lower reaches of Changjiang River	Upland	15.8 ± 6.7	0.93 ± 0.33	10.0 ± 3.0
	Tea plantations	14.5 ± 5.4	0.81 ± 0.25	10.4 ± 1.6
	Paddy	22.7 ± 9.2	1.34 ± 0.47	9.8 ± 1.5
South of Changjiang River	Upland	15.7 ± 6.1	0.90 ± 0.29	10.2 ± 2.2
	Tea and citrus	18.3 ± 3.4	0.97 ± 0.24	11.3 ± 2.2
Loess Plateau	Upland	10.4 ± 4.2	0.70 ± 0.28	8.8 ± 1.6
Huang-Huai-Hai Plain	Upland	9.7 ± 4.8	0.63 ± 0.29	9.0 ± 1.8
	Paddy	15.1 ± 6.3	0.93 ± 0.29	9.4 ± 0.8

[1] Zhu (1988).

Differences can also be found among the cultivated soils from the remaining regions. The amounts of organic matter and N increase in the order: the middle and lower reaches of Changjiang River and south of Changjiang River < Inner Mongolia, Xinjiang, Yunnan-Guizhou Plateau and Sichuan < South China, South Yunnan, Qinghai and Tibet. Although this variation can be explained largely by the difference in climatic conditions, some other factors may also be responsible. For instance, the high organic matter and N concentrations of the cultivated soils in the black soil region of northeast China may be due to the high concentrations of these entities in the virgin soils and the short history of cultivation. The low organic matter and N concentrations in the cultivated soils of the Loess Plateau and the Huang-Huai-Hai Plain are the result of severe erosion and frequent flooding, respectively.

Human activities affect the concentrations of organic matter and N in soil in a number of ways:

(i) The input of organic material in cultivated soils is usually much lower than that in soils under natural vegetation, because a large portion of the biomass production is removed through harvesting. In addition, the rate of decomposition of organic matter in cultivated soils is much faster than that in virgin soil, because of the frequent drying–wetting cycles and the improved aeration in the plow layer of cultivated soil through plowing. Consequently, the concentrations of organic matter and N in cultivated soil are always lower than those in the virgin soils.

(ii) The pattern of soil utilization is one of the most important factors controlling the concentrations of organic matter and N in cultivated soils. For example, in rice growing areas, the amount of organic manure applied to flooded fields is greater than that applied to upland fields. In addition, the flat land surface of rice fields greatly inhibits erosion, and submergence during the growth of rice retards the decomposition of organic matter. Consequently, the concentrations of organic matter and N in paddy soils are usually greater than those in upland and plantation soils (Table 1.2).

This discussion relates only to the general trends in organic matter and N concentrations in the soils under natural vegetation or cultivated soils. The concentrations may vary considerably between soils of the same type or in the same region because of differences in natural conditions and human activities. For instance, the concentrations of organic matter and N in the soils of the middle and lower reaches of the Changjiang River in Jiangsu Province vary greatly: the highest concentrations are found in the soils of the Taihu Lake region (28.2 g/kg and 1.47 g/kg, respectively), followed by the soils in the depressions of the Lixiahe region, with the soils in the hilly area of the Zhengjiang-Yizhen-Luohe region being the lowest (about 16.4 g/kg and 1.03 g/kg, respectively) (Wen 1979; Zhu 1979).

Furthermore, even within a small area, the concentrations of organic matter and N in soil may vary due to differences in topography and parent material. Generally, in hilly areas, the soils on the slopes have lower concentrations of organic matter and N than the soils in the valleys. On the plain, soil organic matter and N vary with distance from the village. Because of ease of transportation, the soils nearer the village receive more organic manure than those far from the village.

Soil water regime and texture are two important factors controlling the concentration of organic matter and N in soil. With excessive water an anaerobic system results and the rate of decomposition of organic matter in soil is slowed. In heavy-textured soil, microbial activity is retarded by poor aeration and the clay may protect organic matter from decomposition. Therefore, within a region the heavy-textured and poorly-drained soils generally have higher organic matter and N concentrations than the light-textured and well-drained soils. For example, with the sandy soils, sandy loam soils and clay loam soils of the North China Plain, the organic matter concentrations vary according to the ratio 1:3:5.3 (ISWC 1961). In the Taihu Lake region, the neutral clay loam hydromorphic paddy soils contain 25.8 g of organic matter/kg and 1.59 g N/kg, compared with 19.2 g organic matter/kg and 1.16 g N/kg in the calcareous light-textured paddy soils (Xu *et al.* 1980).

1.2.3. Distribution of nitrogen in the soil profile

Types of soil differ not only in the contents of organic matter and N in the surface layer, but also in the distribution down the soil profile. The mode of addition of organic material to soil, which mainly depends on the type of vegetation, is the most important factor affecting the distribution of organic matter and N in the profile. In the case of forest soil, organic matter and N tend to accumulate in the surface layer, and their concentrations decline sharply with depth, because litter, the major source of soil organic matter, is added to the soil surface. This is particularly true of forest soils in the cool temperate zone, where the existence of a frozen layer in the profile slows the extension of plant roots to deeper horizons. On the other hand, with herbaceous plants, in addition to the residues on the soil surface, roots provide considerable organic material to the deeper horizons. Consequently, there is little variation in the concentrations of organic matter and N with depth. Moreover, in the steppe soils in different bioclimatic zones the distribution of organic matter and N in the soil profile may also differ due to differences in the type of plant and soil-forming conditions. As shown in Fig. 1.1, the humus horizon in the black soil is 80–100 cm deep, while in the castanozem it is much shallower, ranging from 22–44 cm. Organic matter and N in the profile of the desert soil are very low, and this can be attributed to the low biological accumulation and intensive decomposition. In alpine and subalpine meadow soils, and alpine and subalpine steppe soils, the distribution is similar to that of the forest soils because of the presence of a frozen layer in the profile.

Manure is one of the major sources of organic matter in cultivated soils. Because it is applied to the plow layer and plant roots concentrate in that layer, especially in

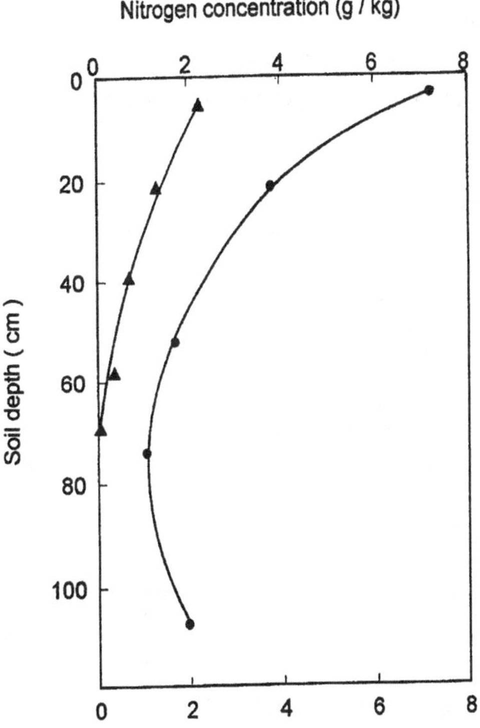

Figure 1.1. The distribution of nitrogen in soil profiles. ●, black soil; ▲, dark castanozem.

paddy soils with a well-developed plow pan, organic matter and N usually decrease sharply with depth. The exceptions are those soils with a deep humus horizon or with a buried humus-rich horizon.

Within the virgin soils the reserves of organic carbon and N are highest in the black soils; in the other soils the amounts decrease in the order: dark brown soil > latosol > red soil > chernozen > castanozem > sierozem > desert soil. Within the agricultural soils, those in the black soil region have the highest organic C and N, followed by those in the Taihu Lake region, while the soils of the Huang-Huai-Hai Plain and Loess Plateau have the lowest amounts of C and N (Table 1.3). For example, the C and N in the 0–100 cm layers of the soils of Ansai County in the Loess Plateau account for only ~1/3 of that in the soils of Taihu Lake region.

1.3. Inorganic nitrogen in soil

Inorganic N occurs in soil as ammonium in the soil solution, adsorbed on cation exchange sites, and fixed in 2:1 type clay minerals, and as nitrite, nitrate and nitrous oxide. These forms of inorganic N account for 1–40% of the total N in surface soils and 1.7–60% of total N in the 0–100 cm soil layer. Clay mineral-fixed ammonium constitutes the major portion of the inorganic N, and its origin, content and availability will be discussed in Chapter 5.

1.3.1. Exchangeable ammonium and nitrate

From the agronomic point of view, exchangeable ammonium and nitrate are considered to be the most important forms due to their ready availability for plant

Table 1.3. Organic matter and total N (t ha^{-1}) in some cultivated soils of China.[1]

Region	Soil	0–20 cm		0–100 cm	
		C	N	C	N
Taihu, Jiangsu	Hydromorphic paddy	35.41	3.48	87.96	10.15
	Bleached paddy	30.28	2.98	69.43	9.19
	Gleyed paddy	42.29	4.18	129.4	12.59
	Calcareous submergenic and hydromorphic paddy	20.79	3.15	66.30	9.02
	Mean[2]	34.71	3.68	94.29	10.93
Huang–Huai–Hai Plain	Mean[2]	16.82	1.57	67.77	5.81
Ansai, Shanxi	Dark loessial	21.24	3.64	88.05	9.84
	Yellow cultivated loessial	9.14	1.41	37.51	4.95
	Sandy loessial	6.76	0.86	29.31	3.86
	Mean[2]	6.93	0.90	29.91	3.93

[1] Unpublished data of Wen *et al.* and Cheng *et al.*
[2] Estimated on area basis.

uptake. The dynamics of ammonium and nitrate in soil are governed by a series of transformation and transport processes, such as mineralization–immobilization, nitrification, denitrification, volatilization, leaching, and plant uptake.

1.3.2. Dynamics in the plow layer of cultivated soils

In cultivated soils without fertilization, exchangeable ammonium and nitrate in the plow layer are usually low due to plant uptake. Mineral N generally varies from ~1 to 10 mg N/kg. It is highest prior to seeding and decreases rapidly at the tillering stage and thereafter. Application of fertilizer N increases the amount of mineral N in the soil at first, but then the amount quickly declines. The extent of the increase and the duration for which the content is maintained at a high level depends on the rate of N applied, the time of application and the type of crop. When N was applied as a basal dressing, exchangeable ammonium and nitrate were as high as 150 mg N/kg, and were sustained at >15 mg N/kg for a long period of time (Li *et al.* 1982). For winter wheat, it lasted for two months (Table 1.4), and for rice about 15–30 days (Figure 1.2). Even at later stages of growth the concentration of mineral N was greater than that in the soil without N application. In the fallow soils, exchangeable ammonium and nitrate vary with season. They are lowest in winter due to low temperatures and hence low mineralization rates, and increase with increasing temperature and rate of mineralization (Zhu *et al.* 1978).

Exchangeable ammonium and nitrate (mineral N) in soil are governed by a number of factors. The concentration varies with soil type and water content. In soils of the same type, the amount of mineral N is positively correlated with the amount

Table 1.4. Nitrate and ammonium (mgN kg soil^{-1}) in different horizons of a fluvo-aquic soil at different growth stages of winter wheat.[1]

Treatment[2]	Date of sampling	0–20 cm		20–40 cm		40–60 cm		60–80 cm		80–100 cm	
		NO_3^-	NH_4^+	NO_3^-	NH_4^+	NO_3^-	NH_4^+	NO_3^-	NH_4^+	NO_3^-	NH_4^+
	10/8, 1986[3]	17	0.8	5.5	0.5	8.7	3.0	16	0.7	7.8	0.7
P	12/8	4.2	0.2	3.5	6.5	5.8	1.0	5.2	0.1	4.5	0.04
NP	12/8	16	0.6	9.8	0.3	14	0.1	15	0.1	14	0.1
P	2/22, 1987	1.9	1.0	4.5	0.5	20	0.3	19	1.5	14	tr
NP	2/22	8.5	2.6	15	1.2	12	1.5	8	0.9	4.3	0.6
P	4/4	3.2	tr	4.1	tr	–	–	9.9	tr	13	tr
NP	4/4	2.6	tr	1.1	tr	3.6	tr	11	tr	7.2	tr
P	5/5	0.9	tr	0.2	tr	1.3	tr	3.5	tr	4.5	tr
NP	5/5	1.9	tr	0.4	tr	0.8	tr	1.6	tr	3.3	tr
P	5/31	1.3	tr	0.2	tr	0.2	tr	2.5	tr	0.4	tr
NP	5/31	3.9	tr	1.9	tr	2.6	tr	5.9	tr	3.5	tr

[1] Unpublished data of Zhang Shao-lin, Zhu Zhao-liang and Xu Yin-hua.
[2] P, Superphosphate (39 kg P ha^{-1}) was banded on October 10, 1986.
NP, Superphosphate (39 kg P ha^{-1}) and urea (75 kgN ha^{-1}) were banded on October 10, 1986.
[3] Prior to fertilization.
tr = trace.

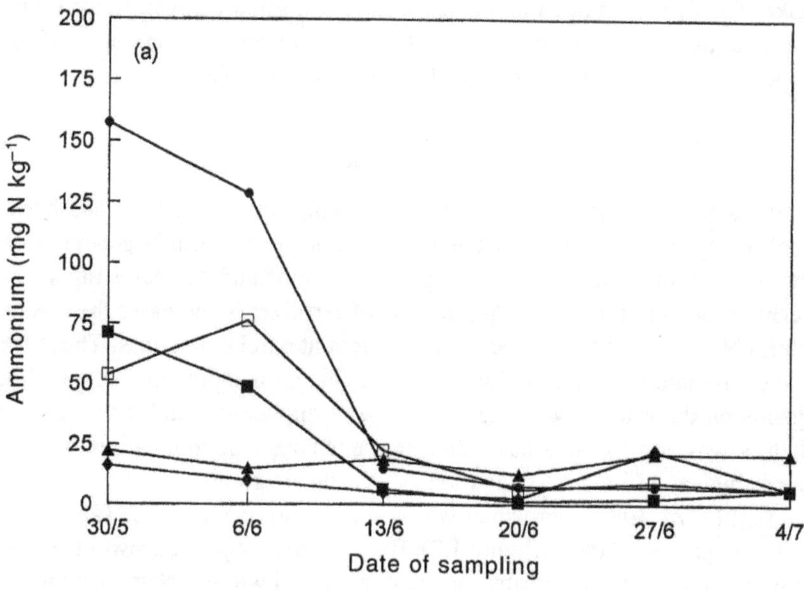

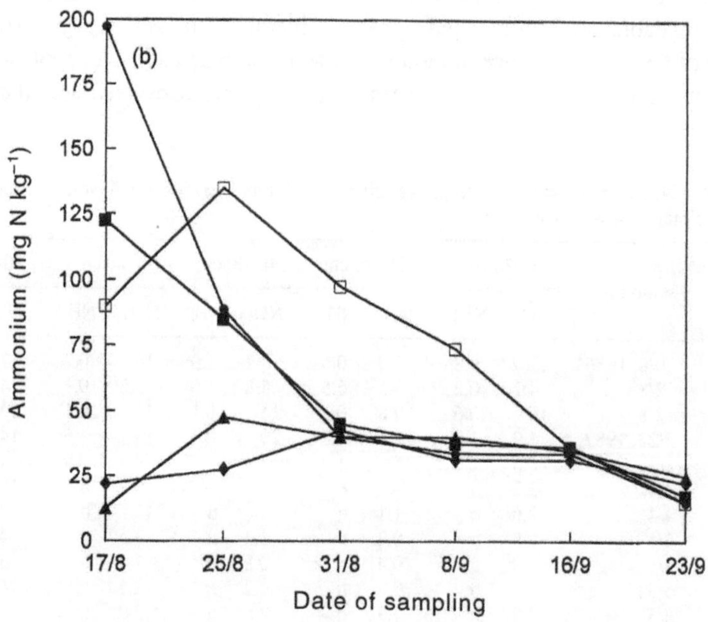

Figure 1.2. Ammonium in the plow layer of a paddy soil during the rice growing season. (a) Early rice, transplanted on May 19, (b) late rice, transplanted on August 7.♦, no fertilizer applied; ▲, no fertilizer or rice; ■, 112 kg N as ammonium sulfate (AS)/ha; □, 112 kg N as AS/ha + 15 × 10^3 kg fresh farmyard manure/ha; ●, 187 kg N as AS/ha.

of soil organic matter (BSSHYA, 1961). Under submerged conditions the amount of exchangeable NH_4^+ is generally greater than the amount of mineral N under aerobic conditions because of the lower N requirement of anaerobic microorganisms. Mineral N in the surface layer of aerobic soils is also greatly affected by the frequency and amount of rainfall, the cycles of drying and wetting and leaching. The drier the soil, the greater the amount of N that mineralizes when the soil is remoistened. In addition, the greater the number of cycles of drying and wetting, the greater the amount of N mineralized. Li and Yang (1965) reported that early deep-plowing of a fallow field increased the amount of mineral N in the soil. Early deep-plowing resulted in the accumulation of twice as much mineral N as occurred with delayed deep-plowing. Peng *et al.* (1981) found a relationship between rainfall (x, mm) during the summer time and depth (y, mm) of nitrate leaching in a fallow field, viz.

$$y = 3.86\,x$$

Furthermore, the soil water regime affects both the amount and the composition of inorganic N in soil. Under aerobic conditions, because the rate of nitrification is usually greater than that of mineralization, the dominant form of inorganic N in soils is nitrate. It generally accounts for more than 80% of the mineral N (Shao *et al.* 1988; Zhang *et al.* 1989a). Under anaerobic conditions, nitrification is retarded due to lack of oxygen, and ammonium is the dominant form.

1.3.3. Spatial variability of mineral nitrogen in the plow layer

The distribution of ammonium and nitrate in the plow layer is highly variable spatially because of the heterogeneity of soil, and the distribution of roots and fertilizer. The enrichment or depletion of inorganic N in the rhizosphere compared with the non-rhizosphere soil depends primarily on the rate of N uptake by the roots relative to the rate of transfer of inorganic N from the soil bulk to the rhizosphere. Qin and Liu (1989) found that nitrate was enriched in the rhizosphere of both upland and flooded soils and attributed this to the faster transfer of nitrate through mass flow from the bulk soil to the rhizosphere than uptake by the roots. The transfer of ammonium from the bulk soil to the rhizosphere is largely through diffusion, and the rate of diffusion of ammonium is lower than its rate of uptake by the roots. Consequently, there was a depletion of ammonium in the rhizosphere and the ammonium concentration decreased towards the rhizosphere. The effect of fertilizer application on the spatial variability of inorganic N varies with the type of fertilizer, and it is greater for chemical fertilizer than organic manure. Ji and Wang (1978) investigated the transfer of supergranular ammonium bicarbonate N in a neutral paddy soil derived from lacustrine deposits in the laboratory. They found that after 5 days at room temperature (28°C) the concentration of ammonium 1 cm from the granule was 36×10^{-3} mol/L, which was 1.56, 8.57, 18.0, 21.2 and 32.7 times the concentrations at 2, 3, 4, 5 and 11 cm from the granule. The concentration may have been even higher in the vicinity of the granule. In addition, root exudates and sloughed-off root hairs may also influence the distribution of inorganic N in the rhizosphere through mineralization, immobilization or denitrification.

1.3.4. Transfer and distribution of nitrate in soil profiles

In upland soils, especially in loamy and sandy soils, the distribution of nitrate in the profile varies with the water regime of the soil and the highest amount of nitrate is not always found in the surface layer. In an investigation on the variation of nitrate with season in an old-manured loessial soil, Peng *et al.* (1981) found that, under the growth of winter wheat–summer maize, the nitrate content of the 0–15 cm layer was the highest in the profile from the end of March to mid-June and from late September to mid-October. However, as a result of rainfall in late June to mid-September and from late November to mid-December, the highest content of nitrate was found in the 15–30 cm and deeper layers. A similar distribution pattern was found in the summer-fallow field (Figure 1.3), but the downward transfer of nitrate due to rainfall was more pronounced under summer-fallow than under the growth of summer maize. Investigations conducted on the North Jiangsu Plain (Yuan *et al.* 1985) showed that in some cases the highest content of nitrate in the soil prior to seeding maize was found in the 15–30 cm layer. When an excess amount of N was applied to the preceding crop, the nitrate content in all soil layers below 15 cm was higher than that in the 0–15 cm soil layer. In extreme cases the highest content was found in the 70–100 cm layer (Table 1.5).

The nitrate accumulated in the soil profile is a source of N for crop growth. Yuan *et al.* (1985) found that when excess amounts of N were applied to a preceding winter crop, the amount of nitrate in the 0–100 cm soil layer prior to seeding maize

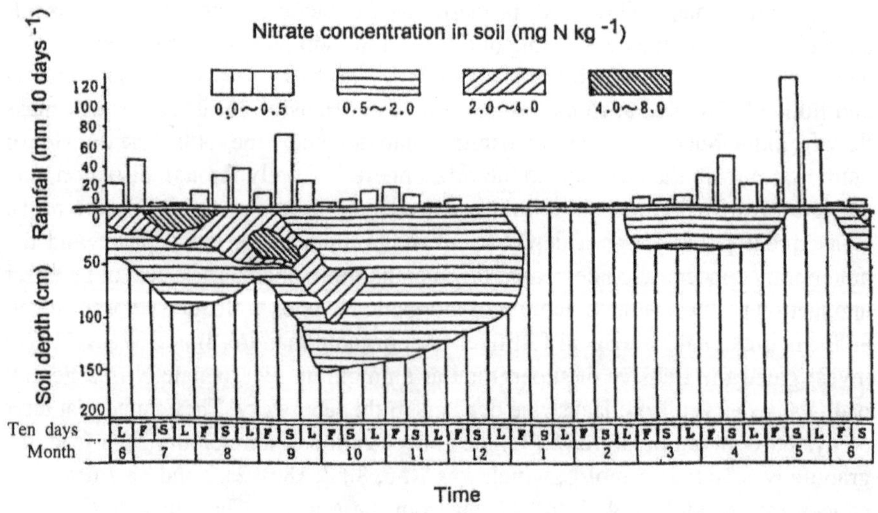

Figure 1.3. Seasonal variation in nitrate in a winter wheat field after a summer fallow. F = first ten days, S = second ten days, L = last ten days.

Table 1.5. Nitrate (kg N ha^{-1}) in sandy loam fluvo-aquic soils of Feng county, Jiangsu Province.[1]

Sampling site	Depth (cm)				
	0–15	15–30	30–50	50–70	70–100
Huashan	0.58	0.23	0.20	0.09	0.22
Sunlou	2.17	2.33	0.03	0.31	0.19
Zhaozhuang	0.59	2.24	2.45	2.61	1.91
Liuwanglou	3.98	2.18	1.49	0.96	2.71
Shanlou	1.53	3.05	3.76	4.00	7.08

[1] Yuan *et al.* (1985).

was as high as 170–290 kg N/ha. This is sufficient to meet the N requirement of the maize crop. Large amounts of nitrate can accumulate in the soil profile in the fallow period before sowing a cereal crop. Thus, Wang Xian Zhong (unpublished data) found that, in a field experiment with maize conducted on a sandy loam very low in organic matter and total N (5.6 g/kg and 0.49 g/kg, respectively), the yield of the control plot (without N application) was comparable with the yield of the treatment receiving 150 kg N/ha as urea, ammonium nitrate, ammonium sulfate or ammonium bicarbonate. The residual nitrate in the profile is an index of the N-supplying capacity of an upland soil and should be taken into account when recommending the rate of N application.

Nitrate is highly mobile in soil. There is no specific adsorption for nitrate ion in soil (Wang *et al.* 1987), and the diffusion coefficient for nitrate is greater than that of Cl^- at low pH (Hu 1989). The upward movement of water in a soil profile facilitates the transfer of nitrate from deep layers to the rooting zone for plant uptake. However, when downward movement of water is dominant in a soil profile, nitrate may be leached to a depth of >2 m, i.e. lost from the soil–plant system, and may result in groundwater pollution. In the Guanzhong and North Jiangsu plains, the groundwater down to 50 m is enriched with nitrate (>10 mg N/L). In some areas the concentration may be as high as 100 mg N/L (NIBSSC 1973; Lei *et al.* 1974). The origin of nitrate in the groundwater is not clear; part may originate from the application of organic manure and fertilizer because some of the groundwater rich in nitrate is found under vegetable gardens with high rates of N application. Epidemiological investigations found that in certain regions, the death rate due to liver cancer was closely related to the nitrate content of soils (Hu *et al.* 1983).

1.3.5. Nitrite

Nitrite is formed during ammonium oxidation and nitrate reduction. However, because nitrite formed during these processes is rapidly oxidized or reduced, its concentration is very low, usually less than 0.3 mg N/kg. In an incubation experiment under aerobic conditions, Hu *et al.* (1983) found that nitrite accounted for only 0.12–0.34% of the nitrate formed.

1.3.6. Nitrogen-containing gases

Inorganic N also occurs in soil, in trace amounts, in the form of N-containing gases, such as NH_3, N_2O, NO and NO_2. Since N_2O is one of the greenhouse gases, its emission from soil is of great concern. Few studies on N_2O emission from soil have been conducted in China. Li *et al.* (1991) measured the rates of emission of N_2O from a flooded soil in a pot experiment. Su *et al.* (1992) reported that the flux of N_2O from a drab soil during the growth of winter wheat was mainly governed by the nitrate level in the soil; it was the highest immediately after irrigation when fertilizer N was applied, amounting to as high as 46 μg N m^{-2} h^{-1}. It then dropped gradually and averaged 6 μg N m^{-2} h^{-1} when the wheat matured.

1.4. Organic nitrogen

A great variety of N compounds have now been isolated from soils and identified. These include amino acids, amino sugars, purine, pyrimidine and trace amount of chlorophyll-type compounds, phospholipids, mono- and poly-amines and vitamins (Bremner 1967; Cortez and Schnitzer 1979; Schnitzer and Spiteller 1986; Stevenson 1982). However, these compounds account for only about 50% of the organic N in soil, and the forms of the remaining N are still unidentified.

The release of considerable amounts of amino acids from soil organic matter during incubation with proteolytic enzymes (Sowden 1970) provides evidence for the existence of proteinaceous compounds in soil. 'Ligno-protein' complexes and 'humoprotein' complexes were isolated using Savag's method (Jenkinson and Tinsley 1960) or extraction with phenolic solvents (Biederbeck and Paul 1973) but the amounts were small. Several scientists have postulated that carbohydrates, N-containing compounds and humic substances exist in soil separately (Schnitzer 1978) but there is currently no evidence to support this hypothesis. In addition, little is known about the mechanism of binding of the identified N-containing compounds to the humic substances and carbohydrates. Current knowledge concerning the nature of the organic N in soil is based largely on studies of the organic N compounds released by acid hydrolysis.

1.4.1. Fractionation of soil nitrogen

Suzuki (1906) was probably the first scientist to use the method of protein hydrolysis to characterize organic N in soil (Bremner 1967). Since then most studies on the forms of organic N in soil have been based on the use of the Van Slyke method of protein analysis, involving separation and estimation of humin-N, ammonium-N, basic and nonbasic nonamino acid-N, and basic and nonbasic amino acid-N. Bremner (1965) developed a procedure for separating and estimating ammonium-N, amino sugar-N, amino acid-N, and hydrolyzable unknown-N in acid hydrolysates and nonhydrolyzable-N.

Ammonium in acid hydrolysates is derived partly from the decomposition of organic N-compounds and partly from the release of clay mineral-fixed ammonium during acid hydrolysis. During acid hydrolysis, tryptophan is decomposed completely. Methionine, serine, threonine, and amino sugars are decomposed to different extents depending upon the conditions of hydrolysis. The percentage of clay mineral-fixed ammonium recovered in acid hydrolysate varies with soil type and was in the range of 31–83% (mean 62.4%). In addition, ammonium in hydrolysates may have originated from the amides, asparagine and glutamine, which are deaminated during acid hydrolysis. It is not possible to quantify the contributions of these sources to the ammonium in acid hydrolysates because little information is available on the content of the amides, asparagine and glutamine, in soil. It is estimated that about 50% of the ammonium in acid hydrolysates is derived from clay mineral-fixed ammonium, amino acids and amino sugars.

Purines and pyrimidines are undoubtedly the constituents of the hydrolyzable unknown N fraction. However, they account for only 0.06–0.88% of the soil N. Recently, Schnitzer and Spiteller (1986) identified 18 heterocyclic-N compounds including hydroxy- and oxy-quinolines and isoquinolines, amino benzofurans, etc. by capillary gas chromatography–mass spectrometry. However, their content was negligible; the N content of the fraction from which these compounds were isolated was only 0.1% of the total N of humic acid. In addition, only α-amino-N was determined by the conventional ninhydrin-NH_3 method. Therefore, the content of amino acid-N was underestimated due to the presence of non-α-amino N in arginine, tryptophan, lysine and proline. It appears that 10–59% (mean 31%) of the hydrolyzable unknown-N of soils occurred as non-α-amino-N (Goh and Edmeades 1979; Greenfield 1972; Griffith *et al.* 1976; Shi *et al.* 1992; Sowden *et al.* 1976). From these facts, it can be roughly estimated that about 60–65% of the hydrolyzable unknown N remains unidentified.

Treatment of the nonhydrolyzable residue with 6 N HCl containing 3% H_2O_2 provides evidence for the presence of N-phenyl amino acids in the residue. According to Aldag (1977), Griffith *et al.* (1976), and Shi *et al.* (1992), this form of N accounts for only 2.3–10.4% of the acid-nonhydrolyzable-N. Ladd and Butler (1966) indicated that N-phenyl amino acids could not be broken down completely by oxidative acid hydrolysis; the percentage of amino acid released from the N-phenyl amino acid derivatives varied with the replacement groups in the benzene ring and ranged from 30 to 70%.

Part of the clay mineral-fixed ammonium in soil is not released by acid hydrolysis and thus is included in the acid-nonhydrolyzable fraction. The amount generally accounts for 1.8–47% of the acid-nonhydrolyzable fraction, and depends on the amount of total-N and the type of clay minerals in the soil. The percentage is lower in soils derived from granite and sandstone in tropical and subtropical regions, particularly in those soils with a higher content of organic matter. It is high in soils derived from purple sandstone and shale and in the recent alluvium of the Changjiang River which contains little organic matter.

Acid-nonhydrolyzable-N can be formed during hydrolysis through the Maillard reaction of carbohydrates or their derivatives with amino acids and ammonium. Asami and Hara (1971) found that addition of glucose to soil immediately prior to acid hydrolysis increased acid nonhydrolyzable-N by 3% and resulted in a decrease in ammonium and amino acid-N. Zhuo Su-neng (unpublished data) also found that clay minerals are likely to facilitate the condensation. In the absence of clay minerals, acid nonhydrolyzable-N produced by reaction of glycine with xylose was equivalent to 0.3% of the added glycine-N. It increased to 4.4% in the presence of clay minerals. However, it is unlikely that the acid nonhydrolyzable-N in soil is an artifact.

1.4.2. Distribution of nitrogen in soil

Since 1965, Bremner's procedure for the fractionation of N has been widely used for characterizing soil N. Results available show that the distribution of the forms of N in the surface layer varies greatly among soils. The percentage of total N as acid-hydrolyzable-N in the surface soils of China ranged from 65 to 92%; the corresponding figures for ammonium-N, amino sugar-N, amino acid-N and acid-hydrolyzable unknown-N are 15–51%, 1.1–15%, 19–45% and 6–26%, respectively (Song 1988; Shen and Shi 1990; Wu 1986; and unpublished data of Ye Wei, Zhuo Su-neng, Huang Dong-mai and Zhu Pei-li, and Zhao Bin-jun). Little is known about the factors affecting the distribution of the forms of soil N, and no consistent trend has been found relating the distribution to climatic conditions and soil properties. For example, there was no significant difference in the distribution of the forms of N between a black soil or a dark brown soil in the cool temperate zone and a latosol in the tropics, in spite of the great differences in climatic conditions, vegetation and parent material (Song 1988; Shi *et al.* 1992). This also applied for a latosol, lateritic red soil, red soil and yellow-red soil developed on the same parent material, granite. There was also no relationship between the type of clay minerals or soil acidity and N distribution. The distribution in a red limestone soil was almost the same as that in a yellow-brown soil, red soil or brown soil from the same region. However, an effect of water regime on N distribution was found under comparable conditions. The percentage of total-N as amino sugar-N and ammonium-N in paddy soils was usually lower than that in upland soils and virgin soils derived from the same parent material (Table 1.6). The results obtained in a model experiment further support this conclusion (Cheng Li-li unpublished data). As shown in Table 1.7, under submerged conditions, the newly formed organic matter had considerably less amino sugar-N and ammonium-N and markedly more amino acid-N than that formed under upland conditions, regardless of the type of plant material used in the incubation (azolla or rice straw), soil parent materials (Xiashu loess or Quaternary Red Clays), presence or absence of calcium carbonate, or length of incubation (3, 5 or 7 y). The effect on amino sugar-N may be attributed to the suppressed growth of fungi in the paddy soils (Hao and Cao 1988). Amino sugar occurs as a constituent of chitin in cell walls and fungal cell walls have a higher content of chitin. The reason for the lower concentration of ammonium-N in acid hydrolysates of paddy soils is still unknown. It may be that the low-molecular-weight N-containing compounds in soil, which are more susceptible to deamination during acid hydrolysis than the high-molecular-weight compounds, are more easily

Table 1.6 Effect of water regime on distribution of nitrogen in soil.[1]

Parent material	Water regime	Total N ($mg\ kg^{-1}$)	Percentage of total soil N					
			Hydrolyzable					Nonhydrolyzable
			Total	NH_4^+	Amino sugar	α-Amino acid	Unidentified	
Granite	Upland	470	82.5	22.5	7.2	38.0	14.8	17.5
	Submerged	1160	86.9	21.5	5.8	42.9	16.7	13.1
Quaternary Red Clays	Upland	1350	77.6	22.1	6.6	32.0	16.9	22.4
	Submerged	1240	87.2	21.4	5.3	34.2	26.3	12.8
Limestone	Upland	2240	82.8	26.0	8.9	34.9	13.0	17.2
	Submerged	1870	81.2	23.9	4.2	33.6	19.5	18.8
Purple sandstone	Upland	930	72.5	26.4	6.3	30.9	8.9	27.5
and purple shale	Submerged	1750	78.0	24.2	6.2	34.2	13.4	22.0
Tertiary red sandstone	Upland	370	82.4	22.9	6.9	30.9	21.7	17.6
	Submerged	1090	90.4	21.0	5.9	36.2	27.3	9.6

[1] Unpublished data of Ye Wei.

Table 1.7. Effect of water regime during incubation on distribution of nitrogen in decayed plant material.[1]

Parent material	Water regime	Organic N ($mg\ kg^{-1}$)	Percentage of organic N					
			Hydrolyzable					Nonhydrolyzable
			Total	NH_4^+	Amino sugar	α-Amino acid	Unidentified	
			Xiashu loess					
Azolla	Upland	816	80.0	17.9	11.4	31.6	19.1	20.0
	Submerged	980	85.2	10.2	8.6	41.3	25.1	14.8
Rice straw	Upland	531	6.0	16.0	21.5	27.3	21.2	14.0
	Submerged	652	87.3	15.7	17.6	28.2	25.8	12.7
			Quaternary Red Clay					
Azolla	Upland	798	75.6	12.4	14.1	26.7	22.5	24.4
	Submerged	958	76.6	14.0	9.6	38.4	14.6	23.4
Rice straw	Upland	550	69.2	15.0	10.3	28.6	15.3	30.8
	Submerged	591	79.4	11.5	10.4	36.3	21.2	20.6

[1] Unpublished data of Cheng Li-li *et al.*

leached from the surface layer of paddy soils, resulted in the low concentration of ammonium in the soil layer. The higher concentration of amino acid-N in paddy soils compared with upland soils may also be explained by the lower degree of humification of organic matter under waterlogged conditions; amino acid-N in marine sediments and peats was found to be higher than that in soils (Isirimah and Keeney 1973; Sowden *et al.* 1978).

It was expected that the different forms of soil organic N may differ in their biological susceptibility, and knowledge concerning the distribution of the forms of soil organic N might provide a basis for evaluating the availability of soil N. However, investigations on the distribution of the forms of N in a cultivated latosol and a cultivated black soil and the corresponding virgin soils (Shi *et al.* 1992) indicated that there was no significant difference between them; the amino acid-N tended to decrease after cultivation whereas ammonium-N and acid-nonhydrolyzable-N tended to increase slightly, even though the total N decreased by 50–70% during cultivation. These results demonstrate that there is little difference in biological susceptibility among the forms of soil N fractionated by acid hydrolysis.

Although surface soils and subsoils are derived from the same parent materials and developed under the same climatic conditions, the hydrothermal conditions, source of plant materials and biological activity differs. Such differences may affect the distribution of N. Song (1988) found more amino sugar-N and amino acid-N, and less hydrolyzable ammonium-N and acid nonhydrolyzable-N in surface soils than in subsoils. In an investigation of the distribution of N forms in 40 Canadian soils, Sowden (1977) also found that the proportion of total N as amino acid-N in the mineral horizons was lower, and that liberated as NH_3 was higher than that in the litter layer. The greater proportion of the total-N present as amino acid-N and amino sugar-N in the surface layer may be attributed to the greater content of newly formed humic substances in that layer. The high percentage of total N liberated as NH_3 may result partly from the increased proportion of soil N as fixed ammonium-N in the deeper horizons. It is generally postulated that soil humic substances are a series of high- and low-molecular-weight compounds formed, with or without the participation of enzymes, from polyphenols or sugars, and amino acids or ammonium through nucleophilic reactions, the Maillard reaction, fixation of ammonia and nitrite, and polymerization. Humification is generally considered to be a continuous process for the formation and decomposition of humic substances. So the greater the degree of humification, the greater the chemical and biological stability of the N. Consequently, the proportions of total N present as amino acid-N and amino sugar-N in newly formed humic substances are greater than those in humic substances of more ancient origin.

The above-mentioned differences in distribution of N forms between newly formed and native humus in the same soil layer are more pronounced. Results obtained by Shi *et al.* (1981) and Huang Dong-mai and Zhu Pei-li (unpublished data) indicate that, in comparison with the indigenous soil humus, the newly formed humus from fertilizer or green manure was higher in α-amino acid-N and lower in acid nonhydrolyzable-N and ammonium-N (Table 1.8).

The composition of organic matter in different particle-size fractions may vary due to the large difference in composition and physical and chemical properties of the minerals. However, no distinct difference in the distribution of N forms has been found between different particle-size fractions with the exception that the

Table 1.8. Comparison of nitrogen in newly formed and indigenous soil organic matter.[1]

Source of newly formed material	Organic matter	Percentage of total N					
		Hydrolyzable					Nonhydrolyzable
		Total	NH_4^+	Amino sugar	α-Amino acid	Unidentified	
After harvesting first crop (three months after incorporation of plant material)							
Azolla	Newly-formed	85.3	19.3	1.2	36.4	28.4	14.7
	Native	79.8	31.2	2.2	28.6	17.7	20.2
Milk vetch	Newly-formed	83.5	20.2	3.7	41.8	17.8	16.5
	Native	79.2	30.9	4.8	35.1	8.5	20.8
Water-hyacinth	Newly-formed	85.3	23.2	4.0	39.3	18.8	14.7
	Native	79.6	32.2	1.2	33.5	12.7	20.4
After harvesting third crop (one year after incorporation of plant material)							
Azolla	Newly-formed	79.4	16.8	tr	23.1	39.5	20.6
	Native	71.9	30.6	3.1	16.6	21.6	28.1
Milk vetch	Newly-formed	76.1	17.6	1.9	32.2	24.4	23.9
	Native	70.9	28.3	2.7	26.6	13.2	29.1

[1] Shi *et al.* (1981).

percentage of total-N as ammonium-N increases as the particle size decreases (Zhang *et al.* 1984).

1.4.3. Nitrogen distribution in humic and fulvic acids

The forms of N in humic and fulvic acids are the same as those in soil N but the distribution pattern is somewhat different. For example, the proportions of total N as ammonium-N, amino sugar-N and hydrolyzable-N were lower, and those of amino acid-N and acid nonhydrolyzable-N were higher in humic acid than in soil. These trends also hold true when the proportions are calculated on an organic-N basis. Generally, for humic acid the percentage of total N as acid hydrolyzable-N varied from 64 to 80%, while that of ammonium-N and amino acid-N ranged from 4–12% and 33–48%, respectively (Table 1.9). The distribution in fulvic acid varied with the method used for preparation. Compared with humic acid, the percentages of total N as nonhydrolyzable-N in fulvic acid were lower, while that of ammonium-N and amino sugar-N were higher.

Little information is available on the distribution of N in humic acid extracted from different soils and it is difficult to compare the data reported by different authors because of the different methods used for extraction and preparation of humic acid. Little is known about the effect of soil environment and soil properties on the distribution of N in humic acids. The percentages of total-N as ammonium-N and amino sugar-N in paddy soils and the humic acid extracted from them tended to

Table 1.9. Distribution of nitrogen in humic and fulvic acids.

Source	Percentage of total N						Reference
	Hydrolyzable					Nonhydrolyzable	
	Total	NH_4^+	Amino sugar	α-Amino acid	Unidentified		
Humic acid							
Red brown soil, Red soil, Paddy soils,	66.4–77.5	6.5–12.0	1.1–3.8	33.0–48.3	17.8–23.6	22.5–33.6	Ye *et al.* (1991)
Black soil, Yellow brown soil, Lateritic red soil	63.6–81.5	3.7–8.2[1]	7.4–11.7[2]	37.6–47.6	6.2–27.4	18.5–36.4	Peng *et al.* (1984)
Dark brown soil	75.8	12.1	4.7	36.1	22.9	24.2	Zhuo and Wen (1994)
Newly formed humus	71.6–79.4	6.4–8.6	1.1–4.2	42.6–45.3	20.7–22.8	20.6–28.4	Ye *et al.* (1991)
Fulvic acid							
Red soil, Lateritic red soil	82.4–86.7	8.5–11.2[1]	12.6–26.8[2]	39.3–47.3	5.3–18.3	13.7–17.6	Peng *et al.* (1984)

[1] 1 N HCl, 100°C, 3 h.
[2] The difference between steam distillation-N with phosphate-borate buffer at pH 11.2 and NH_4^+ N.

be lower than those for the corresponding upland soils (Table 1.10; Ye *et al.* 1991). Similar results were obtained in a model experiment (Table 1.11; Ye *et al.* 1991).

Humic acid can be split into fractions with different molecular weights by gel permeation chromatography. As shown in Table 1.12, the proportion of total N as hydrolyzable-N, amino acid-N and amino sugar-N decreased, whereas that of nonhydrolyzable-N increased as the molecular weight decreased; the nonhydrolyzable form of N was as high as 66% of the total N. Possible explanations for the decrease in the proportion of amino acid-N with the decrease in molecular weight are as follows. In the humic acids of high molecular weight, the peptide side-chain is longer than that in the low-molecular-weight humic acid. Another explanation is that much of the amino acid-N present is not a component of the humic acid; rather, it is present as a contaminant or is attached to the humic acid by hydrogen bonding, and during gel permeation the polypeptides with high molecular weights enter the high-molecular-weight humic acids fraction. Results obtained in fractional precipitation of humic acid by NaOH–alcohol seem to support the latter explanation. It is well established that in a series of high polymer homologues, the larger particles will be precipitated before the smaller ones upon the addition of alcohol. Using this

Table 1.10. Nitrogen distribution in humic acids of paddy soils and corresponding upland soils.[1]

Parent material	Soil	Percentage of total					
		Hydrolyzable					Nonhydrolyzable
		Total	NH_4^+	Amino sugar	α-Amino acid	Unidentified	
Granite	Red brown	73.1	7.6	2.4	45.2	17.8	26.9
	Paddy	77.5	6.5	2.3	48.3	20.4	22.5
Quaternary Red Clays	Red	66.7	9.8	3.8	31.1	22.1	33.3
	Paddy	72.4	8.3	2.2	38.4	23.6	27.6
Limestone	Terra rossa	66.4	12.0	2.0	33.0	19.4	33.6
	Paddy	66.6	9.0	1.1	34.7	21.9	33.4

[1] Ye *et al.* (1991).

Table 1.11. Effect of calcium carbonate and water regime on nitrogen distribution in humic acids newly-formed from rice straw.[1]

Treatment		Percentage of total N					
$CaCO_3$ ($g\ kg^{-1}$)	Water regime	Hydrolyzable					Nonhydrolyzable
		Total	NH_4^+	Amino sugar	α-Amino acid	Unidentified	
0	Upland	76.0	8.3	4.2	42.8	20.7	24.0
	Submerged	71.6	6.7	1.1	42.5	21.3	28.4
10	Upland	77.7	8.3	3.6	44.5	21.3	22.3
	Submerged	73.4	6.4	1.2	42.2	22.6	26.6
70	Upland	79.4	8.6	2.8	45.2	22.8	20.6
	Submerged	76.9	7.2	2.0	45.3	22.4	23.1

[1] Ye *et al.* (1991).

technique, Zhuo and Wen (1994) fractionated the humic acid extracted from a dark brown soil into 12 fractions and found that the amount of total N and the relative proportion of total N as amino acid-N increased consistently, from 25.1% for fraction 2 to 46.5% for fraction 12, while the E_4 value decreased gradually from 1.42 to 0.19 with the sequence of precipitation. Grey-white substances appeared in the precipitates after the 8th fraction. This suggested that the prepared humic acid was comprised of two kinds of polymer homologues with different properties. The high-molecular-weight humic acid did not contain more amino acid-N. In addition, co-existing polypeptides were also found in the fulvic acid fraction. As shown by Sequi *et al.* (1975), the amount of total N and the percentage of total N as amino acid-N in fulvic acid was reduced considerably by passing it through a H^+-saturated cation exchange resin.

Table 1.12. Nitrogen distribution in humic acids of different molecular weight.[1]

Source	Molecular weight	Percentage of total N					
		Hydrolyzable					Nonhydrolyzable
		Total	NH_4^+	Amino sugar	α-Amino acid	Unidentified	
Black soil	>150,000	70.5	15.5	3.1	40.9	13.0	29.5
	150,000–30,000	62.8	11.7	1.0	28.6	21.5	37.2
	<30,000	33.5	9.2	0.5	10.0	13.8	66.5
Lateritic red soil	>150,000	82.8	16.9	5.8	45.1	15.0	17.3
	150,000–30,000	77.0	12.2	3.0	37.1	24.7	23.0
	<30,000	71.7	21.1	1.2	27.0	22.4	28.3

[1] Peng *et al.* (1984).

Most of the proposed hypotheses concerning the formation of humic substances consider amino acid or ammonium as materials indispensable for their formation. These hypotheses not only explain why humic substances contain N, but also help to elucidate the forms of unknown-N in humic and fulvic acids. Humic acid produced by reaction of phenols with peptides contains not only amino acid and N-phenyl amino acid, but also acid hydrolyzable and nonhydrolyzable unknown-N. Similarly, the 'humic acid' formed by reaction of amino acid or ammonium with sugar also contains acid hydrolyzable and nonhydrolyzable unknown-N. Some scientists using the [^{15}N]NMR technique, an effective approach for studying the forms of N, investigated the forms of N in the 'melanoidins' synthesized by reaction of sugar and amino acid or ammonium. Results showed that the N in the melanoidin formed from the reaction of xylose with amino acid, like that in soil humic acid, was mainly in the form of secondary amides, and only a small proportion of the N was present as aliphatic amine, pyrrole and/or indole. However, more than half of the N in the melanoidin formed from the reaction of xylose with ammonium was in the form of heterocyclic-N. Most of this was indole- and pyrrole-like N and the remainder was pyridine-like N. These authors considered that the unknown-N in humic acid is largely in the form of sterically hindered secondary amides (Benzing-Purdie *et al.* 1983, 1986).

1.4.4. Amino acids

As much as 56% of total N in soil can be present as amino acid-N, including α-amino-N, non-α-amino-N and N-phenyl amino acid (Griffith *et al.* 1976).

The occurrence of amino acids in soil was first reported in 1906. At present more than 30 amino acids have been identified, 10 amino acids are commonly found and some, such as diaminopimelic acid, are usually not found in animals and plants. Only trace amounts are present in soil as free amino acids, and the majority exist in soil in combined forms, such as constituents of soil microorganisms or components of humic substances. They are mainly bound to external and internal surfaces of clay minerals or bound to the surface of humus colloids (Stevenson 1982).

1.4.5. Distribution of amino acids in soil

Considerable information concerning the amino acid composition in soil has been obtained since chromatography was used for the identification. Unfortunately, the relevant data are difficult to compare because of analytical errors which have arisen from incomplete extraction, and loss of certain amino acids during hydrolysis, desalting and decolorization.

In some early studies, Carles and Decau (1960) and Wang *et al.* (1967) concluded that climatic conditions affected the amino acid composition of soil. They found that the relative proportion of N in basic amino acids was higher in soils of the tropical region where decomposition was more intensive than in other regions. Presumably this was due to the preservation of the basic compounds as a result of interaction with other soil components. However, results obtained by Sowden *et al.* (1977) failed to confirm these findings. The type of cropping system is another factor affecting the distribution of amino acids in soil. Stevenson (1956) found that the percentage of basic amino acids in a soil under continuous maize was higher than that in a soil in a maize–oats–clover rotation. However, this was not found in a similar investigation on the effect of cropping systems (Khan 1971). Zhang (1989) investigated the effect of continuous application of pig manure on the amino acid composition of a brown soil and a meadow soil and found that the content of total N was markedly higher, and the absolute and relative contents of amino acids were higher in the soil of the manured plots than in the control treatment or where fertilizer (NPK) was applied. However, no effect was found on the amino acid composition (Table 1.13). Table 1.13 also shows that there was a great similarity in the amino acid distribution between the brown soil and the meadow soil, and that of soils in other climatic conditions. All soils were comparatively rich in glutamic acid, glycine, arginine, aspartic acid, alanine and lysine (Sowden *et al.*, 1977).

The amino acid composition of the acid hydrolysates of some plant residues and pig manure is similar to that of soil (Lowe 1973; Zhang *et al.* 1989b). However, the protein constituents in the plant residues and pig manure will not remain intact after addition to soil due to their greater susceptibility to microbial decomposition. The great similarity in the amino acid composition of different soils and in different cropping systems suggests that the same factors govern the distribution of amino acids in different types of soils. The similarity in amino acid composition of different microorganisms is reflected in the amino acid composition of soils. Small variations in the distribution of amino acids in different soils are found, and this can be attributed to the different reactions of amino acids with phenols and reducing sugars, and the different effect of soil inorganic components on the decomposition of the amino acids. It can be expected that amino acids derived from microorganisms will resist microbial decomposition due to their retention by inorganic components in

Table 1.13. Percent molar distribution of amino acids in soils from experimental plots at the Shenyang Agricultural University (brown) and Liaoning Academy of Agricultural Sciences (meadow).[1]

Amino acid	Brown soil			Meadow soil		
	Control	Organic manure	NPK	Control	Organic manure	NPK
Aspartic acid	8.58	9.62	8.89	9.01	8.73	10.20
Glutamic acid	10.69	9.93	10.84	11.40	10.96	11.42
Total acidic	19.27	19.55	19.73	20.41	19.69	21.62
Threonine	4.52	4.31	4.80	4.36	4.58	4.46
Serine	7.58	5.65	6.04	5.35	5.31	5.80
Glycine	10.16	10.85	10.51	11.10	11.42	10.72
Alanine	7.23	7.35	8.25	8.31	7.61	8.70
Valine	4.17	4.35	4.64	5.00	3.92	5.74
Isoleucine	7.64	6.35	6.04	6.98	6.50	7.13
Leucine	6.17	6.27	5.93	6.63	6.81	6.38
Tyrosine	4.17	4.81	4.91	5.29	5.11	3.13
Phenylalanine	2.82	3.85	2.91	2.79	3.34	2.32
Proline	5.46	4.73	4.64	4.94	5.73	4.70
Total neutral	59.92	58.42	58.67	60.75	60.33	59.08
Lysine	7.76	7.08	9.81	7.09	6.54	6.96
Histidine	0.88	1.50	0.97	0.70	1.27	1.22
Arginine	11.05	12.00	9.70	9.77	10.92	9.74
Total basic	19.69	20.85	20.46	17.56	18.73	17.92
Cystine	0.71	0.96	0.92	0.87	0.81	0.99
Methionine	0.18	0.38	0.22	0.41	0.46	0.41
Total S	0.89	1.34	1.14	1.28	1.27	1.40

[1] Zhang (1989).

soil, or they will be preserved through chemical or enzymatic reactions with polyphenols or reducing sugars forming non-amino acid compounds. Shi *et al.* (unpublished data) found that the amino acid composition of calcareous soils was different from that of noncalcareous soils; the former were characterized by higher contents of the acidic amino acids, aspartic and glutamic acids.

1.4.6. Distribution of amino acids in humic and fulvic acids

Humic and fulvic acids contain the same amino acids found in soil hydrolysates. Generally, aspartic acid, glutamic acid, alanine, leucine and valine are the dominant amino acids in humic and fulvic acids. However, in comparison with soil, the content of basic amino acids is considerably lower and that of acidic amino acids is markedly higher in humic and fulvic acids. The amino acid composition of fulvic acids varied greatly with the method of preparation. In general the content of neutral amino acids is lower, and that of acidic amino acids is slightly higher, in fulvic acid than in humic acid (Table 1.14).

Table 1.14. Percent molar distribution of amino acids in hydrolysates of humic and fulvic acid fractions of soil organic matter.

Soil	Acidic	Neutral	Basic	Sulfur-containing	Reference
Humic acids					
Black, Lateritic red (2)[1]	28.4–29.4	61.7	8.9–10.0	0	Peng *et al.* (1984)
Desert (7)	29.1–34.1	55.8–60.1	8–10.2	0.4–1.3	Li and Wang (1988)
Dark brown forest (1)	28.5	63.1	8.2	0.2	Zhuo and Wen (1994)
Yellow red, Red, Paddy (6)	25.6–32.1	59.9–65.5	7.5–9.1	0.2–0.9	Cheng *et al.* (1994)
Paddy (6)	25.3–27.3	59.8–62.3	10.9–13.4	0.6–1.6	Hou and Yuan (1986)
Newly-formed humus (6)	23.4–26.0	65.4–68.5	6.1–7.3	0.9–1.4	Cheng *et al.* (1994)
Fulvic acids					
Lateritic red (1)	54.7	41.0	4.3	0	Peng *et al.* (1984)
Paddy (6)	28.3–34.2	52.8–59.6	7.2–12.7	1.7–2.8	Hou and Yuan (1986)

[1] Values in parentheses are the numbers of soils analyzed.

Small differences in amino acid composition have been found between the humic acids separated from different soils. As shown in Table 1.14, the content of acidic amino acids was higher, and that of neutral amino acids was lower, in humic acid from a desert soil than in humic acid derived from other soils. The acidic amino acid content of the humic acid from calcareous soils (32%) was also higher, and the neutral amino acids content was lower, than that of the humic acid from noncalcareous soils (26%; Cheng *et al.* 1995). These results suggest that the presence of large amounts of calcium carbonate may facilitate the preservation of acidic amino acids in soil. The results presented in Table 1.14 also show that the levels of basic and acidic amino acids were lower, and the neutral amino acids, particularly proline, were higher in the newly formed humic acids than in the older humic acids, regardless of the conditions (submerged or upland, calcareous or non-calcareous) under which the humic acids were formed. The relative proportion of proline in newly formed humic acid was ~2–4 times that in the matured humic acids. The differences in amino acid composition between newly formed and matured humic acids may be due to the properties of the individual amino acids. Basic amino acids, for example, which bond directly to phenolic rings at considerably higher rates than neutral and acidic amino acids, will be selectively protected from decomposition. Thus, the relative proportion of amino acids as basic amino acids increased with increasing humification. However, the reason for the wide difference in the proportion of proline is unknown.

Results obtained by further hydrolysis of the acid nonhydrolyzable residue with 2.5 N NaOH or with 6 N HCl containing 3% H_2O_2 support the explanation given above. Compared with the acid hydrolysate of humic acid, the proportion of acidic amino acids in the hydrolysates of the residue was significantly lower, while that of basic amino acids, particularly lysine, was considerably higher (Figure 1.4). The greater proportion of N-phenylamino acids as N-phenyl-lysine seems to indicate that lysine had more opportunities to bind directly with phenolic rings than other amino acids or it combines more strongly with phenols by cross-linkage; these reactions make it more resistant to acid hydrolysis.

Few data are available concerning the amino acid composition of humic acids with different molecular weights. The results obtained by Zhuo and Wen (1994) indicate that the proportions of basic and acidic amino acids were higher, and that of neutral amino acids was lower, in the high-molecular-weight fractions than those of the low-molecular-weight fractions. For example, the proportions of acidic and neutral amino acids in the high-molecular-weight fraction were 31.7% and 60%, respectively, whereas the corresponding values for the low-molecular-weight fraction were 26.1% and 69.4%

1.4.7. Free amino acids

The term free amino acid refers to those amino acids in soil that do not exist in peptide linkages. Free amino acids can be absorbed directly by crops (Zhang and

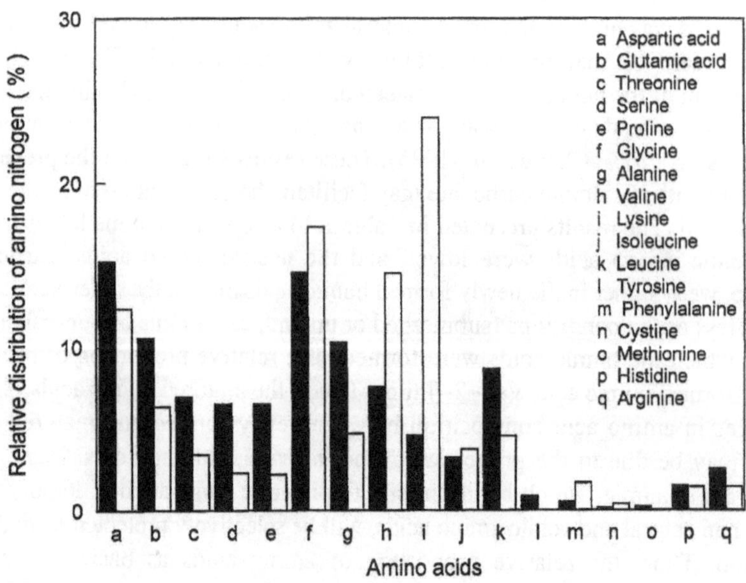

Figure 1.4. Distribution of amino acids in humic acid ■ and the nonhydrolyzable residue □ in red soil.

Sun 1984). Some free amino acids affect rock weathering, and biological availability of micronutrients through complex formation with metal ions, and thus may play an important role in soil fertility and soil formation processes.

Free amino acids may occur in soil solution (Anderson and Vaughan, 1985) or are adsorbed on clay minerals or the surface of humus colloids in such a manner that they are not easily extracted by the solvents commonly used for extraction. Solvents commonly used to extract free amino acids include distilled water, 20% ethyl alcohol, 80% ethyl alcohol, a mixture of water and CCl_4, 0.05 M $Ba(OH)_2$, 0.5 M NH_4OAc, 10% $(NH_4)_2CO_3$, 0.2 N HCl, and alkaline alcohol (pH 11). The extractants differ greatly in the efficiency of removal of free amino acids from soils. Barium hydroxide appears to be the most efficient extractant but part of the organic N may be hydrolyzed during extraction (Kato 1974; Paul and Schmidt 1960; Yamamuro 1980; Zhang 1989). The extraction efficiency of 10% $(NH_4)_2CO_3$, 0.2 N HCl and alkaline ethyl alcohol are close to that of NH_4OAc, which is better than water. However, the recovery by NH_4OAc of amino acids added to soils was only 31–83%. Even for the same extractant, the extraction efficiency differs for different amino acids, and thus the results obtained by extraction may not give the real distribution pattern for free amino acids in soil.

Free amino acids comprise only a small proportion of the total N in soil. The distribution pattern of free amino acids differs greatly from that of combined amino acids. Results obtained by Zhang (1989) using alkaline ethyl alcohol as an extractant show that the dominant amino acids are glutamic acid and alanine, accounting for over 40% of the total. The proportion of basic amino acids is much less than that found in soil hydrolysates (Table 1.15). The ratio of the individual free amino acid to the corresponding combined form varies widely, usually in the range of 0.08 to 1.33% (Tables 1.13, 1.15). Regression analysis indicates that the variation partly arises from differences in the extent of adsorption of the individual amino acid by soil components (Zhang 1989).

Organic manures differ in their content of readily decomposable carbohydrates as well as the content and distribution of free amino acids (Table 1.16). Therefore, the effect of organic manure addition on the content and distribution of free amino acids in soil depends on the kind of manure added. In general, the total amount of free amino acids in soil after addition of manure decreased during the first 2 weeks, then increased sharply, followed by another decrease (Zhang 1989).

Plant roots exert pronounced effects on the content and distribution of free amino acids in soil through the excretion of carbon compounds and amino acids. The content and distribution of free amino acids in soil varied with the stage of plant growth (Yamamuro 1980) and the amount of free amino acid was higher in the rhizosphere than in the remainder of the soil. In addition the composition of the free amino acids fraction in the rhizosphere soil differed from that in the bulk of the soil (Anderson and Vaughan 1985). Shi *et al.* (1988) found that mugineic acid excreted from the roots of barley plants could only be detected within 2 mm of the root surface.

Table 1.15. Percent molar distribution of free amino acids in soils.[1]

Amino acid	Brown soil			Meadow soil		
	Control	Organic manure	NPK	Control	Organic manure	NPK
Aspartic acid	4.43	4.50	4.87	5.15	5.06	5.33
Glutamic acid	26.17	30.20	25.90	27.70	28.79	27.60
Total acidic	30.60	34.70	30.77	32.85	33.86	32.93
Threonine	4.55	4.70	4.37	4.95	4.64	5.23
Serine	4.98	4.81	4.74	6.20	5.85	6.42
Glycine	8.87	8.29	8.80	8.09	7.71	7.78
Alanine	14.74	11.51	15.08	12.57	13.73	11.75
Valine	6.73	4.84	6.30	4.86	5.43	5.08
Isoleucine	1.95	2.09	2.40	2.38	2.20	2.41
Leucine	2.18	2.22	2.30	2.14	2.40	2.02
Tyrosine	2.27	1.90	2.08	1.52	1.55	1.75
Phenylalanine	7.54	6.44	7.73	7.76	6.67	7.61
Proline	1.65	2.75	1.69	2.72	2.03	1.95
Total neutral	55.46	49.56	55.11	53.18	52.21	51.99
Lysine	3.49	3.73	3.53	4.50	4.57	5.11
Histidine	1.71	1.94	2.20	2.07	2.04	2.19
Arginine	5.53	7.41	5.51	5.44	4.63	5.25
Total basic	10.72	13.08	11.24	12.01	11.24	12.55
Cystine	2.61	2.14	1.91	1.29	2.04	2.02
Methionine	0.61	0.52	0.97	0.67	0.65	0.51
Total S	3.22	2.66	2.88	1.96	2.69	2.53

[1] Zhang (1989).

1.5. Conclusions

Despite considerable research on the characterization of soil N information is still required in many important areas. For example, little is known on how ammonium and nitrate are transformed into organic N complexes of high resistance to microbial attack.

Before further progress can be achieved detailed identification and characterization of the unknown-N in soil is required. Although a series of N compounds have been isolated and identified, a considerable portion of the soil N has not been identified. Even though we can account for much of the soil N as non-α-amino-N in basic amino acids, N-phenyl amino acid-N and clay mineral-fixed ammonium, the nature of about 50% of the soil N still remains unknown. Furthermore, the extent of decomposition of amino acids and amides during extraction is unknown.

The majority of the unknown N in soil exists as acid nonhydrolyzable-N. Attempts have been made to isolate nitrogenous substances from the nonhydrolyzable residue with various degradation methods but they have not been successful. We know only that a small amount of the acid nonhydrolyzable-N occurs as

Table 1.16. Percent molar distribution of free amino acids in livestock excrement[1].

Amino acid	Pig manure	Cattle manure	Sheep manure	Chicken manure
Aspartic acid	4.86	6.34	4.84	6.45
Glutamic acid	11.46	24.56	11.55	24.27
Total acidic	16.32	30.90	16.39	30.72
Threonine	5.70	3.98	5.66	5.54
Serine	3.90	8.54	2.75	4.06
Glycine	10.74	8.27	10.41	9.54
Alanine	19.10	17.35	25.56	18.45
Valine	6.99	5.65	12.17	9.54
Isoleucine	6.88	6.67	4.67	5.60
Leucine	12.38	5.74	4.59	7.79
Tyrosine	7.80	1.57	0.76	1.21
Phenylalanine	3.85	1.98	1.21	2.35
Proline	3.22	5.92	13.18	3.70
Total neutral	80.56	66.02	80.97	67.79
Lysine	1.09	1.97	0.73	0.35
Histidine	0.75	1.41	0.60	0.26
Arginine	0.03	ND	0.18	ND
Total basic	1.87	3.38	1.51	0.61
Cystine	0.24	ND	ND	0.49
Methionine	1.01	0.04	1.13	0.39
Total S	1.25	0.04	1.13	0.88
Total (mg N kg^{-1})	2404	1815	736	3895

[1] α-Amino-N/total α-amino-N.
[2] Zhang and Sun (1983) and Sun *et al.* (1986).

N-phenyl amino acids. It seems that the introduction of new separation and analytical techniques is required to solve the problem.

Further work is also required on the mechanism of binding of N-containing compounds to other soil constituents and the factors controlling the resistance of these compounds to microbial attack.

1.6. References

Aldag, R W 1977. Relations between pseudo-amide N and humic acid N released under different hydrolytic conditions. Soil Organic Matter Studies. Proceedings Symp. Braunschweig. IAEA, Vienna. 1:293–299.

Anderson, H A and Vaughan, D 1985. Soil nitrogen: its extraction, distribution and dynamics. In: Vaughan D and Malcolm R E (eds.), Soil Organic Matter and Biological Activity. pp. 290–319. Martinus Nijhoff/Dr W Junk Publishers, The Hague.

Asami, T and Hara, M 1971. On the fractionation of soil organic nitrogen after hydrolysis using hydrochloric acid. Soil Sci. Plant Nutr. 17:222.

Benzing-Purdie, L, Ripmeester, J A and Preston, C M 1983. Elucidation of the nitrogen forms in melanoidins and humic acid by nitrogen-15 cross polarization-magic angle spinning nuclear magnetic resonance spectroscopy. J. Agric. Food Chem. 31:913–915.

Benzing-Purdie, L, Cheshire, M V, Williams, B L, Sparling, G P, Ratcliffe, C I and Ripmeester, J A 1986. Fate of glycine in peat as determined by ^{13}C and ^{15}N CP-MAS NMR spectroscopy. J. Agric. Food Chem. 34:170–176.

Biederbeck, V O and Paul, E A 1973. Fractionation of soil humate with phenolic solvents and purification of the nitrogen rich portion with polyvinylpyrrolidone. Soil Sci. 115:357–366.

Bremner, J M 1965. Organic forms of nitrogen. In: Black, C A (ed.), Methods of Soil Analysis. (Part 2). pp. 1148–1178. Amer. Soc. Agron., Madison ,Wisconsin.

Bremner, J M 1967. Nitrogenous compounds. In: McLaren, A D and Petersen, G H (eds.), Soil Biochemistry. Vol. 1. pp. 19–66. Marcel Dekker, Inc. New York.

Carles, J and Decau, J 1960. Variations in the amino acids of soil hydrolysates. Sci. Proc. R. Dublin Soc. Ser. A. 1:177–182.

Cheng, L L, Ye, W and Wen, Q X 1995. Amino acid composition of humic acids in paddy soils. (in Chinese). Soils 27:195–198.

Cortez, J and Schnitzer, M 1979. Nucleic acid bases in soil and their association with organic and inorganic soil compounds. Can. J. Soil Sci. 59:277–286.

EBSSHYA 1961. (Editorial Board of the Series for Studies on High Yield Agriculture, Academia Sinica). Soil Environments and High Yield of Rice. (in Chinese). pp. 240–288. Science Press. Beijing.

Goh, K M and Edmeades, D C 1979. Distribution and partial characterization of acid hydrolysable organic nitrogen in six New Zealand soils. Soil Biol. Biochem. 11:127–132.

Greenfield, L G 1972. The nature of the organic nitrogen of soils. Plant Soil. 36:191–198.

Griffith, S M, Sowden, F J and Schnitzer, M 1976. The alkaline hydrolysis of acid-resistant soil and humic residues. Soil Biol. Biochem. 8:529–531.

Hao, W Y and Cao, Z B 1988. Soil microorganisms. In: Hseung, Y and Li, C K (eds.), Soils of China. (in Chinese). pp. 537–558. Science Press. Beijing.

Hou, H Z and Yuan, K N 1986. Studies on organo-mineral complexes in soil III. Distribution of amino acids and forms of nitrogen in organo-mineral complexes. (in Chinese). Acta Pedol. Sin. 23:228–235.

Hu, G S 1989. Anion adsorption by variable charge soils in relation to the surface characteristics of the soil. Ph.D. Dissertation. Institute of Soil Science, Academia Sinica.

Hu, Y M, Ma, L S and Xu, C Y 1983. Studies on the relationship between soil nitrogen, nitrate, amine and liver cancer. (in Chinese). Chinese. J. Environ. Sci. (Beijing) 5:275–282.

Isirimah, N O and Keeney, D R 1973. Nitrogen transformations in aerobic and waterlogged Histosols. Soil Sci. 115:123–129.

ISWC 1961. (Edited by Institute of Soil and Water Conservation, Academia Sinica, and Soil Survey Staff of Institute of Beijing Survey and Design under the Ministry of Water Conservancy and Electric Power). Soils in North China Plain. (in Chinese). 291–317.

Jenkinson, D S and Tinsley, J 1960. A comparison of ligno-protein isolated from a mineral soil and from a straw compost. Sci. Proc. R. Dublin Soc. Ser. A. 1:141–147.

Ji, K L and Wang, C H 1978. Application of micro-electrodes for studying the diffusion of ammonium from granulated ammonium bicarbonate in paddy soils. (in Chinese). Acta Pedol. Sin. 15:182–186.

Kato, T 1974. On the water-soluble organic matter in paddy soils. Part 4. Behavior of free amino acids under the submerged condition. J. Sci. Soil Manure, Japan. 45:459–462.

Khan, S U 1971. Nitrogen fractions in a grey wooded soil as influenced by long-term cropping systems and fertilizers. Can. J. Soil Sci. 51:431–437.

Ladd, J N and Butler, J H A 1966. Comparison of some properties of soil humic acids and synthetic phenolic polymers incorporating amino derivatives. Aust. J. Soil Res. 4:41–54.

Lei, W J, Xiao, Z H, Cai, F Q, Zhu, H G and Wang, H L 1974. The geographical distribution of groundwater rich in nutrients in Xuzhou region, Jiangsu province. (in Chinese). Soils. 2:56–63.

Li, H E and Yang, Y L 1965. Effect of fallow ploughing in summer on the yield of succeeding crop. (in Chinese). Acta Pedol. Sin. 13:404–409.

Li, S G and Wang, Z Q 1988. Alkali Soils of Desert Area. (in Chinese). Xinjiang People's Publishing House, Urumqi, China. 92–107.

Li, S Y, Wang, J Y and Kong, V G 1982. Studies on the characteristics of nitrogen supply of paddy soils. II. Effect of fertilization on the soil nitrogen supply and grain yield of double cropping-rice. (in Chinese). Acta Pedol. Sin. 19:13–21.

Li, L M, Wu, Q T, Li, Z G and Pan, Y H 1991. Fluxes of nitrous oxide from different soils. (in Chinese). Soils 23:24–27.

Lowe, L E 1973. Amino acid distribution in forest humus layers in British Columbia. Soil Sci. Soc. Amer. Proc. 37:569–572.

NIBSSC 1973. (Northwest Institute of Biology, Soils and Soil Conservation, Academia Sinica). Groundwater Rich in Nutrients. (in Chinese). Science Press. Beijing.

Paul, E A and Schmidt, E L 1960. Extraction of free amino acids from soil. Soil Sci. Soc. Amer. Proc. 24:195–198.

Peng, L, Peng, X L and Lu, Z F 1981. The seasonal variation of soil NO_3-N and the effect of summer fallow on the fertility of manured loessial soil. (in Chinese). Acta Pedol. Sin. 18:212–222.

Peng, F Q, Gao, Q L and Che, Y P 1984. Distribution of N-containing compounds and carbohydrates in humic substances of divergent origin. (in Chinese). Symp. on Chemistry of Humic Substances. Lusan, China. Chemistry Soc. China. 316–321.

Qin, S W and Liu, Z Y 1989. The nutrient status of rhizosphere soil VI. Distribution of different forms of fertilizer-N in rhizosphere soil. (in Chinese). Acta Pedol. Sin. 26:117–123.

Schnitzer, M 1978. Humic substances: Chemistry and reactions. In: Schnitzer, M and Khan, S U (eds.), Soil Organic Matter. pp. 1–64. Elsevier Scientific Publishing Company, Amsterdam, Oxford, New York.

Schnitzer, M and Spiteller, M 1986. The chemistry of the 'unknown' soil nitrogen. Trans. Int. Conf. Soil Sci. 13th Congr. Hamburg. Vol. 2:473–474.

Sequi, P, Guidi, G and Peteruzzelli, G 1975. Distribution of amino acid and carbohydrate components in fulvic acid fractionated on polyamide. Can. J. Soil Sci. 55:439–445.

Shao, Z Y, Yang, H S, Zhu, Z Y and Liu, L G 1988. Studies on the dynamics of inorganic nitrogen in root zone of chao soil during growth period of winter wheat. (in Chinese). Soils 20:97–98.

Shen, Q Y and Shi, R H 1990. Studies on chemical constituents of organic nitrogen and their availabilities in different soils. (in Chinese). Chinese J. Soil Sci. 21:54–57.

Shi, S L, Lin, X X and Wen, Q X 1981. Decomposition of plant materials in relation to their chemical composition in paddy soil. Inst. Soil Sci., Academia Sinica (ed.), Proc. of Symp. on Paddy Soil. Science Press. Springer-Verlag, Heidelberg. pp. 306–310.

Shi, W M, Chino, M, Youssef, R A, Mori, S and Takagi, S 1988. The occurrence of mugineic acid in the rhizosphere soil of barley plant. Soil Sci. Plant Nutr. 34:584–592.

Shi, S L, Wen, Q X, Liao, H Q and Zhou, K Y 1992. Influence of cultivation on distribution of nitrogen forms and amino acid composition in soils. (in Chinese). Soils 24:14–18.

Song, Q 1988. Studies on the distribution and characteristics of soil organic nitrogen in some soils of China. (in Chinese). Acta Pedol. Sin. 25:95–100.

Sowden, F J 1970. Action of proteolytic enzymes on soil organic matter. Can. J. Soil Sci. 50:233–241.

Sowden, F J 1977. Distribution of nitrogen in representative Canadian soils. Can. J. Soil Sci. 57:445–457.

Sowden, F J, Griffith, S M and Schnitzer, M 1976. The distribution of nitrogen in some highly organic tropical soils. Soil Biol. Biochem. 8:55–60.

Sowden, F J, Chen, Y and Schnitzer, M 1977. The nitrogen distribution in soils formed under widely differing climate conditions. Geochim. Cosmochim. Acta. 41:1524–1526.

Sowden, F J, Morita, H and Levesque, M 1978. Organic nitrogen distribution in selected peats and peat fractions. Can. J. Soil Sci. 58:237–249.

Stevenson, F J 1956. Effect of some long-time rotations on the amino acid composition of the soil. Soil Sci. Soc. Amer. Proc. 20:204–208.

Stevenson, F J 1982. Organic forms of soil nitrogen. In: Stevenson, F J (ed.), Nitrogen in Agricultural Soils. pp. 67–122. Amer. Soc. Agron., Madison, Wisconsin.

Su, W H, Song, W Z, Zhang, H, Cao, M Q, Lu, H R, Zhou, Q, Zeng, J H and Zhang, Y M 1992. Flux of nitrous oxide on typical winterwheat field in Northern China. Environ. Chem. 11:26–32.

Sun, X, Zang, Y S, Ying, Q Z and Tang, C X 1986. Effects of organic manure on soil fertility and crop production. Soil Sci. Soc. China (ed.). Current Progress in Soil Research in People's Republic of China. pp. 197–206. Jiangsu Science and Technology Publishing House.

Wang, T S C, Yang, T K and Cheng, S Y 1967. Amino acids in subtropical soil hydrolysates. Soil Sci. 103:67–74.

Wang, P G, Ji, G L and Yu, T R 1987. Adsorption of chloride and nitrate by variable charged soils in relation to the electric charge of the soil. Z. Pflanzenernaehr. Dueng. Bodenkd. 150:17–23.

Wen, Q X 1979. Organic matter status in soils of Jiangsu province and its relation to soil fertility. Bureau of Agriculture and Forestry, pp. 89–94. Jiangsu Province (ed.). Soil Survey and Agricultural Development. (in Chinese).

Wen, Q X and Lin, X X 1983. Content and characteristics of organic matter in red soil region of China. In: Li, C K (ed.), Red Soils in China. pp. 119–127. Science Press, Beijing.

Wu, G Y 1986. The effect of organic manure on the nitrogen fertility of meadow black soil. (in Chinese). Chin. J. Soil Sci. 17:85–88.

Xu, Q, Lu, Y C, Liu, Y C and Zhu, H G 1980. The Paddy Soil of Taihu Region in China. (in Chinese). Shanghai Scientific and Technical Publishers, Shanghai, China.

Yamamuro, S 1980. Proposal of a method for analysis of free amino acids in paddy soils and investigation on behavior of them in rice growth period. J. Sci. Soil Manure. Japan. 51:131–135.

Ye, W, Cheng, L L and Wen, Q X 1991. Content and partial characterization of unknown N in humic acids. Pedosphere. 1:283–287.

Yuan, L Z, Sui, G Y and Huang, D M 1985. Residual nitrate nitrogen as the basis for predicting fertilizer nitrogen requirements of corn. (in Chinese). Jiangsu Agric. Sci. 6:25–27.

Zhang, X D 1989. Amino acids in soils: distribution, dynamics and their relation to soil fertility. (in Chinese). Ph.D. Dissertation. Shenyang Agricultural University.

Zhang, F D and Sun, X 1983. Determination of free amino acids in organic manures. (in Chinese). Chin. J. Soil Sci. 1:42–43.

Zhang, F D and Sun, X 1984. Studies on the utilization of amino acids by rice seedlings. (in Chinese). Sci. Agric. Sin. 5:61–66.

Zhang, X H, Du, L J and Wen, Q X 1984. Organic matter distribution in particle-size fractions in plow layer of some paddy soils. (in Chinese). Acta Pedol. Sin. 21:418–425.

Zhang, S L, Zhu, Z L and Xu, Y H 1989a. Fate of fertilizer nitrogen and transformation of urea in Chao soil under wheat system. (in Chinese). Acta Agric. Nucl. Sin. 3:9–15.

Zhang, X D, Xu, X C and Chen, E F 1989b. Change in amino acid content of soils after application of pig manure. (in Chinese). Chinese J. Soil Sci.. 20:260–262.

Zhu, Z L 1979. Nitrogen supplying capability of soils in Jiangsu province. Bureau of Agriculture and Forestry, Jiangsu Province (ed.). pp. 95–98. Soil Survey and Agricultural Development. (in Chinese).

Zhu, Z L 1988. Soil nitrogen. In: Hseung, Y and Li, C K (eds.) Soils of China. (in Chinese). pp. 464–482. Science Press, Beijing.

Zhuo, S N and Wen, Q X 1994. Fractionation of humic acid by fractional precipitation technique. (in Chinese). Acta Pedol. Sin. 31:251–258.

Zhu, Z L, Liao, H L, Tsai, K S and Yu, C C 1978. Soil nutrition status under 'rice-rice-wheat' rotation and the response of rice to fertilizers in Suzhow district. (in Chinese). Acta Pedol. Sin. 15:126–137.

2
Natural ^{15}N abundance in soils

XING GUANG-XI, CAO YA-CHENG AND SUN GUO-QING

2.1. Introduction

According to the classical definition, the isotopic composition or relative abundance of an element is a fixed value, and the behaviour of the light and heavy isotopes in chemical reactions is strictly identical. These are two basic assumptions for conducting tracer experiments with stable isotopes. However, it is now known that the variation in natural abundance of the isotopes of elements is a biogeochemical phenomenon existing universally in nature (Hoering 1955; Hauck 1973). For instance, variations are found in the natural abundance of the stable isotopes of C, H, O, N and S in organisms. Moreover, the behaviour of the light and heavy isotopes involved in various reactions is not strictly identical. The variation in the natural isotopic abundance is caused by isotopic fractionation, also called the mass discriminatory effect, which is brought about by a series of biological, chemical and physical processes in nature and, especially, in living organisms. The so-called mass discriminatory effect refers to the preference of the light isotope of an element ^{14}N, for example over the heavy one, ^{15}N, to take part in a reaction. Consequently, in a reaction the newly-formed reaction product is relatively enriched in ^{14}N, while the remaining reaction substrate becomes enriched in ^{15}N. The deviation in natural ^{15}N abundance of a substance from that of the standard (atmospheric N_2) is referred to as $\delta^{15}N$ which is calculated by the equation:

$$\delta\,^{15}N = \frac{R(sample) - R(standard)}{R(standard)} 1000\%$$

where R is the ratio of the ion current intensity of m/e 29 [$^{15}N^{14}N$] to that of m/e 28 [$^{14}N^{14}N$]. R_{sample} may be greater or less than that of the standard. When R_{sample} is greater than that of the standard, the $\delta^{15}N$ measured is positive, showing that the ^{15}N content is greater than that of the standard; when R_{sample} is less than that of the standard, $\delta^{15}N$ is negative, indicating that the ^{15}N content of the sample is less than that of the standard. Cheng *et al.* (1964) were the first to note that variation existed in the natural ^{15}N abundance of various forms of soil N. This attracted the attention of many scientists. They were eager to learn how large the variation was, whether this would cause difficulties in interpreting the results of experiments

Zhu Zhao-liang et al. *(eds.): Nitrogen in Soils of China, 31–41.*

with ^{15}N enriched or depleted materials, and whether the variation in natural abundance of ^{15}N in soil could be used to evaluate N cycle processes in the soil-plant system. It has now been determined that in the soil-plant system, unlike other natural systems, the variation in the natural abundance of ^{15}N is too small to influence the results of tracer studies with ^{15}N enriched or depleted material. As to the second question, some scientists have been trying to use the variation in natural ^{15}N abundance for more than 20 years to determine the extent of N cycle processes such as the N-suppling capacity of soils (Shearer and Legg 1975), the amount of symbiotic N_2 fixation (Amarger *et al.* 1979), the fate of fertilizer-N in soil (Karamanos and Rennie 1981) and the transformation of soil organic matter (Tiessen *et al.* 1984). However, results so far are not satisfactory, and further investigations are required. In China, little has been done on the variation in natural ^{15}N abundance in the soil-plant system. In this chapter, a brief account of the recently obtained results on the variation in the natural abundance of ^{15}N in different soils of China will be given.

2.2. Variation in natural ^{15}N abundance in surface soils

2.2.1. *Forest soils*

Many studies have been made on the variation in the natural abundance of ^{15}N in forest soils. The $\delta^{15}N$ value of the total N was usually low, and mostly negative (Mariotti *et al.* 1980; Natelhoffer and Fry 1988; Rennie *et al.* 1976; Riga *et al.* 1971; Wada *et al.* 1984). This was confirmed by the $\delta^{15}N$ data for the total N in the surface layers of 6 different forest soils in China (Table 2.1). In spite of the differences in soil type and plant species, the $\delta^{15}N$ value of the surface soils was low, ranging between −1.03 and +2.27 (average +0.47). The podzolic soil, the dark brown soil from Da Hinggan Mountains, and the brown soil from Tai'an in Shandong Province had negative $\delta^{15}N$ values in the surface soil N.

The main reason given for the low $\delta^{15}N$ values is that the surface soil contains considerable partially decomposed plant residue, and the plants from which these residues are derived are often depleted in ^{15}N relative to the atmospheric N_2; i.e. the low $\delta^{15}N$ values for the total N in the surface soil is due to the continuous addition

Table 2.1. Isotopic composition of the total nitrogen ($\delta^{15}N$) in the surface layers of forest soils.[1]

Soil	Location	Depth (cm)	Total N (g kg^{-1})	OM (g kg^{-1})	$\delta^{15}N$
Podzolic	Mangui, Da Hinggan Mtns	0–4	14.3	563	−1.03
Brown coniferous forest	Mangui, Da Hinggan Mtns	0–4	10.2	633	+0.15
Dark brown	Zalantun, Inner Mongolia	0–1	15.4	441	−0.10
Gray forest	Yakeshi, Inner Mongolia	0–5	15.9	423	+2.07
Brown	Tai'an, Shandong	0–1	2.47	58	−0.55
Acid purple	Leshan, Sichuan	0–7	1.68	38	+2.27

[1] Sun *et al.* (1991), and unpublished data of Xing G X *et al.*

of plant residues with extremely low $\delta^{15}N$ values. The low $\delta^{15}N$ value of forest litter may be attributed to the fact that N in the plants grown in mountainous areas is derived primarily from biologically fixed N, and N from precipitation (Wada *et al.* 1984). It is known that the $\delta^{15}N$ values of both the biologically fixed N (Delwiche and Steyn 1970) and ammonium or nitrate in precipitation (Moore 1977) are low. Investigations on several representative forest tree species in China confirmed that the leaves of most had negative $\delta^{15}N$ values, except that of *Betula platyphylla* grown on a gray forest soil (Table 2.2).

2.2.2. *Steppe soils*

In China, black, chernozem and chestnut soils are the major uncultivated steppe soils. The $\delta^{15}N$ values for the total N in their surface layers were +6.69, +6.08 and +6.23, respectively (mean +6.33; Table 2.3), which is 5.86 units higher than the average value obtained for forest soils (+0.47) and is close to the value for soils in the Canadian chernozem region (+8.8; Rennie *et al.* 1976). The surface layers of the meadow bog soils on the Zalantun steppes in Inner Mongolia, had a $\delta^{15}N$ value of +4.63, which is lower than the $\delta^{15}N$ value for the neighbouring black soils, but much higher than that of forest soils. It was found that plants representative of the chernozem, chestnut and meadow bog soils all had positive $\delta^{15}N$ values. Although the $\delta^{15}N$ value of *Ancurolepidium Chinensis* grown on a chestnut soil was somewhat lower than that of *Betula platyphylla* grown on a gray forest soil, it was considerably higher than the $\delta^{15}N$ values of other plants grown on forest soils (Table 2.3). The

Table 2.2. Isotopic composition of the total nitrogen ($\delta^{15}N$) in leaves of different plants.[1]

Soil	Plant	Location	$\delta^{15}N$
Podzolic	*Pinus sylvestris*	Mangui, Da Hinggan Mtns	−3.86
Brown forest	Hinggan deciduous pine	Mangui, Da Hinggan Mtns	−2.41
Dark brown	*Quercus mongolica*	Zalantun, Inner Mongolia	−0.34
Gray forest	*Betula platyphylla*	Yakeshi, Inner Mongolia	+2.94

[1] Sun *et al.* (1991).

Table 2.3. Isotopic composition of the total nitrogen ($\delta^{15}N$) in steppe soils and plants grown on them.[1]

Location	Soil				Plant	
	Name	Depth (cm)	Total N (g kg^{-1})	$\delta^{15}N$	Name	$\delta^{15}N$
Zalantun, Inner Monglia	Black	0–38	2.83	+6.69	*Aneurolepidium chinenseis*	+4.45
Yakeshi, Inner Mongolia	Chernozem	0–12	2.78	+6.08	*Stipa subsessiliflora*	+5.62
Zalantun, Inner Mongolia	Meadow bog	0–16	3.04	+4.63	*Cyperus*	+6.79
Xilinhot, Inner Mongolia	Chestnut	0–20	1.60	+6.23	*Ancurolepidium chinenseis*	+1.28

[1] Sun *et al.* (1991) and unpublished data of Xing G X *et al.*

relatively high $\delta^{15}N$ values of the plants grown on steppe soils is undoubtedly one reason for the fairly high $\delta^{15}N$ values in the surface layers. Further, although the three steppe soils and the forest soils are located in the cool temperate zone the pH of the steppe soils was higher than that of the forest soils, and the ammonia formed from decomposition may be lost through volatilization. The volatilization of ammonia would also result in ^{15}N enrichment of the N in the surface layers of the steppe soils.

2.2.3 Desert soils

The $\delta^{15}N$ values of the total N in the surface layers of desert soils was higher than that of the forest soils and the steppe soils (Table 2.4). For the three desert soils studied it averaged +10.85, which is ~4.5 units and ~10 units higher than the values for the steppe and forest soils, respectively. The reasons for this may be related to the soil-forming conditions and soil properties. In the Xinjiang Uygur Autonomous Region, especially in the Turpan area with its extremely arid climate (annual precipitation of <100 mm) and very low vegetation cover, the surface soil layers (except for the residual solonchak) had very low organic matter contents and pH values above 8. Under these conditions only a small amount of N, low in ^{15}N, including that fixed biologically and added in precipitation is added to the soil each year. Thus the growing plants get their N mainly from soil, parent material and the ^{15}N-enriched N released from plants by decomposition. After plant residues (including roots) are returned to the soil, the total N remaining in the soil is enriched with ^{15}N by decomposition and ammonia volatilization. Repeated cycling has enriched the soil and plants with ^{15}N.

2.2.4. Agricultural soils

The $\delta^{15}N$ values of the total N in the surface layers of cultivated soils in northern China (e.g. fluvo-aquic soils, stratified old manured loessial soils, dark loessial soils and yellow cultivated loessial soils), as well as those in southern China (e.g. latosols, lateritic red soils, acid purple soils, and alkaline purple soils) ranged from +3.74 to +7.05 (mean +5.51) which is higher than the average value for forest soils, close to that for the uncultivated steppe soils (+6.33) and lower than that for the

Table 2.4. Isotopic composition of the total nitrogen ($\delta^{15}N$) in desert soils and plants grown on them.[1]

Location	Soil					Plant	
	Name	Depth (cm)	Total N ($g\ kg^{-1}$)	OM ($g\ kg^{-1}$)	$\delta^{15}N$	Name	$\delta^{15}N$
Fukang, Xinjiang	Gray desert	0–12	0.38	5.8	+ 9.37	*Keanmuria*	+9.95
Turpan, Xinjiang	Brown desert	0–10	0.25	3.2	+10.22	–	–
Turpan, Xinjiang	Residual solonchak	0–1	0.64	11.6	+12.97	*Alhagi camelorum*	+4.72

[1] Sun *et al.* (1991) and unpublished data of Xing G X *et al.*

desert soils (Table 2.5). According to Tiessen *et al.* (1984), cultivation had no marked influence on the $\delta^{15}N$ value for total N in the surface layer of steppe soils. It is not known whether these variations originated from the preceding soil or resulted from other factors.

The $\delta^{15}N$ values for the total N in the plow layers of paddy soils varied considerably, depending on the pH value of the soil. For acid paddy soils, the $\delta^{15}N$ values were rather low, commonly ranging from +1.23 to +2.21 (mean +1.53). The calcareous paddy soils, or the acid paddy soils limed continuously for a long time, have high $\delta^{15}N$ values (range +4.19 to +8.22, mean +5.5; Table 2.6). Further studies are required to determine the reasons for the difference.

Land use patterns have noticeable effects on the $\delta^{15}N$ values for the total N in the surface layers of soils. Investigations were made on cultivated soils developed from the same parent material, on neighbouring fields, but on different terrain and subjected to different land management. It was found that the $\delta^{15}N$ values of paddy soils were considerably lower than those in the corresponding upland soils (Table 2.7). Under submerged conditions the processes and rate of N transformations differ from those under upland conditions; e.g. nitrification is greatly inhibited and N_2-fixation is enhanced. Presumably, these processes are responsible for the lower $\delta^{15}N$ values of the paddy soils.

2.3. Variation in $\delta^{15}N$ within the soil profile

Different soils vary not only in the $\delta^{15}N$ values of the total N in the surface layers, but also in the distribution of $\delta^{15}N$ in the soil profile and its correlation with total N.

2.3.1. *Forest soils*

The $\delta^{15}N$ values of the total N in forest soils increased with depth as the total N content of the soil decreased. For example, in the brown coniferous forest soil and

Table 2.5. Isotopic composition of the total nitrogen ($\delta^{15}N$) in upland agricultural soils.[1]

Soil	Location	Depth (cm)	Total N (g kg^{-1})	OM (g kg^{-1})	$\delta^{15}N$
Fluvo-aquic	Fengqiu, Henan	0–19	0.79	11.3	+7.05
Stratified old manured loessial	Wugong, Shaanxi	0–18	0.93	13.1	+3.74
Dark loessial	Xifeng, Gansu	0–20	1.01	15.7	+5.49
Yellow cultivated loessial	Yan'an, Shaanxi	0–17	0.49	7.5	+4.73
Latosol	Xuwen, Guangdong	0–20	2.27	42.2	+5.49
Red	Liujia, Jiangxi	0–15	0.82	5.8	+6.36
Lateritic red	Guangzhou, Guangdong	0–14	1.02	20.2	+6.60
Acid purple	Leshan, Sichuan	0–15	0.94	15.4	+5.84
Alkaline purple	Jianyang, Sichuan	0–20	0.75	10.2	+4.29

[1] Unpublished data of Xing G X *et al.*

Table 2.6. Isotopic composition of the total nitrogen ($\delta^{15}N$) in paddy soils.[1]

Soil	Parent material	Location	Depth (cm)	Total N (g kg^{-1})	OM (g kg^{-1})	pH	$CaCO_4$ (g kg^{-1})	$\delta^{15}N$
Acid	Quaternary Red Clay	Liujia, Jiangxi	0–15	2.16	37.6	5.39	0	+1.26
Acid	Granite	Guangzhou, Gdong	0–20	1.76	38.1	5.10	0	+2.21
Acid	Acid purple sandstone	Leshan, Sichuan	0–15	1.65	32.7	6.10	0	+1.42
Acid sulphate	Neritic deposit	Xuwen, Guangdong	0–14	0.99	25.8	4.10	0	+1.23
Calcareous	Alkaline purple sandstone	Jianyang, Sichuan	0–16	0.87	11.2	7.97	91.8	+4.83
Acid (limed)	Quaternary Red Clay	Taoyuan, Hunan	0–15	1.21	25.8	6.65	0	+4.83
Acid (limed)	Basalt	Xuwen, Guangdong	0–15	2.04	33.9	7.51	3.7	+8.72
Calcareous	Recent alluvium	Changshu, Jiangxi	0–16	1.92	30.5	7.87	3.7	+4.19
Calcareous	Recent alluvium	Nantong, Jiangsu	0–17	2.48	27.6	7.92	11.8	+5.66

[1] Unpublished data of Xing G X *et al.*

Table 2.7. Effect of land use patterns on the isotopic composition of the total N in soil.[1]

Location	Parent material	Land use	Depth (cm)	Total N (g kg^{-1})	OM (g kg^{-1})	pH	$\delta^{15}N$
Yujiang, Jiangxi	Quaternary Red Clay	Upland	0–15	0.82	5.8	5.08	+6.36
		Paddy field	0–15	2.16	37.6	5.39	+1.26
Luogang, Guangdong	Granite	Upland	0–14	1.02	20.2	4.92	+6.06
		Paddy field	0–20	1.76	38.1	5.10	+2.21
Leshan, Sichuan	Acid purple sandstone	Upland	0–15	0.94	15.4	5.30	+5.84
		Paddy field	0–15	1.65	32.7	6.10	+1.41

[1] Unpublished data of Xing G X *et al.*

dark brown soil the $\delta^{15}N$ value gradually increased from +0.15 and –0.10, respectively, in the surface horizons to +7.55 and +6.50 in the deepest horizons, as the total N content decreased, respectively, from 10.2 and 15.4 g kg^{-1} in the surface horizon to 0.23 and 0.54 g kg^{-1} (Table 2.8). This is consistent with results obtained by other scientists (Mariotti *et al.* 1980; Natelhoffer and Fry 1988; Rennie *et al.* 1976; Riga *et al.* 1971; Wada *et al.* 1984).

In incubation and in situ long-term experiments Natelhoffer and Fry (1988) investigated the variation in natural abundance of ^{15}N in the profile of forest soils. They concluded that the comparatively low $\delta^{15}N$ value of the surface layer was due to the continuous incorporation of the ^{15}N depleted forest plant litter. On the other hand, as all decomposition processes discriminate against ^{15}N, the highly-decomposed organic matter eluviated from the surface layers was comparatively enriched with ^{15}N, thus resulting in higher ^{15}N enrichments in the subsurface layers. Ledgard *et al.* (1984) reported that the $\delta^{15}N$ value increased with depth down the profile as the total N decreased, in natural and artificial pasture soils in the Crookwell area of N.S.W.,

Table 2.8. Concentration, distribution and isotopic composition of the total N in profiles of forest soils.[1]

Soil	Location	Depth (cm)	Total N (g kg^{-1})	OM (g kg^{-1})	$\delta^{15}N$
Brown coniferous forest	Mangui, Da Hinggan Mountains	0–4	10.2	633	+0.15
		4–10	2.48	122	+5.32
		10–20	1.03	50.8	+6.24
		20–34	0.23	8.6	+7.63
		34–50	0.23	9.4	+7.55
Dark brown	Zalantun, Inner Mongolia	0–1	15.4	441	−0.10
		1–9	4.61	108	+3.55
		9–26	1.40	25.5	+7.65
		26–54	0.59	9.4	+7.02
		54–89	0.54	6.8	+6.50
Red	Liujia, Jiangxi	0–4	1.42	45.7	+2.02
		4–21	0.44	7.0	+6.23
		21–50	0.32	4.4	+8.29
		50–125	0.37	3.5	+8.19
		125–170	0.36	2.6	+7.38
		170–190	0.27	1.5	+7.86
		190–245	0.32	1.6	+4.00
		245–350	0.21	2.0	+4.02
		350–420	0.33	1.2	+4.27
		420–500	0.25	1.3	+4.58
Yellow-brown	Xiashu, Jiangsu	0–4	2.03	32.3	+1.40
		4–10	0.87	10.2	+4.76
		10–35	0.61	6.2	+4.92
		35–60	0.40	2.9	+7.24
		60–130	0.33	2.4	+7.33
		130–300	0.31	1.7	+7.00
		300–400	0.28	1.4	+6.23
		400–700	0.26	1.1	+6.93

[1] Sun *et al.* (1991).

Australia. This was attributed to the downward movement of the ^{15}N-rich fine particles (organo-mineral complexes).

2.3.2. *Steppe soils*

In steppe soils the $\delta^{15}N$ value of the total N varied with soil depth, but not as much as that in forest soil. For example, in a chernozem from the Hulumbeir area and a black soil in Changchun, although the total N content decreased with depth down the profile, the $\delta^{15}N$ value increased only slightly with depth. In addition, there were differences of only +1.75 and +0.66, respectively, between the $\delta^{15}N$ values for the surface soils and deepest horizons (Table 2.9). Rennie *et al.* (1976) also noted that, in the profile of a chernozem from Saskatchewan, Canada, the $\delta^{15}N$ value tended to increase slightly with depth.

Table 2.9. Concentration, distribution and isotopic composition of the total N in profiles of steppe soils.[1]

Soil	Location	Depth (cm)	Total N ($g\ kg^{-1}$)	$\delta^{15}N$
Chernozem	Yakeshi, Inner Mongolia	0–12	2.78	+5.98
		12–25	2.72	+5.80
		25–65	2.04	+6.97
		65–100	0.90	+7.71
		100–130	0.36	+7.73
Black	Changchun, Jilin	0–15	1.61	+6.27
		15–40	1.03	+7.15
		40–72	1.03	+7.15
		72–92	0.57	+6.96
		92–120	0.54	+7.22
		120–140	0.48	+6.93

[1] Unpublished data of Xing G X *et al.*

2.3.3. *Desert soils*

The distribution patterns for the $\delta^{15}N$ values in desert soil profiles were quite different from those of the forest, acid paddy and other soils. Like the total N content, the $\delta^{15}N$ value decreased considerably with depth (Table 2.10). This distribution may be due to the arid climate (annual precipitation <100 mm) and extremely sparse vegetation which results in little ^{15}N-rich humic substances moving downwards.

2.3.4. *Agricultural soils*

Only the acid paddy soils derived from the Quaternary Red Clays and granite were investigated. The distribution pattern in the acid paddy soil profiles was similar to

Table 2.10. Concentration, distribution and isotopic composition of the total N in the profiles of desert soils.[1]

Soil	Location	Depth (cm)	Total N ($g\ kg^{-1}$)	$\delta^{15}N$
Gray desert	Miquan, Xingjiang	0–4	0.80	+11.40
		4–15	0.40	+10.05
		15–31	0.35	+ 8.36
		31–46	0.31	+ 6.46
		46–72	0.28	+ 3.50
		72–99	0.23	+ 2.29
Residual solonchak	Dry delta area, Aydingkol lake, Xingjiang	0–1	0.64	+12.97
		1–13	1.00	+15.64
		13–27	0.72	+15.31
		27–38	0.47	+12.80
		38–60	0.35	+10.81
		60–75	0.31	+ 9.86
		75–125	0.30	+ 8.72

[1] Unpublished data of Xing G X *et al.*

that in the forest soils. The $\delta^{15}N$ values in surface soil and subsurface soils were low, but they increased greatly below the hydromorphic horizon (Table 2.11).

The reasons for the low $\delta^{15}N$ values in the surface layers of acid paddy soils is uncertain. In acid paddy soils it is easier for humic substances to move downwards, so the increase in $\delta^{15}N$ with depth may be due to the downward movement of these substances.

2.3.5. *$\delta^{15}N$ in fixed ammonium*

As shown in Table 2.12 the $\delta^{15}N$ value of fixed ammonium in 10 surface soils was higher than that of the total N. This is because the surface soil usually contains semi-decayed organic materials with low $\delta^{15}N$ values. Similar results were also found for the deeper horizons (Table 2.13). The results for two soils (Table 2.13) show that the $\delta^{15}N$ values for fixed ammonium in different horizons were usually greater than those for total N in the corresponding horizons.

Table 2.11. Concentration, distribution and isotopic composition of the total N in acid paddy soil profiles.[1]

Location	Depth (cm)	pH	Total N (g kg^{-1})	$\delta^{15}N$
Yujiang, Jiangxi	0–15	5.53	2.16	+1.26
	15–22	5.30	1.45	+1.49
	22–45	5.66	0.74	+6.41
	45–72	5.56	0.37	+6.40
Guangzhou, Guangdong	0–20	5.71	1.76	+2.12
	20–30	6.21	0.78	+6.21
	30–60	6.36	0.59	+6.36
	60–100	6.29	0.38	+6.29

[1] Unpublished data of Xing G X *et al.*

Table 2.12. Isotopic composition of fixed ammonium nitrogen in surface soils.[1]

Soil	Location	Parent material	$\delta^{15}N$	
			Fixed ammonium	Total N
Podzolic	Mangui, Da Hinggan Mtns	Granite	+5.64	+0.34
Dark brown	Zalantun, Inner Mongolia	Granite	+5.05	+3.55
Gray forest	Yakeshi, Inner Mongolia	Basalt	+6.46	+5.90
Brown	Tai'an Shangdong	Gneise	+3.43	+2.39
Meadow bog	Zalantun, Inner Mongolia	Granite	+7.43	+4.53
Chernozem	Yakeshi, Inner Mongolia	Acid purplish sandstone	+7.32	+5.98
Chestnut	Xilinhot, Inner Mongolia	Quaternary red clay	+8.55	+6.23
Paddy	Liujia, Jiangxi	Acid purplish sandstone	+6.11	+1.26
Paddy	Leshan, Sichuan	Alluvium	+3.60	+0.70
Paddy	Changshu, Jiangsu	Alluvium	+5.21	+4.19

[1] Shi S L *et al.* (1992).

Table 2.13. Isotopic composition of nitrogen in fixed ammonium of two soils.[1]

Soil	Location	Parent material	Depth (cm)	$\delta^{15}N$	
				Fixed ammonium	Total N
Paddy	Liujia, Jiangxi	Quaternary Red Clay	0–15	6.11	+0.26
			15–22	7.10	+1.49
			22–45	6.62	+5.41
			45–72	6.29	+6.40
Brown coniferous forest	Mangui, Da Hinggan Mountains	Granite	4–10	5.74	+5.32
			10–20	10.06	+6.24
			20–34	11.98	+7.63
			34–50	8.81	+7.55

[1] Shi S L *et al.* (1992).

The increase in $\delta^{15}N$ of the total N with depth may be partly attributed to the increase in the proportion of fixed ammonium to total N with depth (Shi *et al.* 1987; Wen *et al.* 1988).

2.4. Conclusions

As shown by the data now available, the natural ^{15}N abundance of soil N in the surface layer and its variation with depth change with soil type, and is appreciably affected by soil-forming properties and management. This means that the $\delta^{15}N$ is the integrated result of transformations and transfers of N in the soil-plant-atmosphere system. Therefore, it is important to study the isotopic discrimination of N during the processes taking place in ecosystems and its effect on the variation in natural ^{15}N abundance of different soils.

So far few studies have been made on the $\delta^{15}N$ values for the different N-bearing components in soils. Detailed studies on the variation may provide a scientific basis for quantitatively investigating the N cycles in the soil-plant-atmosphere system.

2.5. References

Amarger, N, Mariotti, A, Mariotti, F, Durr, J C, Bourguignon, C and Lagacherie, B 1979. Estimate of symbiotically fixed nitrogen in field grown soybeans using variations in ^{15}N natural abundance. Plant Soil. 52:269–280.

Cheng, H H, Bremner, J M and Edwards, A P 1964. Variations of nitrogen-15 abundance in soils. Science 146:1574–1575.

Delwiche, C C and Steyn, P L 1970. Nitrogen isotope fractionation in soils and microbial reactions. Environ. Sci. Technol. 4:929–935.

Hoering, T 1955. Variations of nitrogen-15 abundance in naturally occurring substances. Science 122:1233–1234.

Hauck, R D 1973. Nitrogen tracers in nitrogen cycle studies-past use and future needs. J. Environ. Qual. 2:317–327.

Karamanos, R E and Rennie, D A 1981. The isotopic composition of residual fertilizer nitrogen in soil columns. Soil Sci. Soc. Am. J. 45:316–322.

Ledgard, S F, Freney, J R and Simpson, J R 1984. Variations in natural enrichment of ^{15}N in the profiles of some Australian pasture soils. Aust. J. Soil Res. 22:155–164.

Mariotti, A, Pierre, D, Vedy, J C, Bruckert, S and Guillemot, J 1980. The abundance of natural nitrogen-15 in the organic matter of soils along an altitudinal gradient. Catena 7:293–300.

Moore, H 1977. The isotopic composition of ammonia, nitrogen dioxide and nitrate in the atmosphere. Atmos. Environ. 11:1239–1243.

Natelhoffer, K J and Fry, B 1988. Controls on natural nitrogen-15 and carbon-13 abundances in forest soil organic matter. Soil Sci. Soc. Am. J. 52:1633–1640.

Rennie, D A, Paul, E A and Johns, L E 1976. Natural nitrogen-15 abundance of soil and plant samples. Can. J. Soil Sci. 56:43–50.

Riga, A, Van Praage, H J and Brigode, N 1971. Natural abundance of nitrogen isotopes in certain Belgium forest and agricultural soils under certain cultural treatments. Geoderma 6:213–222.

Shearer, G and Legg, J O 1975. Variations in the natural abundance of ^{15}N of wheat plants in relation to fertilizer nitrogen applications. Soil Sci. Soc. Am. Proc. 39:896–901.

Shi, S L, Wen, Q X and Liao, H Q 1987. Contents of fixed ammonium in the main soils of China. (in Chinese). Soils 19:79–83.

Shi, S L, Xing, G X, Zhou, K Y, Cao, Y C and Yang, W X 1992. Natural nitrogen-15 abundance of ammonium nitrogen and fixed ammonium in soils. Pedosphere 2:265–272.

Sun, G Q, Xing, G X, Cao, Y C and Xu, H 1991. Study on variation in natural ^{15}N abundance. I. Characteristics of variations in natural ^{15}N abundance of forest soils. Pedosphere 1:277–281.

Tiessen, H, Karamanos, R E, Stewart, J W B and Selles, F 1984. Natural nitrogen-15 abundance as an indicator of soil organic matter transformation in native and cultivated soils. Soil Sci. Soc. Am. J. 48:312–315.

Wada, E, Imaizum, R and Takai, Y 1984. Natural abundance of ^{15}N in soil organic matter with special reference to paddy soils in Japan: biogeochemical implications on the nitrogen cycle. Geochem. J. 18:109–123.

Wen, Q X, Zhang, X H, Du, L J and Wu, S L 1988. Fixed ammonium and its availability in the main soils of the Taihu region (in Chinese). Acta Pedol. Sin. 25:24–30.

3
Mineralization of soil nitrogen

ZHU ZHAO-LIANG

3.1. Introduction

Mineralization of organic N and immobilization of mineral N are two opposing processes taking place simultaneously in soils. Their relative rates are influenced by a number of factors, especially the quantity and the microbial susceptibility of existing carbonaceous compounds which act as a source of energy. When there is an excess of readily decomposable carbon, immobilization of mineral N proceeds to a greater extent than the mineralization of organic N, which results in net immobilization. Immobilization rates decrease gradually with the decomposition of the carbon compounds. At equilibrium, the rates of mineralization and immobilization are equal and there is neither net mineralization nor net immobilization. With the consumption of further carbonaceous material the rate of mineralization surpasses the rate of immobilization and net mineralization results (Zhu 1963). The organic and mineral N which take part in the transformations can be either soil N or added N. This chapter is devoted to the discussion of net mineralization of soil N and soil N supply. Nitrogen mineralization and immobilization in soil after the addition of organic materials will be discussed in Chapter 12.

3.2. Mineralization of soil nitrogen

Both the amount and pattern of soil N mineralization should be considered in evaluating the capacity of soils to mineralize N. Research on this topic in China has focused mainly on paddy soils. Soil N mineralization capacity can be evaluated from the amount and proportion of the total N mineralized.

3.2.1. Effect of soil type and properties

The amount of soil N mineralized (N_{min}) is a function of the amount of soil organic matter and its biodegradability, moisture, temperature and time of incubation. N_{min} is a capacity factor, while the proportion mineralized (% of soil organic N) is an index of the biodegradability of organic N in soil. As the amount of ammonium fixed by clay minerals is seldom measured in mineralization studies, it is not possible to calculate the amount of organic N in soil, and therefore the mineralization

Zhu Zhao-liang et al. *(eds.): Nitrogen in Soils of China, 43–66.*

percentage is mostly calculated on the basis of the amount of total N in soil. Hence the result derived this way is not strictly the mineralization percentage.

Of the factors controlling mineralization, total N is the most intensively studied. The correlation between N_{min} and total N varies substantially as seen from the numerous results obtained in China (Zhu *et al.* 1984a, b; 1986). Some of the results are presented in Table 3.1. The results show that although N_{min} (measured by the increase in exchangeable ammonium after incubation) correlated significantly with total N in all soils except the gleyed paddy soils, the relationships were not strong and the coefficients differed greatly. This seems to be due largely to the difference in biodegradability of the organic N in the soils tested.

The N mineralization percentage for paddy soils varied substantially, not only for soils from different regions, but also for soils of the same type in the same region (Table 3.2). Thus it is not surprising that the N_{min} and total N are not strongly related.

Table 3.1. Relationship between nitrogen mineralized (N_{min}) and total or organic nitrogen in paddy soils.[1]

Site	Soil	Number	Correlation coefficient.		Reference
			N_{min} vs TN	N_{min} vs ON	
Taihu Region	paddy	68	0.689**		Zhu *et al.* (1984a)
CC, SC & SWC	paddy	67	0.693**		
Taihu Region	I	8	0.790**		Zhu *et al.* (1984b)
	II	19	0.816*		
	II	12	0.611*		
	III	10	0.277		
	III	9	0.598		
	paddy	9	0.579	0.644**	Chen (1985)

[1] Air dried soil incubated anaerobically at 30°C for 2 weeks: CC = Central China; SC = South China SWC = Southwest China; I = calcareous submergic and periodically submergic paddy soil; II = non-calcareous periodically submergic paddy soil; III = gleyed paddy soil; TN = Total nitrogen; ON = Organic nitrogen; * and ** significant at P = 0.05 and P = 0.01, respectively.

Table 3.2. Nitrogen mineralized (% total soil N) in the plow layer of paddy soils.[1]

Site	Soil	Number	Nitrogen mineralized			Reference
			Range	Mean	SD	
Taihu Region	paddy	68	0.82–7.60	4.38	1.48	Zhu (1986)
CC, SC & SWC	paddy	67	0.08–7.75	4.86	2.36	
	I	8	1.81–5.35	3.51	1.18	Zhu *et al.* (1984b)
	II	19	0.82–7.26	4.72	1.91	
	II	12	2.68–5.60	4.42	1.14	
	III	10	2.30–7.60	4.52	1.82	
	II	9	2.84–5.71	4.45	0.75	

[1] Air-dried soil incubated anaerobically at 30°C for 2 weeks.
Abbreviations as in Table 3.1.

Numerous studies have been carried out to determine the reasons for the large differences in the biodegradability of organic N in different soils. Unfortunately, the reasons have not been determined (Zhu, 1963 1979). The biodegradability of organic N in soils has been investigated in relation to the chemical forms of soil organic N (Ivarson and Schnitzer 1979; Keeney and Bremner 1964; 1966), the mineral-organic matter complex (Harada 1959; Shen 1987), soil chemical properties (Zhu *et al.* 1984b), and physical factors (Adu and Oades 1978; Waring and Bremner 1964; Xu and Xu 1981). For instance, the same air-dried soil ground differently has different N_{min} values (Waring and Bremner 1964) and well structured paddy soils easily soften when submerged and their N_{min} values are significantly higher than those of paddy soils with poor structure (Xu and Xu 1981).

3.2.2. *Effect of air-drying*

Air-drying can greatly increase the mineralization of soil N. This phenomenon is known as the drying effect or Birch effect (Huang and Zhang 1957; Shen *et al.* 1959; Birch 1958). Hence, it is preferable to use fresh soil samples in the study of N mineralization. Under most circumstances this is difficult to achieve, especially when a large number of soils are being investigated. The effect of air-drying paddy soils on mineralization of organic N is illustrated in Table 3.3.

The samples for study were taken from the Taihu region and as the mineralization rate for field moist soils were low long incubation times, up to 10 weeks, were required to produce sufficient mineral N for measurement. The results show that air-drying increased N_{min} for all three soil types. The greatest effect occurred in the noncalcareous paddy soils, followed by the gleyed paddy soils, while the smallest effect occurred in the calcareous paddy soils. The mineralization rates increased in the same order with the air dried soils as with field moist soils. Chen (1985) found that N_{min} values for field moist soils after 10 weeks incubation were strongly correlated with those for air-dried soils incubated for 2 weeks ($r = 0.873$ **, $n = 9$). Similar results were obtained by other authors (Yoshino and Dei 1974). The conclusion from these results is that the ranking of soils according to their capacity to mineralize N is not affected by air drying. This conclusion, however, seemed to be

Table 3.3. Effect of air-drying on mineralization of nitrogen (mg N kg^{-1}) in paddy soils. [1]

Soil	Number	Mean N_{min}	
		Air-dried soil (2 wk)	Field moist soil (10 wk)
I	3	35.7 c[2]	15.9 e
II	3	81.6 a	28.1 d
III	3	50.7 b	20.6 e

[1] Air-dried soil incubated anaerobically at 30°C for 2 weeks (Chen 1985). Abbreviations as in Table 3.1.

[2] Duncan's test.

incongruous with the belief that gleyed paddy soils had a stronger drying effect (Shen *et al.* 1959). A possible reason for the conflict is that the noncalcareous paddy soils which are more fertile, contain more microbial biomass, and hence partial sterilization, which is the main mechanism of the drying effect, produced more mineral N in this type of paddy soil.

It is worth noting that mineralization during incubation is affected by the period of storage of air-dried soil (Birch 1960; Cornfield 1964; Gasser 1961). A study on 9 soil samples from the Taihu region stored for two weeks and 5 months showed that there was no change in soil water content, but the mineralization of N in the samples stored for 5 months was 20% higher than that in the soils stored for two weeks (Chen 1985). It follows that the time of storage of soil samples under air-dried conditions should be standardized in order to obtain sensible results for the comparison of the amount of N mineralized by different soils.

3.2.3. *Significance of fixation of mineralized ammonium*

In most mineralization studies little attention has been paid to other processes occurring at the same time, especially fixation of mineralized ammonium by clay minerals. Recent research indicates that this process should not be ignored. Keeney and Bremner (1966) found that ammonium fixation increased significantly after incubation, but they did not study the implications of this finding. Chen (1985) found that misleading conclusions could be drawn from the results of anaerobic incubation studies when the soil had the ability to fix ammonium, if the increased fixed ammonium was not taken into consideration (Tables 3.4 and 3.5).

The results in Table 3.4 show that fixed ammonium increased from –3 to 41 mg N kg^{-1} (mean 17 mg N kg^{-1}) after incubation, implying that part of the mineralized ammonium was fixed by clay minerals. Therefore, the amount of N mineralized, N_{min}, should be the total increase in exchangeable and fixed ammonium after

Table 3.4. Increase in exchangeable and fixed ammonium (mg N kg^{-1}) during incubation under flooded conditions.[1]

Soil	Exchangeable NH_4^+	Fixed NH_4^+	Total
I	59	26	85
	47	20	67
	45	26	71
II	95	11	106
	79	41	120
	83	7	90
III	61	20	81
	67	8	75
	56	–3	3
Mean	66	17	83

[1] Air-dried soil incubated anaerobically at 30°C for 2 weeks (Chen 1985). Abbreviations as in Table 3.1.

Table 3.5. Mineralization of soil nitrogen calculated from the increase in exchangeable and fixed ammonium during incubation.[1]

Soil	Calculated from exch NH_4^+			Calculated from exch + fixed NH_4^+	
	N_{min} (mg N kg^{-1})	(% TN)	(% ON)	N_{min} (mg N kg^{-1})	(% ON)
I	50c	3.7b	4.5b	74b	6.7a
II	86a	4.6a	5.4a	105a	6.7a
III	61b	3.4b	4.0b	69b	4.5b

[1] Mean value for three soil samples; air-dried soil incubated anaerobically at 30°C for 2 weeks (Chen 1985).
Abbreviations as in Table 3.1.

incubation. Otherwise not only would the actual amount of N mineralized be underestimated, but also the ranking of soils according to their capacity to mineralize N would be changed due to differences in the extent of fixation of mineralized ammonium (Table 3.5). The ranking for the 3 paddy soils would be changed from noncalcareous > gleyed > calcareous to noncalcareous > calcareous > gleyed. Similarly, if soils were ranked according to the proportion of organic N mineralized they would again have a different order. If exchangeable ammonium only is regarded as N_{min} then the order for the paddy soils from the Taihu region is noncalcareous > calcareous > gleyed paddy soil. However, if the increase in both exchangeable and fixed ammonium is taken as the amount of N mineralized, the order is, calcareous ≑ noncalcareous > gleyed paddy soil. Thus both exchangeable and fixed N need to be measured to obtain N_{min} (Chen 1985).

It is probably more important to take ammonium fixation into account when comparing mineralization in soil from different regions, because the type of clay minerals in soils from different regions may vary greatly. As a result, native fixed-ammonium may also vary substantially. Thus soils from different regions should not be compared on the basis of the proportion of total N mineralized. Ranking on the proportion of organic N mineralized will also be incorrect if the increase in exchangeable ammonium on incubation is regarded as N_{min}, due to the large differences in the capacities of soils to fix the mineralized ammonium. A large proportion of the soils from Central, South and Southwest China referred to in Table 3.2 contain mainly 1:1 type clay minerals, which have low native fixed ammonium and low capacity to fix ammonium (Wen and Zhang 1986). Thus there would be little difference in the rankings irrespective of whether organic N or total N is used to calculate the proportion mineralized. However, this would not be the case for the soils from the Taihu region, which contain primarily 2:1 type clay minerals, and have high native fixed ammonium and a high capacity to fix mineralized ammonium. Consequently, the proportion of the organic N mineralized in the soils of the Taihu region is probably higher than that of the soils in Central, South and Southwest China. This conclusion seems to contradict the general belief that 2:1 type clay minerals can protect organic matter against microbial decomposition better than 1:1 type clay minerals.

3.2.4. *Mineralization pattern in anaerobic soil*

The pattern of mineralization of soil N is an important aspect that should be taken into consideration in the management soil N. The cumulative effective temperature equation is usually used to describe the pattern of mineralization of soil N during incubation under flooded conditions (Cai *et al.* 1979; Yoshino and Dei 1974, 1977). The general cumulative effective temperature equation is

$$Y = k\,(T - To)^n,$$

where Y is N_{min}, T is soil temperature (°C), and To = 15°C. (T – To) represents the cumulative effective temperature and k is a coefficient. n not only affects $N_{min,}$ but the mineralization pattern as well. (T – To) could be substituted by (T – To)D when soil is incubated at a constant temperature (D and T are incubation time and temperature, respectively). Theoretically, for soils with n values <1, N mineralized per unit of cumulative effective temperature decreases with time of incubation and the cumulative effective temperature. The reverse is true for the soils whose n value is >1.

n values obtained from incubation of Taihu soils under flooded conditions are given in Table 3.6. Although n values were quite different for paddy soils of the same type, the mean values for the calcareous soils were the lowest (0.70 and 0.72), while most of the n values for the gleyed soils were higher than 1. There was great variation in n values for the noncalcareous soils with a mean value around 1. The conclusion from these results is that the biodegradability of organic N in paddy soils formed under different water regimes are significantly different. Of course it is only one of the factors controlling the pattern of mineralization. The type of clay

Table 3.6. Effect of pretreatment of paddy soils from the Taihu Region on the n value in the cumulative effective temperature equation $Y = k\,(T–To)^n$,where $Y = N_{min}$ and T = soil temperature.[1]

Soil	Pretreatment	n value	Samples	Reference
I	air-dried, pre-incu. 2 wk	0.57	2	Gao *et al.* (1984b)
	field moist soil	0.78	3	Chen (1985)
	mean	0.70	5	
II	air-dried, pre-incu. 2 wk	0.86	5	Gao *et al.* (1984b)
	field moist soil	1.36	3	Chen (1985)
	field moist soil	0.76	2	Cai *et al.* (1979)
	field moist soil	0.72	2	Cai and Zhu, unpublished
	mean	0.95	12	
II	air-dried, pre-incu. 2 wk	0.74	3	Gao *et al.* (1984b)
	field moist soil	0.70	2	Cai and Zhu, unpublished
	mean	0.72	5	
III	air-dried, pre-incu. 2 wk	1.96	4	Gao *et al.* (1984b)
	field moist soil	0.83	3	Cai and Zhu, unpublished
	field moist soil	1.63	3	Chen (1985)
	mean	1.52	10	

[1] Soil incubated anaerobically at 30°C for 2 weeks (Chen 1985).
Abbreviations as in Table 3.1.

mineral, as pointed out by Inubushi *et al.* (1985), is probably another factor which results in different mineralization patterns.

The effect of soil drying on N mineralization needs to be considered in studies on the pattern of mineralization. As shown in Table 3.7, air drying greatly increased the k value and significantly decreased the n value. The effect of air drying on mineralization in the initial stages of incubation was to decrease n values, but the further increase in N_{min} during the later stages of incubation is similar to that of field moist soils when the cumulative effective temperature reached a certain level. Therefore, in mineralization studies using air-dried soils, the mineralization pattern representative of the field moist soil could be obtained by subtracting the N mineralized during preincubation. However, the preincubation required may be different for different soils. It has been suggested that the calcareous paddy soils need to be incubated at 30°C for 2 weeks (cumulative effective temperature 210°C. day) and gleyed paddy soil at 30°C for 4 weeks (cumulative effective temperature of 420°C. day) (Cai *et al.* 1979).

3.2.5. *Mineralization pattern in aerobic soils*

First order reaction kinetics have also been used to describe mineralization patterns under aerobic conditions (Zhu and Huang 1983). The results in Table 3.8 indicate that the mineralization potential (N_0) is highly correlated with total N (r = 0.786**, n = 20, Bai and Zhao 1981; r = 0.813**, n = 16, Wang *et al.* 1986; and r = 0.981**, n = 22, Shen 1987). N_0 was about 11–19% of the total N.

The rate constant k is temperature dependent (Campbell *et al.* 1981; Stanford *et al.* 1973a; Stanford *et al.* 1973b, Zhang *et al.* 1987). As shown in Table 3.8, the mean k value was only 0.036 wk^{-1} when soils were incubated at 30°C in the experiment of Shen (1987); this k value was considerably lower than that obtained in the other two incubations at 35°C (0.082 wk^{-1} and 0.045 wk^{-1}). Although it is commonly assumed that k does not vary greatly between soils incubated at the same temperature (Bai and Zhao 1981; Wang *et al.* 1986; Shen 1987; Stanford and Smith 1972), the mean k values for the two experiments conducted at 35°C by Bai and Zhao (1981) and Wang *et al.* (1986) differed greatly (Table 3.8). The reason for the

Table 3.7. Effect of air-drying paddy soils[1] on coefficients of the cumulative effective temperature equation $Y = k\,[T{-}To]^n$.

		k value		n value	
Soil	Samples	Dried	Field moist	Dried	Field moist
I	3	0.51	0.041	0.20	0.78
II	3	2.69	0.0004	0.22	1.36
III	3	0.78	0.00002	0.17	1.63

[1] Soil incubated anaerobically at 30°C for 2 weeks (Chen 1985).
Abbreviations as in Table 3.1.

Table 3.8. Mineralization potential (N_0) and mineralization rate constant (k) under aerobic conditions.[1]

Site	Incubation temperature	Soil[2]	N_0 (mg N kg^{-1})	N_0/TN (%)	k (wk^{-1})	Reference
Shaanxi	35°C	Huangmiantu	73 ± 19	16.8 ± 2.7	0.072 ± 0.017	Bai & Zhao (1981)
		Hailutu	85 ± 12	11.2 ± 0.7	0.087 ± 0.008	
		Loutu	97 ± 23	11.9 ± 1.4	0.083 ± 0.007	
		Yellow paddy	69	11.1	0.082	
		Mean	83 ± 20	13.1 ± 3.0	0.082 ± 0.008	
Beijing & Shanxi	35°C	Loess	324 ± 93	11.8 ± 1.0	0.037 ± 0.008	Wang *et al.* (1986)
Beijing		Cinnamon	650 ± 153	19.2 ± 2.9	0.045 ± 0.008	
Shandong		Fluvo-aquic	318 ± 137	13.7 ± 1.9	0.052 ± 0.007	
		Mean	445 ± 204	15.0 ± 3.9	0.045 ± 0.004	
Jiangsu,	30°C	Paddy	268 ± 75	18.6 ± 1.6	0.033 ± 0.004	Shen (1987)
Zhejiang		Upland	212 ± 104	19.0 ± 1.9	0.038 ± 0.005	
& Anhui		Mean[3]	233 ± 99	18.8 ± 1.6	0.036 ± 0.007	

[1] Data in the table are mean ± SD;
[2] Huangmiantu = yellow cultivated loessial soil; Hailutu = dark loessial soil; Loutu = old manured loessial soil.
[3] 3 uncropped soils were included.

difference is not known, but incubation experiments have shown that the k values for soils with 2:1 type clay minerals are usually lower than those for soils with 1:1 type clay minerals (Inubushi *et al.* 1985).

It was suggested recently that two first order kinetic equations should be used to describe the mineralization of quickly mineralizable and slowly mineralizable organic N, before combining them into one equation (Beauchamp *et al.* 1986; Inubushi *et al.* 1985; Molina and Clapp 1980; Ritcher *et al.* 1982). The coefficients assigned to the mineralization potentials for the quickly and slowly mineralizable organic N are N_{0q} and N_{0s}, and the corresponding rate constants are k_q and k_s. Mineralization of N in soils of North China was studied with air-dried samples under aerobic conditions and the results are given in Table 3.9 (Wang *et al.* 1986). The mineralization potential for the quickly mineralizable N was lower than that for the slowly mineralizable N in all soils, while the rate constants for the quickly mineralizable fraction were significantly higher than those for the slowly mineralizable fraction. According to other studies, N_{0q} is strongly affected by air-drying (Inubushi *et al.* 1985), indicating that N_{0q} is partly of microbial biomass origin.

The fraction of organic N which is readily susceptible to mineralization might be removed when sequential leaching treatments are imposed in the study of mineralization in open incubation systems. It was suggested that the fraction removed should be taken into account in calculating N_0 (Xiong *et al.* 1986; Smith *et al.* 1980; Smith 1987). This approach, however, is not necessarily correct as the fraction leached might not be completely mineralizable. Shen (1987) put forward an improved method of mixing a combined anion and cation exchange resin with the

Table 3.9. Mineralization potential and rate constants for quickly and slowly mineralizable nitrogen under aerobic conditions.[1]

Site	Soil	Quickly mineralizable		Slowly mineralizable	
		N_{0q} (mg N kg^{-1})	k_q (week^{-1})	N_{0s} (mg N kg^{-1})	k_s (week^{-1})
Beijing & Shanxi	Loess	26.5 ± 11.7	0.408 ± 0.386	308 ± 62	0.029 ± 0.008
Beijing	Cinnamon	36.2 ± 14.6	0.367 ± 0.389	562 ± 109	0.046 ± 0.009
Shandong	Fluvo-aquic	43.0 ± 29.1	0.370 ± 0.093	302 ± 157	0.031 ± 0.003

[1] Wang *et al.* (1986). Data in table are mean ± SD.

soil sample to allow of all the mineralized ammonium and nitrate to be adsorbed and determined.

3.3. Effect of fertilizer nitrogen and plant growth on mineralization

3.3.1. Effect of added nitrogen

It is commonly observed in incubation experiments or planted pot studies with labelled N that mineralization and plant uptake of nonlabelled soil N increases after the addition of labelled fertilizer. This phenomenon is known as a positive priming effect, but sometimes a negative priming effect can be observed. Jenkinson *et al.* (1985) suggested that the term Added Nitrogen Interaction (ANI) should be used in place of priming effect.

Studies demonstrate that the priming effect of fertilizer N is, in most cases, only an apparent phenomenon. It is the result of pool substitution between added labelled N and nonlabelled soil N through immobilization (Zhu 1986; Cai *et al.* 1981; Jenkinson *et al.* 1985; Zhu *et al.* 1984a). The added labelled N and non-labelled soil N coexist in the mineral N pool. During immobilization labelled mineral N is transformed into organic N partly in place of some nonlabelled soil mineral N, and hence there is less immobilization of soil mineral N than occurred in the absence of labelled N; the apparent priming effect therefore results. The driving force for immobilization is the energy released by the decomposition of soil organic matter and the sloughing off of plant roots. However, there could be a real priming effect under some circumstances (Jenkinson *et al.* 1985). For instance, the addition of fertilizer N to infertile soils can stimulate the extension of plant roots and therefore increase uptake of soil N.

Different methods have been used to evaluate the priming effect. In order to inhibit nitrification, and its effect on immobilization, and avoid the subsequent loss of N by denitrification, Shen (1986) first sterilized the soil with chloroform and then added a small amount of fresh soil before adding the ^{15}N-labelled fertilizer for incubation. The results (Table 3.10) show that mineralization increased with increasing N addition. However, the increase in mineral N was almost equivalent to the amount of labelled N immobilized. As a result there was no net priming effect.

Table 3.10. Effect of nitrogen addition on mineralization of nitrogen and carbon.[1]

N addition ($\mu g\ g^{-1}$)	Immobilized N ($\mu g\ g^{-1}$)	Increase in mineral N ($\mu g\ g^{-1}$)	Emission of CO_2 ($\mu g\ C\ g^{-1}$)
0	–	–	231.6
5.65	1.41	1.29	218.4
58	6.02	5.85	228.7

[1] Aerobic incubation for 20 days (Shen 1986).

Table 3.10 also shows that carbon dioxide emission did not increase after the addition of N; this provides additional evidence that the rate of mineralization did not increase as a result of the addition of N.

Results of pot and field experiments relevant to the priming effect have been summarized previously (Zhu 1986) and the recent results are summarized here. The net priming effect (the difference between soil N mineralized and labelled N immobilized) and its proportion of the labelled N added were calculated in order to evaluate the priming effects. The so-called net priming effect gives an indication of the net mineral N increased through the priming effect. Of the 29 data sets collected relevant to flooded rice, the range of net priming effects was within ± 10% of the added N except for three sets (–16.0%, +10.9% and +12.3%, respectively) (Ge *et al.* 1987; Zhu 1986; Zhu and Wu 1983; Zhu *et al.* 1986; Huang *et al.* 1981; Cai *et al.* 1981; Yoshida and Padre 1977). These results show that most of the net priming effects were within the allowed measurement error.

In addition, 66 data sets were collected from experiments conducted with upland crops. In 27 of these the net priming effect exceeded ± 10%. However, 14 of the 27 sets had a negative net priming effect and only 13 sets had a positive net priming effect (Jin *et al.* 1983, 1984; Li *et al.* 1984; Riga *et al.* 1980; Shen 1987; Xiong *et al.* 1986; Zhou 1983; Zhu 1986; Zuo *et al.* 1983). The reason for the greater number of upland soils with net priming effects than flooded rice has not been determined, but from the number with negative net priming effects, it could be assumed that more fertilizer N is immobilized under aerobic conditions than under flooded conditions.

It can be concluded that under most circumstances the priming effect is an apparent phenomenon only. It will neither increase the level of N nutrition nor will it increase the consumption of soil N.

3.3.2 *Effect of plants*

Plant growth will probably affect both mineralization and immobilization (see Zhu 1963). On one hand, H^+ and organic acids excreted by plant roots will promote mineralization, and high biological activity in the rhizosphere will stimulate N mineralization. On the other hand, the high C:N ratio of the material sloughing off roots could increase immobilization. Consequently, whether soil N mineralization is depressed or promoted by plant growth is determined by the relative extent of these two effects. In addition, the uptake of mineralized N by plant roots could reduce N loss, which would result in higher apparent mineralization in the presence of plants than in their absence (Broadbent 1978).

Usually N mineralization is estimated by subtraction of the N in the seeds or seedlings from the N in plants grown without N addition. Recent studies (Zhu *et al.* 1986), however, have shown that this method of estimation overestimates soil N mineralization, particularly with flooded rice, because it has included N derived from nonsymbiotic N fixation (Ndfa) as mineralized N. As shown in a pot experiment with 3 soils labelled with ^{15}N, Ndfa could amount to 19.6–23.0% of the total uptake in the rice plants at maturity (Zhu *et al.* 1986; Table 3.11). However, it appears to be appropriate to subtract the N derived from non-soil sources, particularly Ndfa, before calculating mineralization of N from a paddy soil. Results calculated by this method are given in Table 3.12.

Mineralization of nitrogen in the absence of plants amounted to 27 mg N kg $soil^{-1}$ whereas the apparent mineralization in the presence of rice (i.e. when Ndfa is not deducted) averaged 43 mg N kg $soil^{-1}$; the real mineralization in the presence of plants (i.e. after deducting Ndfa) averaged 34 mg N kg $soil^{-1}$. Thus, the apparent promotion of mineralization by plant growth could be attributed largely to nonsymbiotic

Table 3.11. The contribution of nonsymbiotic nitrogen fixation (Ndfa) to total nitrogen uptake by rice.[1]

	Tops			Roots			Whole plant
Soil[2]	N uptake ($\mu g\ pot^{-1}$)	Ndfa ($\mu g\ pot^{-1}$)	Ndfa (%)	N uptake ($\mu g\ pot^{-1}$)	Ndfa ($\mu g\ pot^{-1}$)	Ndfa (%)	Ndfa (%)
I	28.5	5.4	19.1	0.74	0.29	39.7	19.6
II	107.5	23.9	22.2	3.15	0.84	26.6	22.5
III	103.4	22.8	22.8	2.94	0.72	24.4	23.0
Mean			21.4			30.3	21.7

[1] Pot experiment with ^{15}N labelled soils without added nitrogen (Zhu *et al.* 1986).
[2] Abbreviations as in Table 3.1.

Table 3.12. Mineralization of nitrogen ($\mu g\ kg^{-1}$) in flooded soils with and without rice.[1]

Soils	I	II	III	Mean
Apparent N min[d] (+ rice)[2]	38.3	51.1	39.9	43.1
Real min[d] N (+ rice)	30.8	39.6	30.7	33.7
Incubation N (–rice)	24.0	35.1	22.1	27.1

[1] Abbreviations as in Table 3.1; there were three soils in each group, and other results for these soils are given in Tables 3.4 and 3.5 (Zhu *et al.* 1986).
[2] Apparent N min[d] = (total N in rice + exchangeable NH_4^+) – (seedling N + initial exchangeable NH_4^+).
Real min[d] N = apparent N mineralized – Ndfa, where Ndfa = (total N in rice – seedling N) × %Ndfa. %Ndfa = 21.7;
Incubation N = exchangeable NH_4^+ after incubation – exchangeable NH_4^+ before incubation.

N fixation. Again as shown in Table 3.4, N mineralization calculated from the increase in exchangeable ammonium was 28% lower than when both exchangeable and fixed ammonium were taken into account. Taking all the above results into consideration, the so-called promotion of mineralization by rice was only an apparent phenomenon which could be attributed to (i) the overestimation of mineralization because Ndfa was not taken into account, and (ii) the underestimation of mineralization because the increase in fixed ammonium after incubation was not taken into account.

The two aspects mentioned above for flooded soils should also be considered in evaluating the effect of plant growth on mineralization in upland soils. However, it should be less important for upland soils as Ndfa for nonleguminous upland crops is less than that for flooded rice. In addition, much of the mineralized ammonium would be oxidized to nitrate under upland conditions resulting in less fixation of mineralized ammonium by clay minerals.

Rice growth can also markedly affect the apparent mineralization pattern (Tables 3.13 and 3.14). The results in Table 3.13 show that in the early stages of plant growth there was no difference between the apparent mineralization in the presence

Table 3.13. Effect of rice growth on mineralization (mg N kg^{-1}) of nitrogen in paddy soils.[1]

Soil	Method	Jun. 12	Jul. 6	Aug. 5	Aug. 31	Sept. 28	Oct. 30
II	Pot exp. early rice	11.1a[2]	34.8a	74.3a	–	–	–
	Pot exp. single rice	9.9a	36.3a	70.4a	90.6a	103a	116a
	Anaerobic incubation	10.3a	28.4a	54.6b	65.2b	76.3b	75.4b
III	Pot exp. early rice	5.7a	20.3a	51.3a	–	–	–
	Pot exp. single rice	5.8a	19.9a	45.3a	67.0a	67.7a	74.3a
	Anaerobic incubation	5.3a	18.7a	31.7b	44.1b	41.0b	47.9b
Cumulative temp. (°C. day)		157	494	969	1328	1578	1684

[1] Pot experiment and anaerobic incubation were conducted under flooded conditions without added nitrogen. The pots were flooded on May 27 and seedlings were transplanted on May 29. (Cai and Zhu 1983). Abbreviations as in Table 3.1. Soil cumulative temperature is the sum of the daily mean temperature 5 cm below soil surface minus 15°C.
[2] Duncan's test was used to compare results.

Table 3.14. Effect of rice growth on the coefficients in the cumulative effective temperature equation used to describe mineralization.[1]

Soil	Method	k	n	r^2
II	Pot exp.early-maturing rice	0.00570	1.04	0.999
	Pot exp.late-maturing rice	0.00600	1.02	0.998
	Anaerobic incubation	0.0138	0.86	0.998
III	Pot exp.early-maturing rice	0.00130	1.20	0.998
	Pot exp.late-maturing rice	0.00231	1.09	0.998
	Anaerobic incubation	0.00575	0.91	0.992

[1] Cai and Zhu (1983); abbreviations as in Table 3.1.
[2] r is the correlation coefficient for the relationship between log Y and $\Sigma(T - T_0)$.

and absence of plants. However, with time the difference became apparent, and this was presumably due to nonsymbiotic N fixation after the elongation stage of the rice (Ito and Watanabe 1981). Hence, the n values in the treatment with rice should be surely greater than those in the incubation treatment without plants when the cumulative effective temperature equation is used to describe the mineralization pattern (Table 3.14). It seems that the k values obtained with plants should be higher than those without plants because the apparent mineralization with plants was greater than that without plants. However, due to the greater n value with plants, the k values with plants were not necessarily higher. In contrast, the k values with rice were appreciably smaller than those without plants (Table 3.14).

3.4. Soil nitrogen supply

The N supply characteristics of soils can be evaluated quantitatively (amount of N taken up by plants during a cropping season), and qualitatively (pattern of supply). The soil N supply (Ns) is the total amount of N available for plant uptake during the growth of the crop. It includes the initial quantity of mineral N in soil prior to fertilization and that mineralized during the cropping season. In the presence of plants Ns is calculated as the sum of the N taken up by plants from plots receiving no fertilizer or manure N (Nop) and the residual mineral N in the soil. The N taken up by plants at maturity (minus seed and seedling N) is usually taken as an approximate estimate of Ns where the residual N is ignored. In fact, it is only an estimate of apparent Ns and would be better to call it apparent N supply (Nas), the reason for which will be discussed later.

3.4.1. Apparent supply for different crops

Nas is one of the main parameters used for estimating the optimal rate of application of fertilizer N. Tables 3.15 and 3.16 present Nas data for different crops and the reliability of obtaining high crop yields using Nas. Nas for crops such as rice and wheat varied from 34.5–126 kg ha^{-1}, accounting for 1.2–3.3% of the total N in the 0–20 cm soil layer (Table 3.15) and the reliability of Nas for obtaining high crop yields was assessed as 45–83% (Table 16). Evidently, a high Nas is one of the main factors for sustained crop production.

As shown in Table 3.15 Nas for the crops grown in the Taihu region were ranked in the following order: single cropping late maturing rice > early rice > late rice >> wheat. Similar results were obtained in other studies (Zhu *et al.* 1978; Zhang *et al.* 1988; Cai and Zhu 1983). The variation in Nas for rice crops grown in different seasons is due to differences in cumulative effective temperature (Zhu *et al.* 1978; Cai and Zhu 1983). The difference in Nas values for rice and wheat may be due to the greater extent of nonsymbiotic N fixation in rice.

As shown in Table 3.16, high yields of upland crops rely, to a large extent on Nas, except for wheat and barley grown in the Taihu region. This difference may result from the fact that the soils of North China (including northern Jiangsu)

Table 3.15. Apparent nitrogen supply (Nas) for the main crops of China.[1]

Crop[2]	Region	Mean Nas[3] (kg N ha^{-1})	Mean (Nas/soil N)[4] (%)
ERT-1	Fujian, Hunan, Jiangsu, Shanghai & Zhejiang	52.5–69.0	1.6–1.9
ERT-2	Guangdong, Jiangsu, Shanghai & Zhejiang	49.5–61.5	1.2–2.0
LRS	Jiangsu, Shanghai & Zhejiang	76.5–108	2.1–3.3
MR	Jiangsu, Sichuan & Yunnan	63.0–126	2.5–3.1
Wheat	Jiangsu & Sichuan	34.5–73.5	1.9–3.1
Barley	Jiangsu	73.5	1.4

[1] Field experiments; Zhu (1986); Zhang *et al.* (1988); unpublished data of Sun G. Y. and Huang D. M.
[2] ERT-1 = early rice; ERT-2 = late rice; LRS = single crop-late maturing rice; MR = single crop-mid maturing rice.
[3] From N uptake in tops at maturity from zero-N plots; the data for Jiangsu, Shanghai and Zhejiang were derived by subtracting the N in seed or seedlings from the N in the plants. The data are the ranges for the mean values of each experiment.
[4] Mean Nas/N content in 0–20 cm soil layer × 100.

Table 3.16. Reliability of high crop yields on apparent supply of soil nitrogen.

Crop[1]	Region	Reliability[2]
Wheat	Tianjin	73.3 ± 6.2
	Henan	78.2
	Taihu	53.0
Barley	Taihu	45.2 – 48.3
	Shanghai	48.3
Maize	Heilongjiang	77.5 ± 4.6
Millet	Shanxi	77.1 ± 6.4
Cotton	Sichuan	82.6 ± 2.4
ERT-1	Taihu	51.9 – 69.8
ERT-2	Taihu	53.7 – 62.9
LRS	Taihu	67.2 – 75.9
MR(1)	Yunnan	64.2 ± 2.2
MR(2)	Yunnan	82.9 ± 8.5

[1] ERT-1, ERT-2, LRS, MR – see Table 3.15. MS(1) = previous crop was wheat; MS(2) = the previous crop was broad bean. Data given are mean SD or range.
[2] Reliability = (N uptake by tops at maturity from zero-N plots)/(N uptake in tops from high yield plots with added N).

contain more available N (especially nitrate) at seeding than the soils of the Taihu region. The high yield crops can rely to a greater extent on soil N supply when a large amount of nitrate accumulates in the soil profile at seeding (Yuan *et al.* 1985; Peng *et al.* 1981).

3.4.2. *Sources of plant nitrogen other than fertilizer and manure*

As discussed previously, Nas is generally estimated from Nop. Sometimes the N in seeds or seedlings is taken into account as part of the apparent soil N supply, but part of the accumulated N comes from non-soil sources, such as nonsymbiotic N fixation, rainfall and irrigation water. It is important that the contributions of the different N sources to Nas be estimated.

The contribution of nonsymbiotic N fixation should be elucidated for two reasons, (i) it is difficult to quantify, and (ii) it may account for a large proportion of the N in rice plants. The contribution was between 19.1 and 22.8% (mean 21.4%, calculated as Ndfa in the tops) for rice, and there was little difference between the soils tested (Table 3.11). Based on the mean Ndfa, the contributions of the different N sources to Nas for rice were calculated and are given in Table 3.17 (Zhu 1988).

The N in the plants derived from rainfall and irrigation water was not included in the above estimates. According to observations conducted at different sites in this region by Liu (unpublished data), the annual N in rainfall is 8.6–16.8 kg ha^{-1} (mean 13.7 kg ha^{-1}) and N in irrigation water is 4.1–5.7 kg ha^{-1} (mean 5.1 kg ha^{-1}). The annual leaching and runoff losses ranged from 4.8 to 8.3 kg N ha^{-1} (mean 6.6 kg N ha^{-1}) giving a net input of 7.4–15.3 kg N ha^{-1} (mean 12.2 kg N ha^{-1}) which would be largely utilized by the rice plants. It is apparent that this source of N should not be overlooked. However, due to the lack of data for each site and crop of rice, it was not possible to allow for this in Table 3.17.

As shown in Table 3.17, an appreciable amount of Nas for rice came from non-symbiotic N fixation. The N actually supplied by the soil (including the N in rain-fall and irrigation water) was 42.0–58.5 kg ha^{-1}, which accounted for 58–72% of the Nas. The reliability of obtaining high rice yields using the data for the actual amount of N supplied by the soil was only 34–46% rather than the 52–76% given in Table 3.16. These values for rice were similar to those obtained for wheat and barley (45–53%). which receive lower contributions of nonsymbiotic N fixation and seed N.

Table 3.17. Contribution of different sources of nitrogen to Nas for rice plants grown on zero-N plots.[1]

		(kg N ha^{-1})		% total Nas	
Rice	N Source	Range	Mean	Range	Mean
ERT-1	Seedling	3.1–9.5	6.2	4–13	9
	Nonsymbiotic	11.3–15.3	13.3	19–21	20
	Soil	41.6–56.0	48.8	69–75	72
	Total Nas	58.5–75.3	68.2	100	
ERT-2	Seedling	12.9–29.8	19.4	16–41	27
	Nonsymbiotic	8.8–16.4	11.5	12–18	16
	Soil	32.4–60.3	42.3	46–66	58
	Total Nas	55.5–91.1	73.1	100	
LRS	Seedling	3.0–13.2	7.2	4–16	9
	Nonsymbiotic	11.3–26.4	16.1	18–21	20
	Soil	41.3–96.8	58.8	66–76	72
	Total Nas	59.6–128	81.8	100	

[1] Field experiments in Taihu Region (Zhu 1988).
Abbreviations as in Table 3.15.

3.4.3 *Pattern of supply*

The pattern of N supply has been studied in China with special reference to cropping systems and the timing of N application (ISSAS 1977, 1978 ; Zhu *et al.* 1978, 1979, 1984a; Cai *et al.* 1979, 1981; Gao *et al.* 1984b; Hseung *et al.* 1980). According to farmers' experience, the pattern of nitrogen supply is characterized by the growth performance of crop plants, and it has been separated into early and delayed development phases. The so-called early development phase relates to the good growth of crop plants early, while the delayed development phase refers to restricted growth in early stages of plant growth and better growth at the later stages. Studies on paddy soils (Zhu *et al.* 1979) showed that these characteristics are determined by (i) the N mineralization pattern, and (ii) the uptake of mineralized N by rice. The mineralization pattern was discussed previously (Table 3.6).

Recent studies indicate that absorption of soil nutrients is determined by their chemical forms and their mobility, namely the so-called bioavailability. The rate of movement of soil nutrients to root surfaces and the extension velocity of plant roots to the locations where the nutrient exists are important factors governing nutrient absorption. Studies in situ in the field (Zhu *et al.* 1979) showed that the extension velocity of rice roots in soil showing early development was significantly faster than that in a soil with delayed development because of better soil structure. The rate of movement of ammonium ions to the root surface was investigated with $^{86}Rb^{+}$, which is similar in mobility to ammonium ions, and the results showed that there was no apparent difference between the two soils of different structure. This suggests that the extension velocity of rice roots is a more important factor than the rate of nutrient movement in controlling the rate of uptake of ammonium by rice plants in the early stages of growth.

3.4.4. *Subsoil supply*

Mineral N, especially nitrate, in deeper soil layers has generally been taken into account in the study of N supply in upland soils because nitrate, the major form of mineral N in upland soils, is highly mobile, and crop roots can usually extend to very deep soil layers (Peng *et al.* 1981). Studies on wheat and corn fields in northern Jiangsu province showed that mineral N or nitrate in the 0–60 cm or 0–70 cm soil layer was highly correlated with N uptake in zero-N plots. However, relatively speaking, nitrate in the upper soil layers contributes more to crop yield than that in the lower soil layers (Yuan *et al.* 1985).

Recent investigations show that it is not sufficient to consider the plow layer only in a study of the N supplying characteristics of paddy soils. Field microplot studies in southern Jiangsu province (Chen 1985) indicated that the subsoil could supply 15–38 kg N ha^{-1} (mean 24 kg N ha^{-1}), which accounted for 16–50% (mean 30%) of the Nas (including the plow layer and subsoil layer). Similar data were reported by Greenland and Watanabe (1982). The contribution from the subsoil will affect the pattern of N supply because the subsoil N is absorbed at the later stages of rice growth.

Further research indicated that the contribution of the subsoil N to the total Nas correlated significantly with the ratio of the total N in the 15–30 cm layer to that in 0–30 cm layer ($r = 0.847$, $n = 9$, $P = 0.01$). However, there was no relationship between subsoil N supply and total N in the 15–30 cm layer. It seems that the subsoil contribution to total Nas comes from the plough-pan (usually the 15–30 cm layer).

3.5. Prediction of supply

One important reason for studying N availability indices is to predict the amount of soil N which will be supplied to plants during the growing season so that the optimum rate of fertilizer N can be applied (Zhu 1982; Gao *et al.* 1984a).

3.5.1. Chemical indices of availability

Biological and chemical indices of N availability have been studied for many years and numerous reviews have been published (Harmsen and van Schreven 1955; Bremner 1965; Zhou 1978a, b; Zhu 1979; Keeney 1982). The reviews indicate that the relationship between the biological availability of soil N and its chemical extractability or chemical forms is very complicated. For example, organic matter in the soil residue after treatment with 6 N HCl can still be degraded by microorganisms (Ivarson and Schnitzer 1979), and the chemical extractability of soil organic N does not always relate to its biodegradability (Juma and Paul 1984). To further complicate the problem, the biodegradability of the same form of organic N in different soils can be very different (Keeney and Bremner 1964, 1966).

In addition, N mineralization is affected by the extent of grinding (Waring and Bremner 1964). Hence, the chemical indices of soil N availability are empirical and different authors recommend different methods of determination. A NaOH microdiffusion method is widely used in China (Zhu 1962; Zhou *et al.* 1976, 1981; Xie 1985). In addition, methods such as alkaline $KMnO_4$ distillation (Xie 1985, Shen 1987), phosphoric acid-sodium borate distillation (Shen 1987) and extraction with boiling KCl (Li 1984) have been proposed.

3.5.2. Evaluation of indices

Incubation, pot and field experiments are often used in the evaluation of availability indices; however, the final judgement should come from the results of field experiments. Table 3.19 provides a summary of the correlations obtained in relevant studies. The results show that NaOH-diffusion N, and total N were generally correlated with mineral N after incubation (N_{min}, the sum of N mineralized during incubation and the initial mineral N). Although the correlation with NaOH-diffusion N was slightly better than that with total N, in fact, NaOH-diffusion N was significantly correlated with total N (Zhu 1986).

Table 3.18. Importance of the subsoil in flooded fields for supplying nitrogen (kg N ha^{-1}) to rice.[1]

	I			II			III			Mean
Source of N	1	2	3	4	5	6	7	8	9	
Plow layer (A)	77.4	52.8	72.1	50.3	75.0	59.3	49.2	42.6	38.8	57.5
Total Nas (B)	92.5	75.1	95.2	73.5	91.7	74.0	81.3	72.5	76.6	81.4
Below plow layer (B – A)	15.1	22.3	23.1	23.2	16.7	14.7	32.1	29.9	37.8	23.9
(B - A)/B, %	16.3	29.7	24.3	31.6	18.2	19.9	39.5	41.2	49.3	30.0
Mean	23.4 ± 6.7			23.2 ± 7.2			43.4 ± 5.3			

[1] Field microplot trial in southern Jiangsu with single-cropped rice (Chen 1985).
a) I, II, III – see Table 3.1.
b) The total Nas was calculated as the total nitrogen accumulation in the aerial parts (minus seedling nitrogen) of the rice plants grown in bottomless cylinders, thus the rice roots could extend freely. In determining the amount of nitrogen supply of plow layer, rice roots were restricted within the plow layer with two layers of 300 mesh nylon cloth at the bottom of the cylinders.

Table 3.19. Relationship between N_{min}, NaOH diffusion N and total N.[1]

Site	Soil	Incubation method	Correlation coefficient for the relationship between		Reference
			N_{min}, TN	N_{min}, NaOH-N	
Jiangsu, Shanghai	I	anaerobic	0.796*	0.753*	Zhu *et al.* (1984b)
	II	anaerobic	0.606*	0.795**	
	III	anaerobic	0.581	0.628	
Zhejiang		anaerobic	0.756**	0.848**	Zhou *et al.* (1976)
Jiangsu		anaerobic	0.653	–	Shen (1987)
		anaerobic	0.600	–	
Inner Mongolia		anaerobic	0.725**	0.783**	Wang (1981)
Jilin		anaerobic	0.813**	–	Zhou *et al.* (1981)
		aerobic	0.766**	–	

[1] Abbreviations as in Table 3.1;
*and ** significant at P = 0.05 and P = 0.01, respectively.

In pot experiments N accumulation in plants without N application is usually measured along with dry weight and grain yield. Table 3.20 shows that N_{min} measured in incubation experiments was strongly correlated with Nas in pot experiments. NaOH-diffusion N and total N were less strongly correlated with Nas.

Strong relationships were not usually found in field experiments. The correlation between Nas and the measured indices was usually weak ($r^2 \cong 0.25$) and seldom significant (Table 3.21), and unsatisfactory for the quantitative prediction of Nas. It should be noted that even the incubation method does not satisfactorily predict Nas in the field.

3.5.3. *Precision of prediction*

The weak relationship between N availability index and Nas in field experiments may result from a number of factors. As indicated previously, the N in plants grown

Table 3.20. Relationship between the apparent supply of soil nitrogen (Nas) and soil nitrogen availability indexes.[1]

Crop	Site	Correlation coefficient for the relationship between			Reference
		Nas, TN	Nas, NaOH-N	Nas, N_{min}	
Rice	Zhejiang	0.929**	0.893**	0.826**	Zhou *et al.* (1976)
	Jiangsu	0.574	0.775*	0.961**–0.969**	Shen (1987)
	Guangzhou	0.514	0.595*	0.956**–0.967**	Xie (1985)
	Guangzhou	0.555	0.669*	0.721*–0.893**	Xie (1985)
	Guangzhou	0.816**	0.924**	0.857**–0.925**	Xie (1985)
Corn	Jilin	–	0.884**–0.938**	0.851**–0.898**	Zhou *et al.* (1981)
Millet	Jiangsu, Zhejiang, Anhui	0.877**	0.795**	0.991**	Shen (1987)

[1] Pot experiment; * and ** – significant at 5% and 1%, respectively.

Table 3.21. Relationship between the apparent supply of soil nitrogen (Nas) and nitrogen availability indexes.[1]

Crop	Site	Soil	Correlation coefficient for the relationship between			Reference
			Nas, TN	Nas, NaOH-N	Nas- N_{min}	
Rice	Zhejiang		0.467*[2]	0.480*	0.504*	Zhou *et al.* (1976)
	Jiangsu & Shanghai	I	0.237	0.207	–0.155	Zhu *et al.* (1984b)
	Jiangsu & Shanghai	II	0.534	–0.409	–0.267	Zhu *et al.* (1984b)
	Jiangsu & Shanghai	III	0.527	0.569	–0.522	Zhu *et al.* (1984b)
Millet	Inner Mongolia				0.939**	Wang (1981)

[1] Field experiment.
[2] * and ** – significant at P = 0.05 and P = 0.01, respectively.

on zero-N plots, which is usually taken as a measure of Nas, is derived from different sources. Therefore, the weak relationships may result not only from errors in predicting the amount of N derived from the soil source, but also from interference with the N derived from non-soil sources.

Available N in soil includes the initial mineral N and the N mineralized throughout the period of plant growth. The initial ammonium has already been taken into account in the NaOH diffusion method or the anaerobic incubation method, and the initial nitrate was also included in the measurements on upland soils using a modified NaOH diffusion method to include nitrate. Therefore the weak correlation between N_{min} or NaOH-N and Nas was not related to the initial mineral N content of the soil.

The prediction of mineralization in a field with availability indices has been extensively studied. Nitrogen mineralized during the period of plant growth is affected by the amount of potentially mineralizable N in the soil, hydrothermal conditions and the duration of plant growth. Management practices such as soil drying after plowing the surface layer and harrowing also profoundly affect N mineralization. Unfortunately, the prevailing incubation or chemical method could at most determine the potentially mineralizable N in the soil or its relative magnitude while the effect of environmental conditions and management practices on N mineralization were usually not quantified. Some suggestions have been proposed in order to improve the precision of prediction (Wang *et al.* 1983; Kafkafi *et al.* 1978; Smith *et al.* 1977; Stanford *et al.* 1973b, 1977). Wang *et al.* (1983) proposed that the n value in the cumulative effective temperature equation for a particular soil type be determined by anaerobic incubation, that the k value be obtained with the NaOH diffusion method, and that the cumulative effective temperature in different fields be estimated according to the regression equation relating soil and air temperature to eliminate errors caused by differences between fields. This method can be used to predict N mineralization, and the mineralization pattern as well. For upland soils, it is suggested that temperature and water content be taken into account in the first order kinetic equation to predict soil N mineralization (Kafkafi *et al.* 1978). It is well known that soil drying after plowing and harrowing strongly influences the mineralization of soil N, but the quantitative determination of the extent of this influence still remains a problem.

Note that in the experiments mentioned above, only the N in the plow layer was used for predicting Nas, and the effect of subsoil supply was ignored. However, as shown in Table 3.18, the contribution from the subsoil to total Nas varied greatly between different fields, even those with the same soil type, or between soils of the same region. This is presumably one important factor for the weak relationship between Nas and availability indices in the plow layer. The problem of how to estimate mineralization of N in different soil layers applies to upland soils also (Hadas *et al.* 1986).

Anaerobic incubation was usually used for the study of paddy soils. It can also be used for determining mineralization in upland soils, because of the strong relationship between mineralization in anaerobic and aerobic incubation. However, the fixation of mineralized ammonium by clay minerals (Zhu 1988) may lead to less precise prediction unless it is taken into account.

The greater the contribution of a non-soil source N to Nas, and the greater the variation in the contribution between the soils studied the weaker the correlation between Nas and soil N availability indices will be.

As discussed earlier, Ndfa accounts for a large proportion of Nas for rice plants. However, it should not have a significant effect on the relationship between Nas and soil N availability indices because it doesn't vary greatly in different soils. In upland soils, it makes a negligible contribution to Nas.

For directly seeded crops, very little N is brought in by seeds and it has very little effect on the prediction of Nas. However, the contribution of seedling N to Nas may be considerable for crops such as transplanted rice, and the amount of seedling N varies greatly due to the variation in the size and quality of seedlings (Table 3.17). In order to minimize prediction error, seedling N should be subtracted from Nas in the correlation study. However, even taking seedling N into account, as in Table 3.21, the relationship was still poor, implying that other factors were involved.

Another factor that will affect the correlation between Nas and availability indices is the N supplied in rainfall and irrigation water. According to the data obtained in the Taihu region, as mentioned earlier, the net input was 7.4–15.3 kg N ha^{-1} yr (unpublished data of Liu). The net amount of N for each crop will be even less. In comparison with the contribution and variation of other non-soil sources, it might not greatly affect the relationship, except for polluted water high in N.

It can be concluded from the above discussion, that in order to improve the prediction of Ns it is important to standardize the minimum sampling depth, and to estimate mineralization and plant uptake in different soil layers. Attention should also be paid to the quantitative estimation of the effect of management practices on soil N mineralization.

3.6. References

Adu, J K, and Oades, J M 1978. Physical factors influencing decomposition of organic materials in soil aggregates. Soil Biol. Biochem. 10:109–116.

Beauchamp, E G, Reynolds, W D, Brashe-Villenueve, O, and Kirby, K 1986. Nitrogen mineralization kinetics with different soil pretreatments and cropping histories. Soil Sci. Soc. Am. J. 50:1478–1483.

Bai, Z J, and Zhao, G S 1981. Nitrogen mineralization potentials of the main arable soils in Shaanxi Province. (in Chinese). J. Soil Sci. 4:26–29.

Birch, H F 1958. The effect of soil drying on humus decomposition and nitrogen availability. Plant Soil 10:9–31.

Birch, H F 1960. Nitrification in soil after different periods of dryness. Plant Soil 12:81–96.

Bremner, J M 1965. Nitrogen availability indexes. In: Black CA (ed.), Method of Soil Analysis, Part 2, pp. 1324–1345. Am. Soc. Agron. Madison, Wisconsin.

Broadbent, F E 1978. Mineralization of organic nitrogen in paddy soils. In: Nitrogen and Rice, pp. 105–118. IRRI, Manila.

Cai, G X, and Zhu, Z L 1983. Effect of rice growth on the mineralization of soil nitrogen (in Chinese). Acta Pedol. Sin. 20:272–278.

Cai, G X, Zhang, S L and Zhu, Z L 1979. Condition trial of sealed incubation under flooded conditions for measuring soil nitrogen mineralization pattern in paddy soils (in Chinese). Soils 6:234–240.

Cai, G X, Zhang, S L and Zhu, Z L 1981. Characteristics of N mineralization of paddy soils and their effect on the efficiency of nitrogen fertilizer. In: Institute of Soil Science, Academia Sinica (ed.), Proc. of Symp. on Paddy Soil, pp. 793–799. Science Press, Beijing.

Campbell, C A, Myers, R J K and Weier, K L 1981. Potentially mineralizable nitrogen, decomposition rates and their relationship to temperature for five Queensland soils. Aust. J. Soil Res. 19:323–332.

Chen, D L 1985. Study on natural nitrogen-supplying abilities of paddy soils. (in Chinese). M. Sc. Thesis. Institute of Soil Science, Academia Sinica. Nanjing.

Cornfield, A H 1964. Effect of air-drying storage of soils on the subsequent accumulation of mineral nitrogen during incubation. Plant Soil 20:260–264.

Gao, J H, Zhang, Y, Huang, D M, Wu, J M and Pan, Z P 1984a. Study on soil nitrogen mineralization parameters to predict fertilization for early rice (in Chinese). Chinese Agric. Sci. 5:67–72.

Gao, J H, Zhang, Y, Huang, D M, Wu, J M and Pan, Z P 1984b. Soil nitrogen mineralization patterns and N fertilizer efficiencies in paddy soils (in Chinese). Acta Pedol. Sin. 21:341–350.

Gasser, J K R 1961. Effect of air-drying and air-dry storage on the mineralizable nitrogen of soils. J. Sci. Food Agric. 12:778–784.

Ge, N F, Zhang, D Y, Ma, S F, Zhang, Y D and Fei, B Y 1987. Study on nitrogen nutrition and the yield response from N fertilizer applications for hybrid rice. II. Fate of urea N applied to soils and its effect on soil N supply. (in Chinese). J. Nanjing Agricultural University 2:69–75.

Greenland, D J and Watanabe, I 1982. The continuing nitrogen enigma. Trans. 12th Inter. Cong. Soil Sci. 5:123–137. New Delhi.

Hadas, A, Feigenbaum, S, Feigin, A and Portnoy, R 1986. Nitrogen mineralization in profiles of different managed soil types. Soil Sci. Soc. Am. J. 50:314–319.

Harada, T 1959. The mineralization of native organic nitrogen in paddy soils and the mechanism of its mineralization (in Japanese). Bull. Natl Inst. Agric. Sci. (Japan), Ser. B, 9:123–199.

Harmsen, G W, and Van Schreven D A 1955. Mineralization of organic nitrogen in soil. Adv. Agron. 7:300–398.

Huang, D M and Zhang, B S 1957. Transformation of nitrogen in paddy soils and yield of rice as affected by ploughing under dry and water-logged conditions (in Chinese). Acta Pedol. Sin. 5:223–233.

Huang, D M, Gao, J H and Zhu, P L 1981. The transformation and distribution of organic and inorganic fertilizer nitrogen in rice-soil system (in Chinese). Acta Pedol. Sin. 18:107–121.

ISSAS 1977. (Institute of Soil Science, Academia Sinica.) The effect of N fertilizers on increasing rice yield under the rotation system of 'rice-rice-wheat' in Suzhou district (in Chinese). Soils 3:127–135.

ISSAS 1978. (Institute of Soil Science, Academia Sinica.) Physical and Chemical Analysis of Soils. (in Chinese). pp. 76–78. Shanghai Science and Technology Press, Shanghai.

Inubushi, K, Wada, H and Takai, Y 1985. Easily decomposable organic matter in paddy soil VI. Kinetics of nitrogen mineralization in submerged soils. Soil Sci. Plant Nutr. 31. 31:563–572.

Ito, O and Watanabe, I 1981. Immobilization, mineralization and availability to rice plants of nitrogen derived from heterotrophic nitrogen fixation in flooded soil. Soil Sci. Plant Nutr. 27:169–176.

Ivarson, K C, and Schnitzer, M 1979. The biodegradability of the 'unknown' soil-nitrogen. Can. J. Soil Sci. 59:59–67.

Jenkinson, D S, Fox, R H and Rayner, J H 1985. Interactions between fertilizer nitrogen and soil nitrogen – the so-called 'priming effect'. J. Soil Sci. 425–444.

Jin, Z Y, Zhou, Z C, Wang, H Y, Sun, W C and Wu, W 1983. Study on loss pathways of solid nitrogen fertilizers applied to upland soils. Soil Science Society of China (ed.), Utilization and Fertility Improvement of Soils of China. (in Chinese). Vol. 3. pp. 190–193.

Jin, Z Y, Zhou, Z C, Wang, H Y, Sun, WC, Zhou, W P, Zou, S F, Yao, W H and Li, L W 1984. Study on loss pathways of different solid nitrogen fertilizers and ways to reduce nitrogen losses in black soils (in Chinese). J. Soil Sci. 4:153–156.

Juma, N G and Paul, E A 1984. Mineralizable soil nitrogen: amounts and extractability ratios. Soil Sci. Soc. Am. J. 48:76–80.

Kafkafi, J, Bar-Yosef, B and Hadas, A 1978. Fertilization decision model – A synthesis of soil and plant parameters in a computerized program. Soil Sci. 125:261–268.

Keeney, D R 1982. Nitrogen – availability indexes. In: Page, A L, Miller, R H, and Keeney, D R (eds.), Methods of Soil Analysis, Part 2, 2nd ed. pp. 711–733. Am. Soc. of Agron. and Soil Sci. Soc. Am., Madison, Wisconsin.

Keeney, D R and Bremner, J M 1964. Effect of cultivation on the nitrogen distribution in soils. Soil Sci. Soc. Am. J. 28:653–656.

Keeney, D R and Bremner, J M 1966. Characterization of mineralizable nitrogen in soils. Soil Sci. Soc. Am. Proc. 30:714–719.

Li, Z L, Li, A R and Cao, Z H 1984. Effect of application methods of nitrogen fertilizers on the recovery of fertilizer nitrogen by spring wheat in calcareous soils (in Chinese). Soils 16:134–137.

Molina, J A E and Clapp, C E 1980. Potentially mineralizable nitrogen in soil: The simple exponential model does not apply to the first 12 weeks of incubation. Soil Sci. Soc. Am. J. 44:442–443.

Peng, L, Peng, X L and Lu, Z F 1981. The seasonal variation of soil NO_3-N and the effect of summer fallow on the fertility of manured loessial soil (in Chinese). Acta Pedol. Sin. 18:212–222.

Richter, J R, Muske, A, Habenicht, W and Bauer, J 1982. Optimized nitrogen-mineralization parameters of loess soils from incubation experiments. Plant Soil 68:379–388.

Riga, A, Fisher, V and Praag, H J Van 1980. Fate of fertilizer nitrogen in microplots through a four-course rotation: 1. Influence of fertilizer splitting on soil and fertilizer nitrogen. Soil Sci. 130:88–99.

Shen, Q R 1987. Studies on the forms and availability of soil nitrogen. (in Chinese). Ph.D Thesis. Nanjing Agricultural University. Nanjing, China.

Shen, S M 1986. The effect of mineral nitrogen on the mineralization and immobilization of soil nitrogen (in Chinese). Acta Pedol. Sin. 23:10–16.

Shen, Z P, Huang, D M, Bai, G Y and Duan, S T 1959. Yield response to drying of paddy soil and its relations with soil properties (in Chinese). Acta Pedol. Sin. 7:124–135.

Smith, J L, Schnabel, R R, Mc Neal, B L and Campbell G S 1980. Potential error in the first-order model for estimating soil nitrogen mineralization potential. Soil Sci. Soc. Am. J. 44:996–1000.

Smith, S J 1987. Soluble organic nitrogen losses associated with recovery of mineralized nitrogen. Soil Sci. Soc. Am. J. 51:1191–1194.

Smith, S J, Young, L B and Miller, G E 1977. Evaluation of soil nitrogen mineralization potentials under modified field conditions. Soil Sci. Soc. Am. J. 41:74–76.

Stanford, G and Smith, SJ 1972. Nitrogen mineralization potentials of soils. Soil Sci. Soc. Am. Proc. 36:465–472.

Stanford, G, Legg, J O and Smith, S J 1973a. Soil nitrogen availability evaluation based on nitrogen mineralization potentials and uptake of labelled and unlabelled nitrogen by plant. Plant Soil 39:113–124.

Stanford, G, Frere, M H and Schwaninger, D E 1973b. Temperature coefficient of soil nitrogen mineralization. Soil Sci. 115:321–323.

Stanford, G, Carter, J N, Westermann, D T and Meisinger, J J 1977. Residual nitrate and mineralizable soil nitrogen in relation to nitrogen uptake by irrigated sugar beets. Agron. J. 69:303–308.

Wang, J Z 1981. A method of measuring soil nitrogen supply in Zhaowuda League, Inner Mongolia (in Chinese). J. Soil Sci. 6:20–22.

Wang, Y H, Jiang, S Z and Gu, Y M 1983. A study on predicting nitrogen supplying capacities of gleyed paddy soil in the suburbs of Shanghai (in Chinese). Acta Pedol. Sin. 20:262–271.

Wang, Y Q, Nordmeyer, H and Richter, J 1986. Nitrogen mineralization of loessial soil, cinnamon soil and fluvo-aquic soil in China (in Chinese). Acta Pedol. Sin. 23:1–9.

Waring, S A and Bremner, J M 1964. Effect of soil mesh-size on the estimation of mineralizable nitrogen in soils. Nature 202:1141.

Wen, Q X and Zhang, X H 1986. Fixed ammonium in soils. In: Soil Agricultural Chemistry and Soil Biology and Biochemistry Committees, Soil Science Society of China (eds.), Advances and Prospects for Soil Nitrogen Research in China (in Chinese). pp. 34–45. Science Press, Beijing.

Xie, L C 1985. Study on methods for measuring available nitrogen in paddy soils. (in Chinese). J. Soil Sci. 1:41–43.

Xiong, Z X, Chen, M H and Meng, Z Y 1986. Study on soil nitrogen supply using 15N labelled technique (in Chinese). Ningxia Agric. Sci. and Technology 4:7–9.

Xu, F A and Xu, S Y 1981. The effect of different structure of paddy soil on the nutrient absorption by early rice plants (in Chinese). Acta Pedol. Sin. 18:199–202.

Yoshida, T and Padre, Jr B C 1977. Transformation of soil and fertilizer nitrogen in paddy soil and their availability to rice plants. Plant Soil 47:113–123.

Yoshino, T and Dei, Y 1974. Patterns of nitrogen release in paddy soils predicted by an incubation methods. JARQ 8:137–141.

Yoshino, T and Dei, Y 1977. Prediction of nitrogen release in paddy soils by means of the concept of effective temperature (in Japanese). J. Cent. Agric. Exp. Stn. 25:1–62.

Yuan, L Z, Sun, G Y and Huang, D M 1985. Study on soil nitrate as an index of fertilization for maize (in Chinese). Jiangsu Agric. Sci. 6:25–27.

Zhang, B Q, Sun, X W and Guan, L Z 1987. Effect of fertilization on soil organic matter and several main fertility properties (in Chinese). J. Soil Sci. 18:156–160.

Zhang, S L, Zhu, Z L and Xu, Y H 1988. On the optimal rate of application of nitrogen fertilizers for rice and wheat in Tai-lake region (in Chinese). Soils 20:5–9.

Zhou, D C 1983. Investigations on the efficiency of ammonium bicarbonate, ammonium sulfate and urea applied to wheat. (in Chinese). Shanghai Academy of Chemical Engineering (ed.), Agrochemical Properties and the Efficiency of Ammonium Bicarbonate, pp. 26–30.

Zhou, M Z 1978a. Methods of measuring soil available nitrogen – Mineralization rate methods (in Chinese). Soils and Agricultural Chemistry 5:25–37.

Zhou, M Z 1978b. Methods of measuring soil available nitrogen – Chemical methods (in Chinese). Soils and Agricultural Chemistry 6:32–44.

Zhou, M Z, Yu, W T and Fang, Z F 1976. Methods of measuring soil available nitrogen (in Chinese). Soils (5–6):316–323.

Zhou, Z C, Wang, H Y, Cai, Y D, Jin, Z Y and Xiang, J 1981. Study on methods for measuring available nitrogen in the main upland soils in Jilin Province (in Chinese). J. Soil Sci. 6:23–26.

Zou, D F and Wei, X M 1983. Investigations on the fate of nitrogen fertilizers applied to calcareous soils. Soil Science Society of China (ed.) Utilization and Fertility Improvement of Soils of China. (in Chinese). Vol. 3. pp. 185–189.

Zhu, P L and Huang, D M 1983. Study on residual nitrogen potential in soils (in Chinese). Jiangsu Agric. Sci. 11:1–7.

Zhu, Z L 1962. Investigation of nitrogen supplying status of soils I. Rate of liberation of ammonia in alkaline hydrolysis as an index for predicting nitrogen supplying status of rice field (in Chinese). Acta Pedol. Sin. 10:55–71.

Zhu, Z L 1963. Transformations of nitrogen in soil – a review of literature (in Chinese). Acta Pedol. Sin. 11:328–338.

Zhu, Z L 1979. Advances in the research of soil nitrogen transformations and movement. (in Chinese). Progress in Soil Sci. 2:1–16.

Zhu, Z L 1982. Parameters for assessing the application rate of nitrogen fertilizer on rice and wheat. (in Chinese). Soils 14:136–140.

Zhu, Z L 1986. Mineralization and supply of soil nitrogen. In: Soil Agricultural and Soil Biology and Biochemistry Committees, Soil Science Society of China (eds.), Advances and Prospects for Soil Nitrogen Research in China. (in Chinese). pp. 14–27. Science Press, Beijing.

Zhu, Z L 1988. Nitrogen mineralization and supply of paddy soil. Proc. 1st Inter. Symp. on Paddy Soil Fertility, Part 1 193–204. December 6–13 1988. Chiangmai.

Zhu, Z L, Liu, C Q and Jiang, B F 1984a. Mineralization of organic nitrogen, phosphorus and sulphur in some paddy soils of China. In: Organic Matter and Rice. pp. 259–272. IRRI, Manila.

Zhu, Z L, Cai, G X, Xu, Y H and Zhang, S L 1984b. Nitrogen mineralization of paddy soils in Tai-lake region and the prediction of soil nitrogen supply (in Chinese). Acta Pedol. Sin. 21:29–36.

Zhu, Z L, Chen, D L, Zhang, S L and Xu, Y H 1986. Contributions of nonsymbiotic nitrogen fixation to the nitrogen uptake by growing rice under flooded conditions (in Chinese). Soils 18:225–229.

Zhu, Z L, Liao, X L, Cai, G X and Yu, J Z 1978. Soil nutrition status under 'rice-rice-wheat' rotation and the response of rice to fertilizers in Suzhou District (in Chinese). Acta Pedol. Sin. 15:126–137.

Zhu, Z L, Chen, Y Y, Yu, Y F, Xu, Y H and Zhang, S L 1979. The effect of forms and methods of placement of nitrogen fertilizer on characteristics of the nitrogen supply in paddy soils (in Chinese). Acta Pedol. Sin. 16:218–233.

Zhu, Z M and Wu, T B 1983. Study on the effect of paddy soil nitrogen supply on the efficiency of nitrogen fertilizers using ^{15}N (in Chinese). Hunan Agric. Sci. 5:36–38.

4
Fixation and release of ammonium

WEN QI-XIAO AND CHENG LI-LI

4.1. Introduction

In 1917 it was found that soil has the ability to fix ammonium (McBeth 1917), but it was not until the 1950s that it was understood that fixed ammonium exists widely in various kinds of soils, layer silicate minerals, and sedimentary and igneous rocks (Bremner and Harada 1959; Rodrigues 1954; Stevenson 1959; Stevenson and Dhariwal 1959). Fixed ammonium may account for as much as 50% of the total N in the top metre of some soils, and the fixation and release of ammonium plays a prominent role in the internal cycle of N in soils.

Li (1938), while studying the adsorption of ammonium by soils, found that soils have the ability to fix ammonium and that the extent of fixation varies from soil to soil. In recent years, investigations have been carried out on the ammonium-fixing capacity, the content and distribution of fixed ammonium in soil profiles, and the dynamics and biological availability of fixed ammonium in soils from various parts of China. In this chapter a brief review will be given on these studies.

4.2. Fixation of ammonium

4.2.1 Mechanism

Clay minerals consist of tetrahedral Si-O sheets and octahedral Al-O-OH sheets. The minerals can be divided into 1:1 and 2:1 types according to the proportion of each unit present. As a result of isomorphous substitution, the lattices of 2:1 type clay minerals possess negative charges, which are balanced by other cations, e.g. Ca^{++}, Mg^{++}, K^+ or NH_4^+, either inside the crystal or outside the structural unit. The size of the ammonium ion is similar to the opening in the hexagonal cavity formed by the six oxygen atoms on the exposed surfaces between the two tetrahedral Si-layers (Xu and Hseung 1983), and the electrostatic force between ammonium and the negative charges in the crystal sheets is greater than the hydration energy of ammonium. Consequently, it is easy for the ammonium ion to shed its hydration water shell and enter the 'lattice void' to be fixed. K^+ and ammonium have nearly identical ionic radii and are fixed by the same mechanism. Other cations cannot be fixed either due to their smaller ion radii or greater hydration energy.

Zhu Zhao-liang et al. *(eds.): Nitrogen in Soils of China, 67–86.*

4.2.2 Factors affecting ammonium fixation

The capacity of soil to fix ammonium is controlled by a series of factors, such as the type of clay mineral, soil texture, soil pH, concentration of ammonium, soil organic matter, and the presence of other cations.

It is generally accepted that 2:1 type clay minerals have the ability to fix ammonium, while 1:1 type clay minerals, such as kaolinite and halloysite, are non-fixing minerals. Among the 2:1 type clay minerals vermiculite has the greatest ability to fix ammonium, followed by montmorillonite. The ability of illite to fix ammonium depends on the degree of weathering and K^+ saturation of the lattice. As shown in Table 4.1, under the experimental conditions used vermiculite did not reach its maximum capacity for fixation, and illite with a high degree of K^+ saturation (7.5% K) fixed little ammonium.

The ammonium-fixing capacity of a soil is closely related to the kind of clay minerals present. Soils derived from the Changjiang River alluvium and purplish sandstone and shale in the subtropics, which are dominated by hydromica and vermiculite in the clay fraction, not only contain a large amount of native fixed ammonium, but also have the capacity to fix considerable added ammonium. However, soils derived from granite and sandstone in the subtropics, and all of the soils in the tropics with kaolinite as the predominant clay mineral contain relatively little native fixed ammonium and fix little fertilizer ammonium (Table 4.2).

In soils of same type, the greater the content of 2:1 type clay minerals, the greater is its ability fix ammonium. Since the clay minerals are present mainly in the clay and fine silt fractions, soils with a high content of clay and silt have a greater capacity to fix ammonium. As shown in Figure 4.1, the ammonium fixation capacity of the coarse clay fraction (1–2 μm) of Xiashu loess was 6 times greater than that of the coarse silt fraction (10–20 μm), and 12 times greater than that of the sand fraction (>20 μm).

Nommik (1957) found that ammonium fixation tends to increase with increasing soil pH; soils with pH values lower than 5.5 generally fix little ammonium. In addition, the effect of soil pH on K^+ fixation may be used to determine the effect of soil

Table 4.1. Ammonium fixation by clay minerals.[1]

	Vermiculite		Montmorillonite		Illite	
$(NH_4)_2SO_4$ added (mg N kg^{-1})	Fixed (mg N kg^{-1})	Newly fixed (% of added N)	Fixed (mg N kg^{-1})	Newly fixed (% of added N)	Fixed (mg N kg^{-1})	Newly fixed (% of added N)
0	90	–	34	–	669	–
250	142	53.2	90	22.4	662	0
500	318	61.8	100	13.1	678	1.8
1000	860	85.0	122	8.8	680	1.2
2000	1806	89.9	171	6.8	689	1.0

[1] Unpublished data of Cheng Li-li *et al.*

Table 4.2. Ammonium fixation by surface soils formed from different parent materials.

Parent material	Soil	Locality	Native fixed NH_4^+ (mg N kg^{-1})	Newly fixed[1] (% of added N)
Changjiang River alluvium	Calcareous hydromorphic paddy	Shanghai, Jiangsu	287 ± 30	64.0 ± 6.69
Loessial sediment	Neutral hydromorphic paddy	Jiangsu	205 ± 33	31.0 ± 10.6
Purple sandstone and shale	Purple, Neutral hydromorphic paddy	Hunan, Zhejiang	357 ± 83	29.8 ± 10.9
Quaternary Red Clay	Acid hydromorphic paddy	Jiangxi	124 ± 21	1.1 ± 1.15
Tertiary red sandstone, granite, sandstone, basalt, etc.	Latosol, Red	Jiangxi, Hainan	54 ± 13	1.5 ± 2.07

[1] 250 mg N as ammonium sulfate were added per kg of soil.

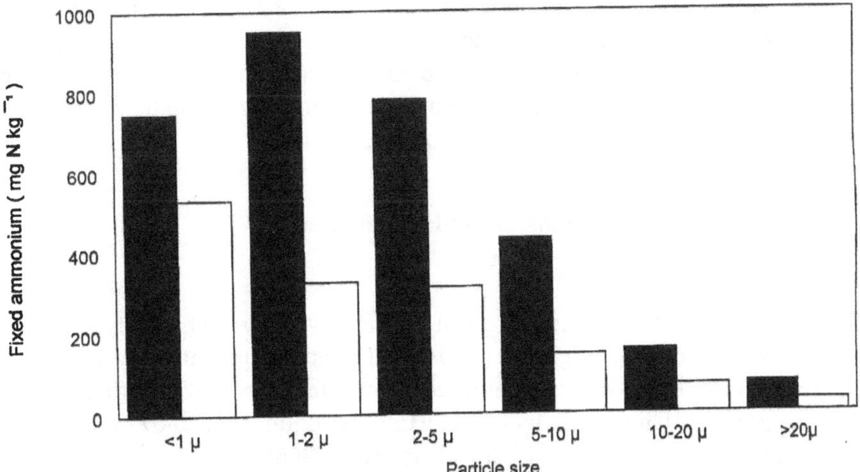

Figure 4.1. Fixed ammonium in different particle size fractions of Xiashu loess ■ and its bleached layer □.

pH on ammonium fixation. Rich (1964) and Thomas and Hipp (1968) in their studies on the fixation of potassium by soils found that in soils with pH values below 5.5, aluminium exists mainly as Al^{3+} and $Al(OH)_x$, which depress the fixation of K^+; between pH 5.5 and 7.0, the amount of K^+ fixed increases gradually with increasing soil pH as a result of the transformation of $Al(OH)_x$ to hydroxy-Al groups.

As with ammonium adsorption, the amount of ammonium fixed by a soil increases, and the percentage fixation decreases as the concentration of ammonium in solution increases. In general, the increase in amount of ammonium fixed by increasing the ammonium concentration is greater in soils with high capacity to fix ammonium than in soils with low capacity. This is also true for different particle-size fractions of a soil. As shown in Table 4.3, particles <10 μm in size, especially those in the 2–5 μm range, fixed more added ammonium than particles >10 μm.

Ammonium fixation tends to increase during drying of a soil after addition of ammonium. This is presumably due to the removal of water and the increase in concentration of ammonium in solution, and the partial contraction of the lattice which leads to the entrapment of ammonium in the interlayers (Black and Waring 1972). The latter may be essential for the fixation of ammonium by soils containing montmorillonite.

Because of the competition between potassium and ammonium for fixing sites potassium has a depressive effect on the fixation of ammonium. However, it is still not known whether the amount of fixed ammonium in a soil will be reduced if large quantities of fertilizer potassium are added to the soil for a long time. Guo *et al.* (1986) found that application of 50–56 kg K ha^{-1} had no effect on the amount of native fixed ammonium.

It was suggested that organic matter can either block the entry of ammonium into fixing sites or prevent collapse of the basal spacing of clay minerals (Hinman 1966), thus reducing ammonium fixation. Guo *et al.* (1986) observed that fixed ammonium in soils which had received organic manure for a long time was lower than that in soils which had not received organic manure.

4.3. Amount and distribution of fixed ammonium

4.3.1. *Amount of fixed ammonium*

The amount of fixed ammonium in a soil is primarily governed by the type of clay mineral, which in turn, largely depends on the type of parent material and its degree of weathering. Due to differences in soil age and parent material, soils in different zones of China vary widely in fixed ammonium, (11 to 677 mg N kg^{-1}; determined by the method of Silva and Bremner 1966). The greatest amount is found in clay soils rich in vermiculite and the smallest amount is found in sandy soils rich in kaolinite. Table 4.4 shows that along a transect from south to north,

Table 4.3. Fixation of ammonium by different particle-size fractions of Xiashu loess.[1]

$(NH_4)_2SO_4$ added (mg N kg^{-1})	<1 μm	1–2 μm	2–5 μm	5–10 μm	10–20 μm	>20 μm
0	100[2]	100	100	100	100	100
676	203	282	324	241	197	144
1014	242	320	381	287	220	181
1352	252	342	413	328	265	252
1690	272	344	419	328	274	235

[1] Unpublished data of Wen Qi-xiao *et al.*
[2] The amount of fixed ammonium in the untreated fraction is set as 100.

Table 4.4. Fixed ammonium (mg N kg^{-1}) in soils of different zones.[1]

Zone	No. of samples	Maximum	Minimum	Mean ± S.D.
Black soil, castanozem	94	303	19	160 ± 58
Brown soil, cinnamon soil	94	279	78	172 ± 38
Yellow-brown soil	86	381	104	238 ± 70
Red soil	258	677	11	144 ± 109
Latosol	52	87	11	53 ± 26

[1] Data from Wen *et al.* (1988), Sun and Wu (1989), Chen and Zhu, (1988), Zhang *et al.*(1989), Shi *et al.* (1987), (1992), Guo *et al.* (1986), Cheng *et al.* (1988), Chu *et al.* (1986), Fan *et al.* (1990), and unpublished data of Department of Soil Biochemistry, Institute of Soil Science, Academia Sinica.

fixed ammonium is lowest in the latosol zone, varying from 11 to 87 mg N kg^{-1} (average 53 mg N kg^{-1}). This is because the dominant clay minerals in these soils are kaolinite, haematite, and gibbsite. Trace amounts only of vermiculite and hydromica are present in these soils because of the high degree of weathering. Soils in the red soil zone which are not weathered to the same extent and contain more hydromica and vermiculite than the latosols, contain more fixed ammonium (144 mg N kg^{-1} on average).

In the yellow-brown soil zone where hydromica and vermiculite are the dominant clay minerals, there is more fixed ammonium than in the soils of the red soil zone (averaging 238 mg N kg^{-1}), which is the highest of the zones studied. North of the yellow-brown soil zone is the brown soil zone. In these soils, which are less weathered, the clay minerals are mainly hydromica or montmorillonite and hydromica, so that fixed ammonium decreases to 172 mg N kg^{-1} (on average). In the zones further north lie the castanozem, chernozem and sierozem soils. Although the predominant clay minerals in these soils are similar to those in the brown soil zone, the lower degree of weathering and increased saturation of the lattice by K^+, Mg^{++} and Ca^{++} etc. result in less fixed ammonium (average 160 mg N kg^{-1}).

Needless to say the discussion above gives a general idea only of the variation in fixed ammonium in soils of the different zones. Even within a zone the soils derived from different parent materials differ greatly in their content of fixed ammonium. For instance, in the Taihu Lake region of the yellow-brown soil zone, the average fixed ammonium in soils formed on loessial sediment is only 198 mg N kg^{-1}, while in soils derived from Changjiang River alluvium fixed ammonium averages 298 mg N kg^{-1}. Similarly, in Shandong Province in the brown soil zone, the average fixed ammonium in brown soils developed on granite is 127 mg N kg^{-1}, while in Chao soils (fluvo-aquic soils) derived from Huanghe River alluvium it is 198 mg N kg^{-1}. In the red soil zone the variation is even greater. The average fixed ammonium in soils developed on weathered granite is the lowest (40 mg N kg^{-1}), while in soils formed on shale, purple sandstone and shale, marl and river alluvium, because of the young soil age and the low degree of weathering, fixed ammonium is usually more than 200 mg N kg^{-1}. The greatest amount is found in soils derived from shale, averaging 332 mg N kg^{-1} (Table 4.5). Because of the presence of such young soils, the

Table 4.5. Fixed ammonium in surface layers of soils derived from different parent material in subtropical China.[1]

Parent material	Locality	Fixed ammonium	
		(mg N kg^{-1})	(% of total N)
Shale	Guizhou, Hunan, Yunnan	332 ± 128	19.8 ± 10.4
Purple sandstone and shale,	Zhejiang, Hunan, Sichuan Yunnan, Guangdong	253 ± 85	30.0 ± 17.5
Marl	Jiangxi, Hunan, Guizhou Guangdong Guangxi, Sichuan	234 ± 124	16.4 ± 11.9
River alluvium	Jiangxi, Anhui, Zhejiang Fujian, Hunan, Sichuan	213 ± 88	20.6 ± 9.6
Quaternary red clay	Jiangxi, Zhejiang, Hunan Guangxi, Fujian, Yunnan, Guizhou, Hubei, Anhui	137 ± 44	13.6 ± 7.5
Phyllite	Jiangxi, Anhui, Zhejiang	107 ± 38	8.0 ± 5.6
Tertiary red sandstone	Jiangxi, Hunan, Guangdong	71 ± 36	12.5 ± 11.0
Basalt	Yunnan, Fujian	74 ± 2.8	5.6 ± 4.0
Siliceous limestone	Jiangxi, Hunan, Guangxi	51 ± 25	4.5 ± 2.6
Granite	Jiangxi, Hunan, Zhejiang Guangdong, Guangxi, Anhui, Fujian	45 ± 24	4.6 ± 3.2
Granite–Gneiss	Guangdong, Sichuan	40 ± 12	5.1 ± 2.4

[1] Shi *et al.* (1987), Cheng *et al.* (1988), and unpublished data of Ye Wei and Zhuo Su-nen.

average fixed ammonium in the red soil zone soils is greater than generally expected.

The amount of fixed ammonium also differs with the degree of weathering of parent material and soil age, even for soils developed on the same parent material in the same soil zone. This is demonstrated by the large standard deviations given in Table 4.5. In the Quarternary Red Clays for example, because of the difference in soil age, relief and parent rock, as well as the variation in climatic conditions during the Pleistocene era (Xu 1987), the amounts of fixed ammonium in the soils derived from this parent material vary from 90 to 270 mg N kg^{-1}. Similar variation is found for soils derived from purple sandstone and shale (162–425 mg N kg^{-1}).

In addition to the composition of clay minerals, soil texture also affects the amount of fixed ammonium in soils developed from the same parent material. A study on soils developed on Huanghe River alluvium in the brown soil zone indicated that the amount of fixed ammonium was positively correlated with the clay content of the soil (Li *et al.* 1992). It should be noted that this relationship was found in the red soil zone and the soil zones to the north of it only. In the latosol zone, due to the high weathering of minerals and the small amounts of hydromica and vermiculite present (Xu 1987), there was no relationship between fixed ammonium and clay content (Shi *et al.* 1987).

The amount of fixed ammonium is also affected by the pattern of land use. For example, fixed ammonium in the surface layer of paddy soils is generally lower than that in adjacent upland soils developed from the same parent material (Table 4.6).

Table 4.6. Fixed ammonium in paddy soils and adjacent upland soils.[1]

Parent material	Locality	Soil pattern	Total N ($g\ kg^{-1}$)	Fixed ammonium-N	
				($mg\ N\ kg^{-1}$)	(% of total N)
Purple sandstone and shale	Jianyang, Sichuan	Upland	0.73	187	25.6
		Paddy	1.10	76.5	7.0
Purple sandstone and shale	Lesan, Sichuan	Forest	1.39	243	17.5
		Paddy	0.88	99	11.2
Purple sandstone and shale	Liujiang, Guangxi	Upland	1.50	181	12.0
		Paddy	1.63	102	6.2
Limestone	Tunxi, Anhui	Upland	2.24	289	12.9
		Paddy	1.87	195	10.4
Quaternary red clay	Liuzhou, Guangxi	Upland	1.41	121	8.6
		Paddy	2.34	90	3.8
Quaternary red clay	Jingxian, Jiangxi	Waste land	1.22	167	13.7
		Paddy	1.66	138	8.3
Quaternary red clay	Jinhua, Zhejiang	Upland	0.99	158	15.9
		Paddy	1.63	129	7.4
Quaternary red clay	Tunxi, Anhui	Upland	1.35	122	9.0
		Paddy	1.24	90	7.3
Granite	Zixi, Jiangxi	Upland	0.85	74.0	8.7
		Paddy	1.78	29.6	1.7

[1] Cheng *et al.* (1988).

This may be attributed to (1) the texture of the surface soil becoming coarser under flooded conditions due to the downward movement of clay particles in percolating water, and (2) alteration of the composition of soil clay minerals due to submergence.

The proportion of fixed ammonium to total N varies greatly depending on the amount of fixed ammonium and organic matter. In the surface layer of sandy soils where organic matter is extremely low, fixed ammonium may account for 54% of the total N, while in soils rich in organic matter, such as black soils, dark brown soils, chernozems, soils formed on granite, sandstone, siliceous limestone in the red soil zone, and most of the soils in the latosol zone, less than 5% is present as fixed ammonium. In other soils the proportion of the total N present as fixed ammonium is usually greater than 13%. For instance, the paddy soils of the Taihu Lake region and Chengdu Plain contain 18.5% and 19.6%, respectively, and the cultivated soils of the Huang-Huai-Hai Plain average 29.2% (Table 4.7). Consequently, for most cultivated soils in China fixed ammonium is an important fraction.

4.3.2. Distribution in profile

In contrast to the sharp decline in organic matter with depth, in most soils fixed ammonium varies little or even increases with depth. In paddy soil profiles, due to the low clay content and change in clay mineral composition in the surface and illuvial horizons, fixed ammonium is generally low.

Table 4.7. Fixed ammonium in agricultural soils.[1]

Region	0–20 cm			0–100 cm		
	Total N (kg ha^{-1})	Fixed N (kg ha^{-1})	Fixed N (% of total)	Total N (kg ha^{-1})	Fixed N (kg ha^{-1})	Fixed N (% of total)
Taihu Lake Plain	3560	659	18.5	11070	3790	34.3
Chengdu Plain	3540	693	19.6	11750	3880	32.6
Huang-Huai-Hai Plain	1570	458	29.2	5810	2560	44.1
Central and South China[2]	2590	355	13.7	8170	2798	34.2

[1] Wen *et al.* (1988), Li *et al.* (1992).
[2] Soils derived from Quaternary red clay.

For most soils, the proportion of fixed ammonium to total N increases with soil depth (Figure 4.2). For example, in the horizons of some cultivated soils below 40 cm, the proportion increased to more than 40%, with the highest being 72%. However, in brown desert soils and some sandy soils which contain little organic matter, and those soils in the latosol zone with extremely low fixed ammonium the proportion present as fixed ammonium does not decrease with depth. Results show that the amount of fixed ammonium in the top 1 m of the soil profile in some agricultural regions of China is high. In soils of the Taihu Lake region, Chengdu Plain, and those derived from Quarternary Red Clays in central and south China, fixed ammonium amounts to 2800–3900 kg N ha^{-1}, accounting for about one-third of the total soil N. In soils of the Huang-Huai-Hai Plain it amounts to 2560 kg N ha^{-1}, accounting for about one-half of the total soil N. The higher proportion in the Huang-Huai-Hai Plain soils results from the lower total N content.

In most published work on the characterization of soil organic matter, fixed ammonium has not been taken into account in the calculation of the C/N ratio; the C/total N ratio has been used rather than the C/organic N ratio. The error may be small for soil horizons with high organic matter content or those with little fixed ammonium. However, misunderstandings will occur for those soils low in organic matter or high in fixed ammonium. For instance, although the C/total N ratios of the plough layers of the soils in the Huang-Huai-Hai Plain are slightly lower than those in the middle or lower regions of the Changjiang River (Zhu 1987), the C/organic N ratios for the two regions are the same (viz. 12). The discrepancy is more obvious for subsoils. The C/total N ratios of most subsoils are relatively low, sometimes <6 (i.e. less than bacteria), but the C/organic N ratios of these subsoils are >7.0. In some sugar cane-growing soils in Taiwan the C/organic N ratios are as low as 2.8 and 4.6 (Wang *et al.* 1967). The reason for this very low ratio is unknown and further research is needed to identify whether the soils contain large amounts of inorganic N.

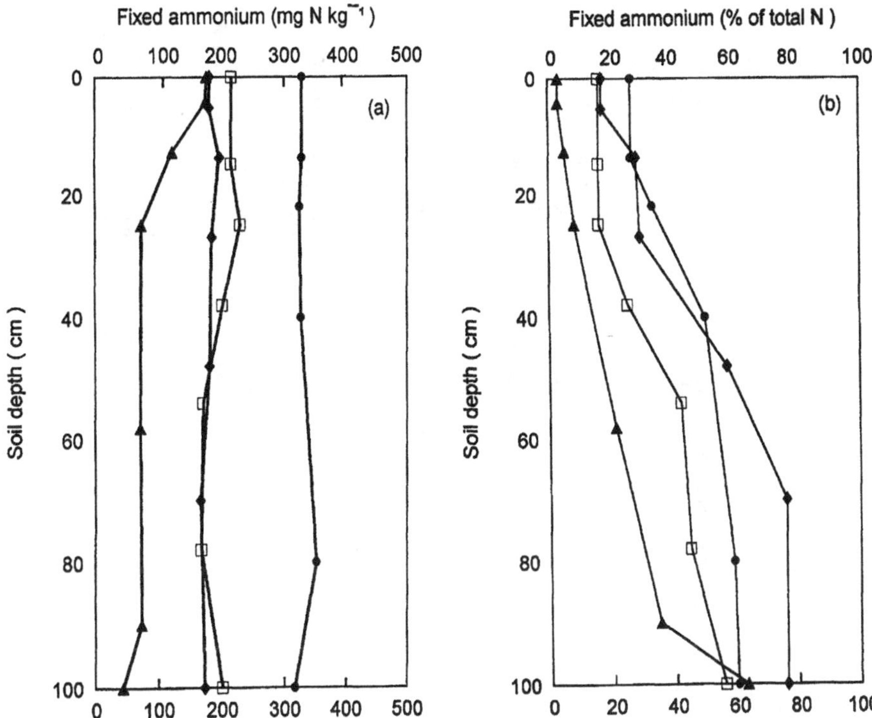

Figure 4.2. Distribution of fixed ammonium in soil profiles. ● calcareous submergenic paddy soil; ◆ sierozem; ▲ dark brown soil; □ neutral hydromorphic paddy soil.

4.4. Biological availability

4.4.1 Availability

Since fixed ammonium can account for a considerable proportion of the total N in soil, particularly in arable soils, its availability for plant uptake has been the subject of much study. However, little is known about the mechanism of release although it is believed that part of the fixed ammonium is released through diffusion.

A number of studies have been conducted on the biological availability of fixed ammonium to the current crop in seedling cultures and in pot and field experiments. The results indicate that the biological availability of fixed ammonium varies greatly. In general, the availability of newly fixed ammonium resulting from the fixation of added ammonium is high. In most seedling cultures and pot experiments the availability was >90%; in field experiments, however, the availability depended on the depth of the soil horizon. Availability was highest in the surface horizon, (>70%), followed by the subsurface layer (20–55 cm), and was lowest in the subsoil (Fan *et al.* 1990; Keerthisinghe *et al.* 1984). On the other hand the availability of native-fixed ammonium was low even in seedling culture (Wen *et al.* 1988; Table 4.8).

Table 4.8 Availability of fixed ammonium to crops.

Soil	Parent material	Fixed NH_4^+ (mg N kg^{-1})	Plant recovery (%)	Experimental method	Test crop	Reference
Newly-fixed ammonium						
Neutral hydromorphic paddy	Loessial sediment	130	100	Seedling culture	Rice	Wen *et al.* (1988)
		130	94	"	Wheat	Wen *et al.* (1988)
		47.1	98	^{15}N pot	Rice	Cheng *et al.* (1989)
Calcareous hydromorphic paddy	Changjiang river alluvium	64.6	92	"	Rice	Cheng *et al.* (1989)
Acid hydromorphic paddy	Quaternary red clay	1.0	67	"	Rice	Cheng *et al.* (1989)
Brown	Loessial sediment	12	100	"	Spring wheat	Zhu (1986)
Meadow	River alluvium	4.8	100	"	"	Zhu (1986)
Lou soil	Loess	275	98	Pot expt	Sudan grass	Sun & Wu (1989)
Lou soil	Loess	379	95	"	"	Sun & Wu (1989)
surface layer	Loess	13.3	70	Field expt	Winter wheat	Fan *et al.* (1990)
20–55 cm layer	Loess	17.1	63	"	"	Fan *et al.* (1990)
subsoil	Loess	19.4	38	"	"	Fan *et al.* (1990)
Native-fixed ammonium						
Calcareous hydromorphic paddy	Changjiang river alluvium	287	3.5	Pot expt	Rice	Cheng *et al.* (1989)
Neutral hydromorphic paddy	Loessial sediment	197	16.8	Pot expt	Rice	Cheng *et al.* (1989)
Acid hydromorphic paddy	Quaternary red clay	99	19.2	Pot expt	Rice	Cheng *et al.* (1989)
Fluvo-aquic	Huanghe river alluvium					
surface layer		128	12.4	Field expt	Winter wheat	Zhang S. L. *et al.* unpub
subsurface layer		149	19.6	Field expt	Winter wheat	Zhang S. L. *et al.* unpub
Lou soil	Loess					
surface layer		235	12.8	Field expt	Winter wheat	Fan *et al.* (1990)
subsurface soil		241	11.2	Field expt	Winter wheat	Fan *et al.* (1990)
subsoil		279	4.3	Field expt	Winter wheat	Fan *et al.* (1990)

Under certain conditions, however, the availability of newly fixed ammonium may also be low. As was demonstrated by Nommik and Vahtras (1982) the availability of newly fixed ammonium was reduced when potassium was added simultaneously with or shortly after the addition of ammonium. Allison *et al.* (1953) reported that in a greenhouse experiment millet recovered only 7% of the newly fixed ammonium in a silty loam soil.

In some cases the availability of the native fixed ammonium may be high. Wen *et al.* (1988) reported that under greenhouse conditions native fixed ammonium in a neutral hydromorphic paddy soil was reduced by 14–16% by cropping with rice or millet, and Mengel and Scherer (1981) found that cropping with barley under field conditions reduced native fixed ammonium in the plow layer and subsoil of loess by 23% and 22%, respectively.

Several factors are responsible for the large variation in biological availability of fixed ammonium, including the type of experiment. It is well known that the density of plant roots per unit volume of soil decreases in the sequence seedling culture > pot and field micro-plots > field plot experiments. It is known that the ability of plants to take up nitrogen is greater when the density of roots is high, and thus the availability of fixed ammonium for a given crop decreases in the same order. Furthermore, because of differences in the ability of crops to take up N, the availability of fixed ammonium also varies with the crop under investigation. For example, Zhu (1986) showed that spring wheat was more effective than pea in absorbing fixed ammonium. Nommik and Vahtras (1982) concluded that cereals were more effective than potato and sugar beet in utilizing fixed ammonium and that mustard was less effective. However, the large variation in biological availability of fixed ammonium cannot be fully explained by differences in the type of experiment or crop grown, and it may be that the variation is related to the ambiguous definitions of newly fixed and native fixed ammonium (Wen and Zhang 1986), how and when they were analyzed and the timing and rate of addition of ammonium or ammonium-producing fertilizer before analysis.

Fixed ammonium is held with different binding strengths at different sites within the interlayers of 2:1 type clay minerals. Under defined conditions, plants can utilize only those fractions of fixed ammonium below a certain binding strength. The blocked release of tightly bound fractions may have arisen from the narrowed interlayer space or the presence of hydroxy-Al (or Fe) 'islands' in the interlayers. Thus, for a given soil the amount of fixed ammonium not available to plants is a fixed value, which can be referred to as the critical value with respect to biological availability. Obviously, the critical value will vary depending on the composition and amount of clay mineral. Although the availability of newly fixed ammonium is generally high some will not be available to plants if the native fixed ammonium is below the critical value of the soil under investigation. From the above discussion, it is not surprising that the availability of native fixed ammonium reported in the literature varies greatly. However, the extremely low availability of newly fixed ammonium reported by Allison *et al.* (1953) can not be explained by the critical value concept. It may result from the repeated leaching of the soil sample with

K_2SO_4, $CaCl_2$, or $MgCl_2$ after addition of ammonium. On the other hand, the high availability of native fixed ammonium found by Mengel and Scherer (1981) is presumably due to the high content of fixed ammonium in the soil (in excess of the critical value) resulting from the high rate of application of sewage sludge to the preceding crop.

The confusion regarding the availability of fixed ammonium in the literature could be solved if fixed ammonium in soil could be fractionated according to its strength of binding to clay minerals.

4.4.2. *Methods for studying availability*

The availability to plants of fixed ammonium, either native or newly fixed, is generally evaluated by the difference method. However, errors may arise from the fluctuation in fixed ammonium during the growth of plants. As shown in a large number of investigations fixed ammonium may increase initially, due to the application of ammonium or ammonium-producing fertilizer and the low rate of uptake of N by seedlings. As plants grow and N uptake increases, fixed ammonium tends to decrease to the original level at the tillering stage, and decreases further until the booting or heading stage when the rate of N uptake is maximal. From this stage onwards, fixed ammonium starts to increase to the initial value (Wen *et al.* 1988; Zhu 1986; Chu *et al.* 1986; Fan *et al.* 1990; Kudeyarov 1981; Figure 4.3); or it may decrease further or increase slightly (Wen *et al.* 1988; Zhu 1986; Cheng *et al.* 1989;

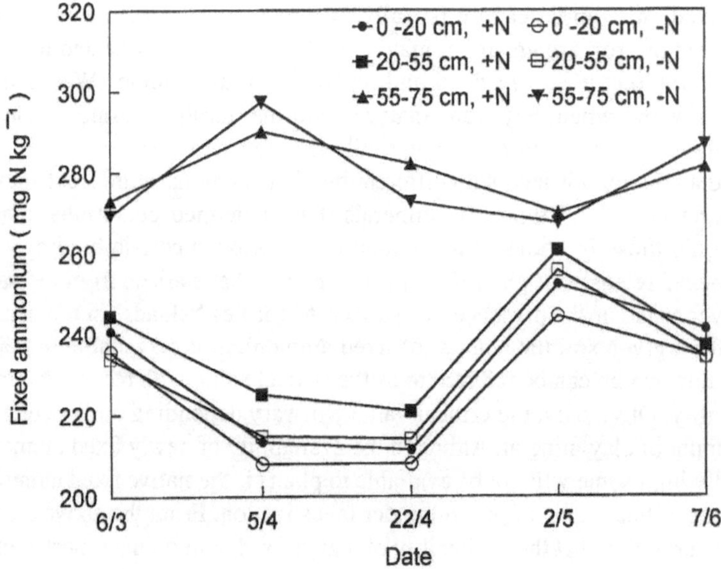

Figure 4.3. Dynamics of fixed ammonium during the wheat growing season (Lou soil, Fan *et al.* 1989).

Kudeyarov 1981; Mengel and Scherer 1981; Figure 4.4). The increase in fixed ammonium in the later stage of growth can be explained by the high mineralization of soil N and the low plant uptake. The decrease in fixed ammonium at the late growth stage may be attributed to insufficient mineralization of soil N to meet the plant's requirements, and the consumption of fixed ammonium by plants. It is obvious that the replenishment of the consumed native fixed ammonium by mineralized ammonium in the later stages of growth may result in an underestimation of the availability of fixed ammonium, particularly for native fixed ammonium. Thus, it can be concluded that the difference method is unsatisfactory for evaluating the availability of fixed ammonium. Furthermore, although the availability of newly fixed ammonium can be evaluated with the ^{15}N tracer method, owing to the biological exchange which exists between soil-N and added N, the recovery of fixed $^{15}NH_4^+$ by plants in the ^{15}N tracer method is not a real measure of the availability of fixed ammonium. Therefore, the development of a simple method to evaluate the availability of fixed ammonium is required.

4.5. Effect on nitrogen fertility and transformations

4.5.1. Effect on soil fertility

Ammonium fixation renders part of the soil N unavailable to plants and thus is unfavourable for the fertility of soils. However, even some native fixed ammonium,

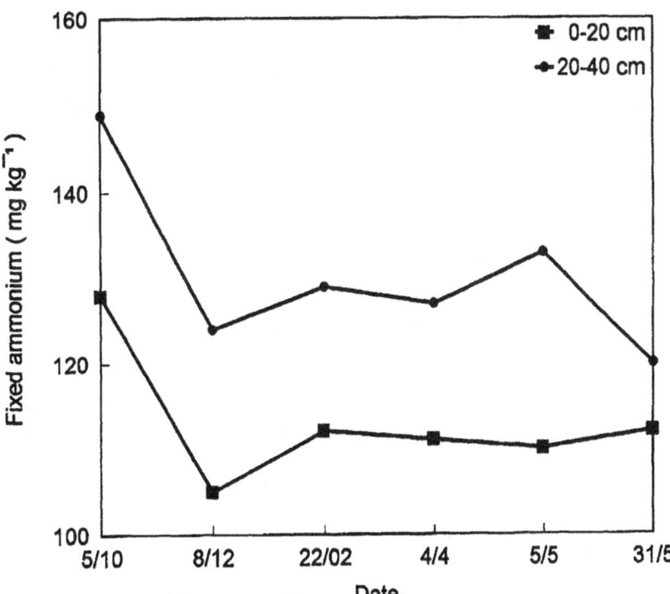

Figure 4.4. Dynamics of fixed ammonium in the +N plot during the growth of wheat (Zhang *et al.* unpublished data).

except that in the bottom layer of some soils, can be utilized by plants at the vigorous growth stage. For instance, in a field experiment with winter wheat, even when applied with N, 9.7%, 8.7% and 3.6% (equivalent to 54, 88 and 28 kg N ha^{-1}) respectively, of the native fixed ammonium in the surface horizon, and the 20–55 cm and subsoil horizons of a Lou soil (old manured loessal soil) was recovered by the crop at the elongation or flowering stage (Fan *et al.* 1990). Similarly, in a field experiment conducted on a Chao soil, with and without applied N, 12.3% and 19.6% of the native fixed ammonium in surface and subsurface layers (equivalent to 40.8 and 79.1 kg N ha^{-1}), respectively, was recovered by winter wheat (unpublished data of Zhang Shao-lin). This also applied to fertile paddy soils. In a calcareous gleyed paddy soil, 7.1% and 8.6% of the native fixed ammonium (equivalent to 19 and 23 mg N kg soil^{-1}) was recovered by early rice from the N and no-N plots, respectively (unpublished data of Shi Shu-lian). Thus it can be concluded that fixed ammonium is an important source of N for crops, and should not be overlooked.

The availability of newly fixed ammonium is usually high and is even higher than that of newly immobilized N. For example, in pot experiments of flooded rice on 2 paddy soils with high capacity to fix ammonium, 56–77% of the fertilizer N was fixed by the soils shortly after fertilization, but most of the newly fixed ammonium (87–93%) was taken up by the rice plants prior to heading (Figures 4.5 and 4.6). Under field conditions, the recovery of newly fixed ammonium by crops although slightly lower, still amounted to at least 70% (Cheng *et al.* 1989; Fan *et al.* 1990). Therefore, fixation of added ammonium is not a negative effect on the efficiency of

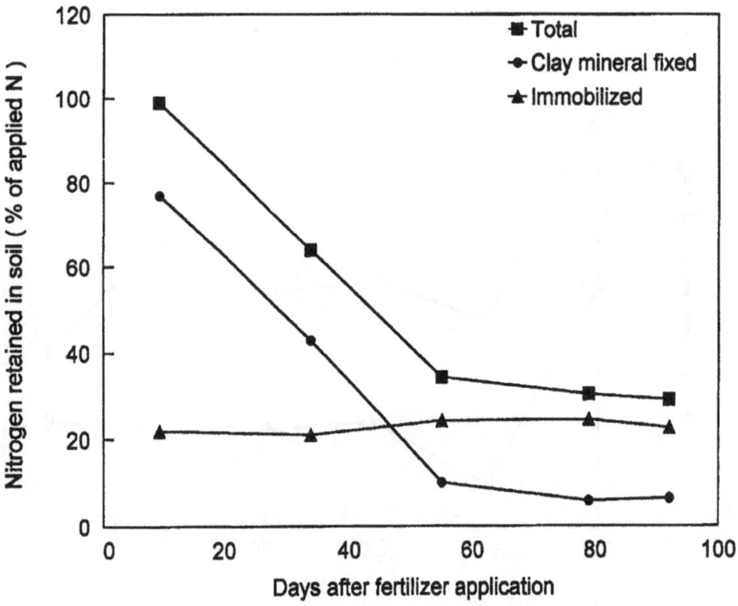

Figure 4.5. Fixation, immobilization and release of fertilizer N during the growth of rice (pot experiment, calcareous hydromorphic paddy soil, 84 mg urea N kg^{-1}, 1.3 g rice straw kg^{-1}; Cheng *et al.* 1989).

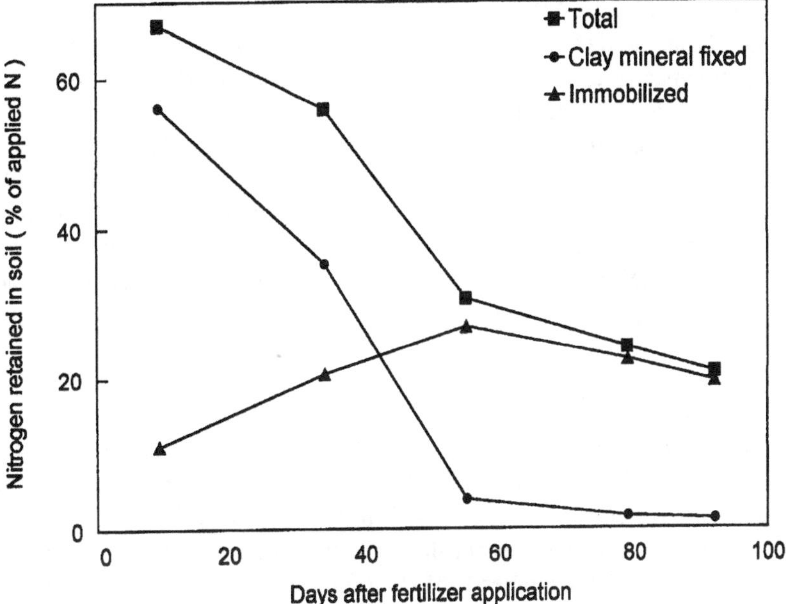

Figure 4.6. Fixation, immobilization and release of fertilizer N during the growth of rice (pot experiment, neutral hydromorphic paddy soil, 84 mg urea N kg^{-1}, 1.3 g rice straw kg^{-1}; Cheng *et al.* 1989).

N fertilizer. Instead, it may limit the concentration of ammonium in the soil solution, resulting in a steady and long-lasting supply of fertilizer-N to crops.

Fixation of ammonium by clay minerals can also improve the quality of composts. It was shown (SMRWAS 1959) that the mud in waterlogged compost can retain part of the available N as exchangeable ammonium and newly fixed ammonium; the relative amounts obviously depend on the composition of the clay minerals in the mud. Mud enriched with 2:1 type clay minerals retained more exchangeable ammonium and an even higher amount of fixed ammonium than mud with less 2:1 type clay minerals. As shown after 3 months in a submerged incubation experiment with azolla added at the rate of 8% w/w, 4.7% and 9.8% of the azolla-N (equivalent to 145 and 305 mg N kg mud^{-1}), was retained by the mud of Xiashu loess in the form of exchangeable and fixed ammonium. This was 25% and 40% higher than that by the Quaternary red clay mud (Wen *et al.* 1987).

It is well known that the rate of N loss from soil is controlled by the concentration of mineral N in the soil solution (Hauck 1981). Thus, fixation of fertilizer-N by clay minerals, resulting in a lower concentration of mineral N in the soil solution, may reduce N losses through volatilization and leaching. However, this has not yet been verified in a field experiment. In an incubation experiment, carried out under upland conditions for 3 and 6 months, N loss from water hyacinth buried 5 cm below the surface of Xiashu loess, with a high capacity to fix ammonium, was 73% and 85%, respectively; it was 79% and 90%, respectively, when the water hyacinth was incubated with a Quaternary red clay having a lower capacity to fix ammonium (Table 4.9). Fischer *et al.* (1981) also reported that 50% of the N, added as ^{15}N-labelled $(NH_4^+)_2SO_4$, was lost when incubated for 127 days in a soil with a low capacity to fix ammonium, while it was only 20% for a soil with a higher capacity to fix ammonium.

The advantage of ammonium fixation in reducing N loss is often taken into consideration in designing fertilizer recommendations. For instance, in the North China Plain it is recommended that farmers irrigate immediately following the surface broadcast of ammonium or ammonium-producing fertilizer to wash the fertilizer to deeper layers, facilitate ammonium fixation and reduce N loss by ammonia volatilization and possibly other mechanisms.

4.5.2. *Effect on nitrogen transformations*

Fixation-release of ammonium is closely linked with N uptake and loss, as well as with other N transformations in soils. Thus, a better understanding of the ammonium fixation-release process is essential for studying other N transformations in soil. For example, it has been established in submerged incubation (Cai *et al.* 1979; Yoshino and Dei 1977; Bremner 1965) that for most soils, part of the mineralized ammonium will be fixed by clay minerals. Therefore, if only the increments in exchangeable and water-soluble ammonium after incubation are taken as a measure of the mineralized N then the mineralization of organic-N will be under-estimated. Wen *et al.* (1988) reported that after 30 days incubation under flooded conditions at room temperature, fixed ammonium in a neutral and a calcareous hydromorphic paddy soil increased by 32 and 52 mg kg^{-1} (equivalent to 2.55% and 4.04% of total

Table 4.9. Fate of water hyacinth nitrogen in two soils.[1,2]

Time (months)	Organic N	Mineralization rate (A)	Fixed NH_4^+	Exchangeable $NH_4^+ + NO_3^-$	Loss rate (B)	B/A
	--------- (% of N added in water hyacinth) ---------					
Xiashu loess (clay content: 345g kg^{-1})						
3	37.8	62.2	8.3	8.8	45.1	72.5
6	38.7	61.3	7.8	1.5	52.0	84.8
12	29.6	70.4	7.4	1.3	61.7	87.6
24	18.7	81.3	7.5	0.7	73.1	89.9
36	19.3	80.7	7.8	0.8	72.1	89.3
60	14.8	85.2	6.6	0.6	78.0	91.5
Quaternary red clay (clay content: 440g kg^{-1})						
3	28.9	71.5	5.4	9.6	56.5	79.0
6	23.3	76.5	5.5	2.2	68.8	89.9
12	21.8	78.2	4.2	0.7	73.3	93.7
24	19.4	80.6	2.4	0.6	77.6	96.3
36	16.8	83.2	2.8	0.5	79.8	95.9
60	10.1	89.9	2.1	0.6	87.2	97.0

[1] Unpublished data of Cheng Li-li *et al.*
[2] Incubated under upland conditions.

soil N), respectively. Furthermore, Cheng and Zhu (1988) demonstrated that after 14 days of incubation under flooded conditions, fixed ammonium in 9 paddy soils changed by –3 to 26 mg kg^{-1} (equivalent to –3.6%–57%) of the increase in exchangeable ammonium (Table 4.10).

When fertilizer-ammonium is added to a soil capable of fixing ammonium, both ammonium fixation by clay minerals and immobilization by microbes can occur. If the fertilizer-N retained in soil, excluding the exchangeable ammonium and nitrate, is regarded as immobilized N, then the maximum rate of immobilization and the remineralization will be overestimated (Broadbent and Nakashima 1970; Yoneyama and Yoshida 1977). However, in soils dominated with 1:1 type clay minerals, the fertilizer-N retained in the soil (excluding exchangeable ammonium and nitrate) can be approximated to the immobilized N (Figure 4.7).

Many investigations have shown that the availability of the residual N from fertilizers is low and it decreases rapidly with time (Huang *et al.* 1982; Jansson 1963; Webster and Dowdell 1985). It has been found that residual N can be differentiated into immobilized N and newly fixed ammonium, and that their availability is different. As demonstrated by Cheng *et al.* (1989) the availability of residual ammonium sulfate N for the second rice crop was significantly greater than that for the fourth rice crop, which is largely the result of the higher availability of the newly fixed ammonium as there was no appreciable difference in the availability of the immobilized N.

The significance of the ammonium fixation-release process differs greatly between soil types, depending on the amount of fixed ammonium present and the

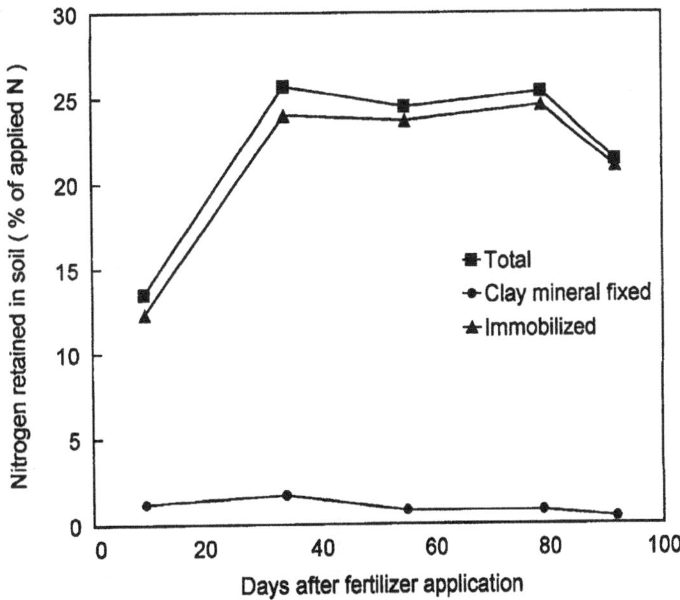

Figure 4.7. Fixation, immobilization and release of fertilizer N during the growth of rice (pot experiment, acid hydromorphic paddy soil, 84 mg urea N kg^{-1}, 1.3 g rice straw kg^{-1}; Cheng *et al.* 1989).

Table 4.10. Increase in fixed ammonium after submerged incubation of air-dried soil samples.[1]

Soil	Locality	Total N ($g\ kg^{-1}$)	Fixed NH_4^+-N			
			Before incubation		Increment after incubation	
			($mg\ kg^{-1}$)	% of total N	($mg\ kg^{-1}$)	Increment in fixed NH_4^+/ Increment in exch. NH_4^+ (%)
Calcareous hydromorphic paddy						
No. 1	Sazhou	1.46	242	16.5	25	42.5
No. 2	"	1.30	246	18.9	20	42.8
No. 3	"	1.27	234	18.4	26	57.4
Neutral hydromorphic paddy						
No. 1	Wuxian	1.84	182	9.9	11	11.5
No. 2	"	1.84	196	10.7	41	51.7
No. 3	"	1.69	247	14.6	8	9.7
Calcareous hydromorphic paddy						
No. 1	Changshu	1.75	274	15.7	19	31.4
No. 2	"	2.15	271	12.6	7	10.5
No. 3	"	1.62	318	19.6	–2	–3.6

[1] Chen and Zhu (1988).

capacity of the soils to fix ammonium. For arable soils in China, the importance of fixed ammonium increases in the following order, soils of the latosol zone, and soils derived from granite, sandstone and siliceous limestone in the red soil zone, which contain very little fixed ammonium and do not have the capacity to fix ammonium < soils derived from Quaternary Red Clays in the red soil zone < soils in the yellowish-brown soil zone < soils derived from shale, purple sandstone and shale, and marl in red soil zone (including the soils of the Chengdu Plain, Dongting Lake Plain, Jianghan Plain and Taihu Lake Plain), and Lou soil in the brown soil zone (Guanzhong Plain). For the high fixing soils, ammonium fixation must be taken into account in studies on N transformations.

4.6. References

Allison, F E, Doestsch, J H and Roller, E M 1953. Availability of fixed ammonium in soils containing different clay minerals. Soil Sci. 75:373–381.

Black, A S and Waring, S A 1972. Ammonium fixation and availability in some cereal producing soils in Queensland. Aust. J. Soil Res. 10:197–297.

Bremner, J M 1965. Nitrogen availability indexes. In: Black, C A (ed.), Methods of Soil Analysis. Part 2. pp. 1324–1345. Am. Soc. Agron. Madison, Wisconsin.

Bremner, J M and Harada, T 1959. Release of ammonium and organic matter from soil by hydrofluoric acid and effect of hydrofluoric acid treatment on extraction of soil organic matter. J. Agric. Sci. 52:137–160.

Broadbent, F E and Nakashima, T 1970. Nitrogen immobilization in flooded soils. Soil Sci. Soc. Am. Proc. 34:218–221.

Cai, G X, Zhang, S L and Zhu, Z L 1979. Experimental conditions for the determination of the mineralization pattern of soil nitrogen under sealed submerged conditions. (in Chinese). Soils :234–240.

Chen, D L and Zhu, Z L 1988. Analytical studies on the nitrogen supplying capacity of paddy soils. (in Chinese). Acta Pedol. Sin. 25:262–268.

Cheng, L L, Wen, Q X and Li, H 1988. Fixed ammonium in soils in tropical and subtropical regions of China. (in Chinese). Soils 20:239–242.

Cheng, L L, Wen, Q X and Li, H 1989. Transformation of ^{15}N-labelled fertilizer N in soils under greenhouse and field conditions. (in Chinese). Acta Pedol. Sin. 26:124–130.

Chu, X Y, Huang, C Y and Mo, H M 1986. Dynamics of soil fixed ammonium during rice growth. (in Chinese). J. Zhejiang Agricultural University 12:233–237.

Fan, X L, Li, C W and Mengel, K 1989. Correlationship between winter wheat growth and release of nonexchangeable ammonium in Lou soil. (in Chinese). Chinese J. Soil Sci. 20:249–251.

Fan, X L , Li, C W and Mengel, K 1990. Availability of nonexchangeable (fixed) ammonium of manured loessial soil in China. (in Chinese). Acta Pedol. Sin. 27:301–308.

Fischer, W R, Pfanneberg, T, Niederbudde, E A and Medina, R 1981. Transformation of ^{15}N-labelled ammonium in two soils differing in NH_4^+-fixing capacity. J. Soil Sci. 32:409–418.

Guo, P C, Zhu, B I and Han, X R 1986. Fixation of ammonium by clay minerals and its release in soils. In: Soil Agricultural Chemistry and Soil Biology and Biochemistry Committees, Soil Science Society of China (eds.), Advances and Prospects for Soil Nitrogen Research in China. (in Chinese). pp. 28–33. Science Press, Beijing.

Hauck, R D 1981. Nitrogen fertilizer effects on nitrogen cycle processes. In: Clark, F E and Rosswall, T (eds.), Terrestrial Nitrogen Cycles. Ecol. Bull. (Stockholm) 33:551–562.

Hinman, W C 1966. Ammonium fixation in relation to exchangeable K and organic matter content of two Saskatchewan soils. Can. J. Soil Sci. 46:223–225.

Huang, D M, Zhu, P L and Gao, J H 1982. Residual effect of organic and inorganic fertilizer nitrogen in paddy soils and upland soils. (in Chinese). Science in China, Ser. B 10:907–912.

Jansson, S L 1963. Balance sheet and residual effects of fertilizer nitrogen in a 6-year study with ^{15}N. Soil Sci. 95:31–37.

Keerthisinghe, G, Mengel, K and De Datta, S K 1984. The release of nonexchangeable ammonium (^{15}N-labelled) in wetland rice soils. Soil Sci. Soc. Am. J. 48:291–294.

Kudeyarov, V N 1981. Mobility of fixed ammonium in soil. In: Clark, F E and Rosswall, T (eds.), Terrestrial Nitrogen Cycles. Ecol. Bull. (Stockholm). 33:281–290.

Li, C K 1938. The extent of ammonium sulfate fixation in main soils of China. (in Chinese). Special Publication on Soils, Ser. B:1–7.

Li, Z P, Cheng, L L and Wen, Q X 1992. The content and availability of nonexchangeable ammonium in soils of the Huang-Huai-Hai Plain. (in Chinese). Chinese J. Soil Sci. 23:200–202.

McBeth, I G 1917. Fixation of ammonia in soils. J. Agric. Res. 8:71–80.

Mengel, K and Scherer, H W 1981. Release of nonexchangeable (fixed) soil ammonium under field conditions during the growing season. Soil Sci. 131:226–232.

Nommik, H 1957. Fixation and defixation of ammonium in soils. Acta Agric. Scand. 7:395–436.

Nommik, H and Vahtras, K 1982. Retention and fixation of ammonium and ammonia in soil. In: Stevenson, F J (ed.), Nitrogen in Agricultural Soils. pp. 123–171. Am. Soc. Agron. Madison, Wisconsin.

Rich, C I 1964. Effect of cation size and pH on potassium exchange in Nason soil. Soil Sci. 98:100–106.

Rodrigues, G 1954. Fixed ammonia in tropical soil. J. Soil Sci. 5:264–274.

SMRWAS. 1959. (Section of Manure Researches, Institute of Soil Science, Academia Sinica, and Wushi Agricultural School, Kiangsu). Investigations on the process of decomposition and nutritive value of Tsao-Tung-Ni, a mixed fertilizer of mud and straw prepared under anaerobic conditions. (in Chinese). Acta Pedol. Sin. 7:190–202.

Shi, S L, Wen, Q X and Liao, H Q 1987. The contents of non-exchangeable ammonium in main soils of China. (in Chinese). Soils 19:79–83.

Shi, S L, Wen, Q X, Liao, H Q and Zhou, K Y 1992. Influence of cultivation on distribution of nitrogen forms and composition of amino acids in soils. (in Chinese). Soils 24:14–18.

Silva, J A and Bremner, J M 1966. Determination and isotope-ratio analysis of different forms of nitrogen in soil. 5. Fixed ammonium. Soil Sci. Soc. Am. Proc. 30:587–594.

Stevenson, F J 1959. On the presence of fixed ammonium in rocks. Science. 130:221–222.

Stevenson, F J and Dhariwal, A P S 1959. Distribution of fixed ammonium in soils. Soil Sci. Soc. Am. Proc. 23:121–125.

Sun, Y and Wu, S R 1989. Fixed ammonium content in Lou soil and its availability for crops. (in Chinese). Chinese J. Soil Sci. 20: 205–207.

Thomas, G W and Hipp, B W 1968. Factors affecting potassium availability. In: Kilmer, V J, Younts, S E and Brady, N C (eds.), The Role of Potassium in Agriculture. pp. 269–291. Am. Soc. Agron. Madison, Wisconsin.

Wang, T S C, Yang, T K and Chen, S Y 1967. Amino acids in subtropic soil hydrolysates. Soil Sci. 103:67–74.

Webster, C P and Dowdell, R J 1985. A lysimeter study of the fate of nitrogen applied to perennial ryegrass swards: Soil analysis and the final balance sheet. J. Soil Sci. 36:605–611.

Wen, Q X, Cheng, L L and Shi, S L 1987. Decomposition of Azolla in the field and availability of Azolla nitrogen to plants. In IRRI (ed.), Azolla Utilization. pp. 241–253. IRRI, Los Banos.

Wen, Q X and Zhang, X H 1986. Fixed ammonium in soils. In: Soil Agricultural Chemistry and Soil Biology and Biochemistry Committees of the Soil Science Society of China (eds.), Advances and Prospects for Soil Nitrogen Research in China. (in Chinese). pp. 34–45. Science Press, Beijing.

Wen, Q X, Zhang, X H, Du, L J and Wu, S L 1988. Fixed ammonium in soils of the Taihu lake region and its availability. (in Chinese). Acta Pedol. Sin. 25:22–30.

Xu, J Q 1987. Soil clay minerals. In: Hseung, Y and Li, C K (eds.), Soils of China. (in Chinese). pp. 374–389. Science Press, Beijing.

Xu, J Q and Hseung, Y 1983. Clay phyllosilicate. In: Hseung, Y. (ed.), Soil Colloids. (in Chinese). Vol. 1. pp. 1–107. Science Press, Beijing.

Yoneyama, T and Yoshida, T 1977. Decomposition of rice residue in tropical soil. III. Nitrogen mineralization and immobilization of rice residue during its decomposition in soil. Soil Sci. Plant Nutr. 23:175–183.

Yoshino, T and Dei, Y 1977. Prediction of nitrogen release in paddy soils by means of the concept of effective temperature. (in Japanese). J. Cent. Agric. Exp. Stn. Japan 25:1–62.

Zhang, S L, Zhu, Z L and Xu, Y H 1989. The transformation of urea and the fate of fertilizer nitrogen in fluvo-aquic soil-winter wheat system in flooded plain of Huanghe River. (in Chinese). Acta Agriculturae Nucleatae Sinica 3:9–15.

Zhu, B J 1986. The release of non-exchangeable NH_4^+ in soil and its relation to plant growth. (in Chinese). Chinese J. Soil Sci. 17:31–33.

Zhu, Z L 1987. Soil nitrogen. In: Hseung, Y and Li, C K (eds.), Soils of China. (in Chinese). pp. 464–482. Science Press, Beijing.

5

Adsorption and diffusion of ammonium in soils

CHEN JIA-FANG

5.1. Introduction

Ammonium in soil serves not only as a major source of N for plant growth, but also as an important product or reactant in the N transformation processes in soil. Ammonium is also the main form in which fertilizer N is applied in agriculture.

Ammonium in soil may be lost through ammonia volatilization under alkaline conditions. Under certain conditions, it can be transformed into nitrate through nitrification, which can be further converted into nitrite, nitric oxide or nitrous oxide through denitrification, thus contaminating underground water or the atmosphere. Moreover, if nitrite accumulates in soil, ammonium can react with it to produce dinitrogen gas.

Obviously, the rates of these processes in soil are all closely associated with the concentration and activity of ammonium in the liquid phase, which in turn are dependent upon the adsorption-desorption characteristics of the soil. In short, ammonium adsorption-desorption directly or indirectly affects the uptake of ammonium by plant roots, the buffering capacity of the soil for ammonium and the transformations of inorganic N in soil.

This chapter deals primarily with ammonium adsorption-desorption by some of the main soils in China, the factors controlling this reaction, and the buffering capacity of the soil. In addition, diffusion which is the rate limiting process controlling ammonium adsorption by soil, and the chief mechanism for the transfer of ammonium in soil, is reviewed briefly.

5.2. Ammonium adsorption

Early on, there was a great deal of interest in the chemical behaviour of ammonium in soil. Most studies regarding ammonium adsorption-desorption by soil and its relation to soil properties were made before the 1950s. However, investigations concerning the distinction between ammonium adsorption and fixation, and the mechanisms involved were carried out 10–20 years later.

Ammonium adsorption is commonly referred to as the net accumulation of ammonium arising from coulombic attraction at the solid phase/bulk solution interface. Ammonium thus adsorbed can be extracted by neutral salt solutions. In some cases, the term 'sorption' (Burchill and Hayes 1981; Fairbridge and Finkl 1979) is

Zhu Zhao-liang et al. *(eds.): Nitrogen in Soils of China, 87–111.*

used instead of 'adsorption', when it is impossible to characterize the mechanism involved. Sorption is one of the most important mechanisms for the retention of ammonium by soils. Ammonium fixation, which mainly occurs in soils with clay minerals of the 2:1 type, is another way for soils to retain ammonium. It is described as the strong absorption of ammonium by soils as a result of NH·O bond formation in hexagonal holes of Si-O sheets, and the balancing of the positive charge deficiency which arises from the isomorphous substitution of Si^{4+} by Al^{3+} (Fairbridge and Finkl 1979; Talibudeen 1981). Unlike adsorbed ammonium, fixed ammonium can only be released through HF treatment (Wen and Zhang 1986; Sun and Wu 1989). However, under some circumstances, they are interchangeable. For instance, drying can change adsorbed ammonium to fixed ammonium, while wetting can result in the lattice expansion of clay minerals, thereby converting fixed to adsorbed ammonium (Bhattacharyya 1971; Kardos 1955). As such, it is possible to distinguish conceptually and methodologically ammonium sorption from fixation. However, since it is difficult to avoid the partial release of nontarget forms of ammonium in soils containing 2:1 type clay minerals during the determination of adsorbed ammonium by selective chemical dissolution, errors will occur in the determination of adsorbed ammonium in NH_4^+ saturated samples, due to the release of fixed ammonium.

5.2.1. *Isothermal adsorption*

As early as in 1938, isothermal ammonium adsorption by soils in China was studied by Li (1938) using 22 samples collected nationwide from Qinghai, Shanxi, Henan, Hebei, Shangdong, Sichuan, Hubei, Jiangsu, Zhejiang, Jiangxi, Yunnan and Guangxi provinces, representing 8 main soil types in China, with molar SiO_2/R_2O_3 ratios ranging from 0.39 to 2.94, and SiO_2/Fe_2O_3 ratios ranging from 1.1 to 18.8. Results fitted well to the Langmuir equation (1), with a correlation coefficient greater than 0.996 ($P < 0.001$–0.01).

$$\frac{1}{Y}=\frac{1}{M}+\frac{K}{M}\cdot\frac{1}{C} \tag{1}$$

where Y is the amount of ammonium adsorbed (cmol kg^{-1}), C is the equilibrium concentration of ammonium in solution (cmol kg^{-1}), M is the maximum adsorption (cmol kg^{-1}) and K is the Langmuir constant relating to the binding energy.

A statistical examination showed that, except for one negative value, M values lay between 4.3 and 54.1, and log M and K, respectively, were related to the molar SiO_2/R_2O_3 and SiO_2/Fe_2O_3 ratios of soil clay fractions (<2 mm) (Figure 5.1; Li 1938). These relationships indicate that the adsorbing capacity and intensity for ammonium are closely related to the mineralogical composition of soil clay fraction. Similar relationships have also been observed for the clay fractions of the other main soils in China, such as phaeozem (black soil), Lou soil (stratified old loessial soil), yellow-brown soil, red soil and latosol, and for montmorillonite and

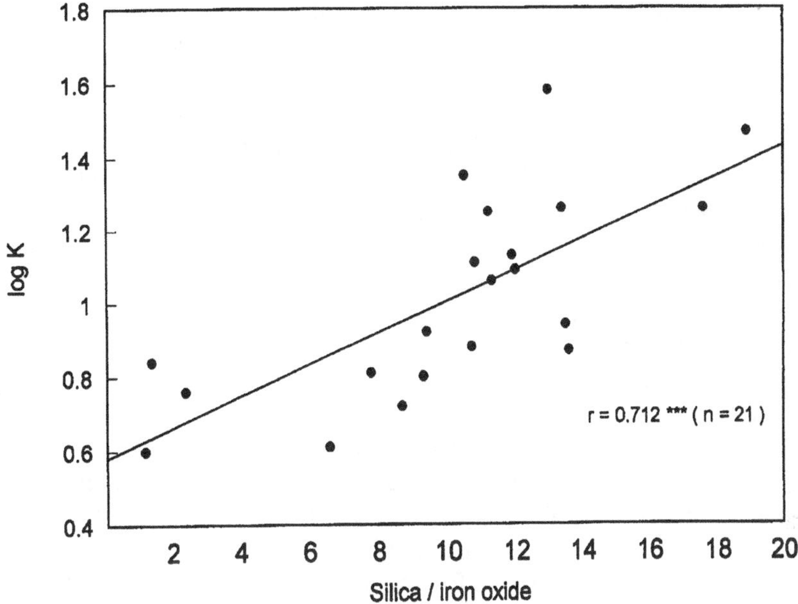

Figure 5.1. Influence of the SiO_2/Fe_2O_3 molar ratio in the clay fraction on the K value obtained from Langmuir adsorption equation (Li 1938).

kaolinite as well. The relationships can all be described by the two surface Langmuir equation (2) (Xie *et al.* 1988).

$$Y = \frac{K_1M_1C}{1+K_1C} + \frac{K_2M_2C}{1+K_1C} \quad (2)$$

where Y, M and K are as in Equation (1), C is the ammonium concentration in the equilibrium solution (mol L^{-1}), M = M1 + M2, is the sum of the maximum adsorption of each surface, and is related significantly to the cation exchange capacity (CEC) (r = 0.94, n = 7; Xie *et al.* 1988).

A preliminary study (Chen and Chiang 1963) on ammonium adsorption by paddy soils derived from Xiashu loess, Taihu lacustrine deposits and Quaternary red earths showed that the results obeyed Freundlich isothermal adsorption (3), although they failed to fit Equation (1), and the M value calculated according to Equation (1) accounted for only 64–89% (mean 77%) of the CEC of the soils.

$$Y = KC^{1/n} \quad (3)$$

or

$$\log Y = \log K + (1/n) * \log C \quad (4)$$

where K and n are constants, and Y and C are the same as in Equation (1).

Statistical examination of the results from isothermal adsorption of ammonium by 29 soil samples (including 10 paddy soils) indicates that apart from two samples, the results can be described by Equation (3) (Chen and Chiang 1963).

In Equation (4), if C = 1, then Y = K. Thus, for an initial ammonium concentration of K + 1, we can assume that, at equilibrium, the amount of ammonium adsorbed from solution should equal or approximate K. This is corroborated by Figure 5.2, which implies that it is possible to obtain the K value in Equation (3) experimentally. In addition, we can see from Figure 5.3 that K is closely related to the CEC.

The studies above suggest that ammonium adsorption by soils can either be described by Langmuir's adsorption equation or by the empirical Freundlich adsorption equation in which the constants are dependent on the clay mineral constituents.

Adsorption of ammonium is affected by the amount of organic matter and the complexes it forms with clay minerals. Table 5.1 shows the relationship between the amount of ammonium adsorbed from solutions of different concentrations before and after removal of organic matter with H_2O_2. From the 'b' values in Table 5.1, we can see that , except for latosol, the amount of adsorbed ammonium increased following the removal of organic matter. On the other hand, Chen and Chiang (1963) found that removal of organic matter by oxidation with H_2O_2 reduced the amount of ammonium adsorbed by the clay fractions of several paddy soils.

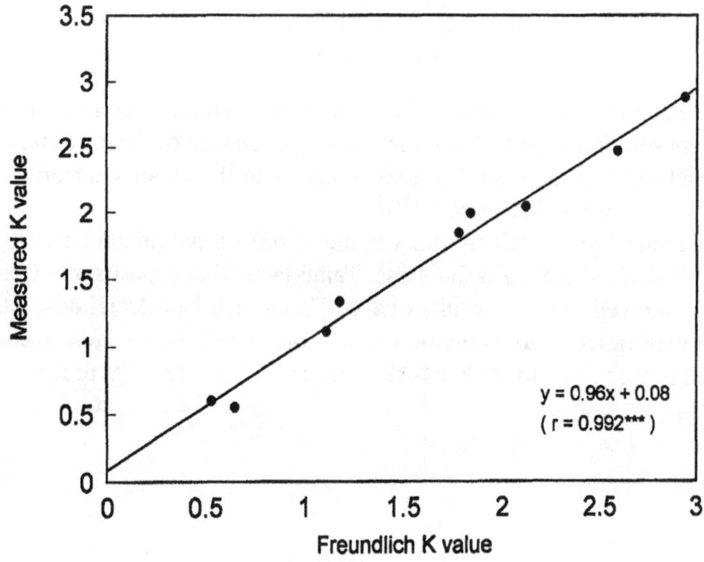

Figure 5.2. Relationship between the Freundlich K value and the measured K value (Chen and Chiang 1963).

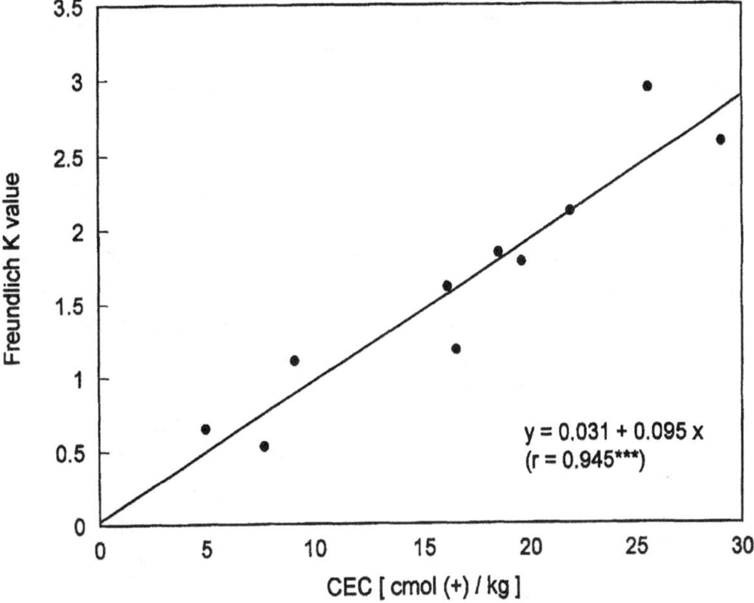

Figure 5.3. Influence of cation exchange capacity (CEC) on the Freundlich K value (Cheng and Chiang 1963).

Table 5.1. Effect of organic matter on ammonium adsorption by clay fractions separated from various soils.[1] (Expressed as y = a + bx, where x and y are the amounts of adsorbed ammonium before and after removal or organic matter, n = 8).

Regression coefficients	Phaeozem (Jiushan)	Phaeozem (Keshan)	Lou soil (stratified old loessial soil)	Yellow brown soil	Red soil	Latosol
r	0.999	1.000	1.000	1.000	0.999	0.995
b	1.04	1.07	1.07	1.08	1.11	0.66
a	0.66	–0.72	–0.06	–0.13	0.00	0.82

[1] Calculated from unpublished data of Xie, P.
Data for adsorption were obtained at 20 ± 1°C with a soil:water ratio of 1:20 by equilibrating Ca^{2+} saturated samples for one hour with pH 6 NH_4Cl solution ranging from 0.002 mol L^{-1} to 0.16 mol L^{-1}.

5.2.2. *Selectivity of soils for ammonium*

Soils may exhibit selectivity for the absorption of cations from soil solution. Cation exchange reactions which occur simultaneously with the isothermal adsorption of ammonium release cations, such as Ca^{2+}, Mg^{2+}, H^+ or Al^{3+} (signified by M) into the soil solution. Thus, at equilibrium, the solid or liquid phases of various soils will have a particular NH_4^+/M ratio. Therefore, according to the law of mass action, at equilibrium, the molar ratio of ammonium in the solid phase to that in the liquid phase can be used to predict the affinity of the soil for ammonium with respect to the other cations. Like the value 'A', which is the ratio of NH_4^+ in the solid phase to that in the liquid phase at a given NH_4^+ concentration in solution (Xie *et al.* 1988); the greater this ratio, the stronger the selectivity for ammonium. An example is given in Table 5.2.

The data given in Table 5.2 show that the ratios for the 4 groups of soil samples differ appreciably from each other, suggesting that the selectivity for ammonium decreases in the order group 1 > group 3 > group 2 > group 4. The samples of group 1 and group 3 contain mainly 2:1 type clay minerals and those of group 2 and group 4 contain mainly 1:1 type clay minerals (ISSAS 1978; Chen and Chiang 1963), thus it seems that the selectivity for ammonium is closely related to the mineralogical composition of the clay fraction. It is believed that the stronger preference of 2:1 type minerals for ammonium arises from the formation of –OH–N–HO– bonds within the lattice space between ammonium ions and the neighbouring O ion in the oxygen ion layer (Russell 1973; Talibudeen 1981). In addition, results obtained with group 3 and group 4 soils show that the molar ratio of ammonium in the solid phase to that in the liquid phase is related to the CEC (Figure 5.4). Since CEC is dependent on mineral type and clay content, this supports the conclusion made above.

The adsorption of Ca^{2+} and NH_4^+ ions from Ca^{2+}–NH_4^+ binary solutions by yellow-brown soils (with hydrous mica as the predominant mineral), latosols (containing mainly kaolinite and gibbsite) and red soils (with kaolinite, together with a certain amount of hydrous mica as the main minerals) was studied after removal of free iron oxides and saturation of the exchange sites with sodium ions. The results show that the NH_4^+/Ca^{2+} ratio of the soils were all less than 2 (Yu and Chen 1982),

Table 5.2. Molar ratio of ammonium in solid phase to liquid phase at equilibrium.[1] (Expressed as $X \pm S_x$).

Soil type	Number of samples	Amount of ammonium added (cmol kg soil^{-1})	
		1	6
Chernozem; Chestnut soil; Gray-brown soil; Purple-brown soil	9	1.42 ± 0.39	1.38 ± 0.56
Red soil; Paddy soils derived from red soil	13	0.64 ± 0.35	0.47 ± 0.18
		Amount of ammonium added (cmol kg soil^{-1})	
		6.05	14.55
Paddy soils derived from Xiashu loess and Taihu lacustrine deposits	7	1.10 ± 0.40	0.69 ± 0.23
Paddy soils derived from red soil	3	0.29 ± 0.12	0.19 ± 0.07

[1] Li (1938) and Chen and Chiang (1963).

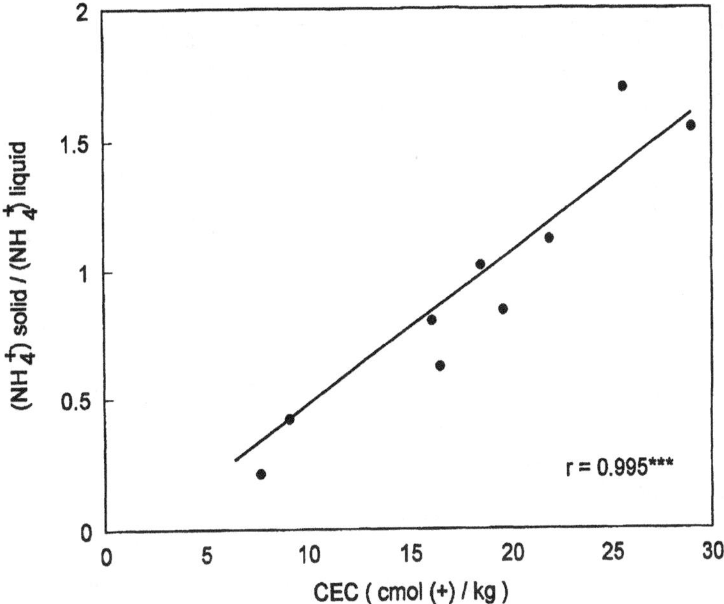

Figure 5.4. Relationship between cation exchange capacity of soils and the molar ratio of ammonium in the solid phase to ammonium in the liquid phase (Chen and Chiang 1963).

except when the initial solution NH_4^+/Ca^{2+} ratios were 8.0 and 3.0 (Figure 5.5). This indicates that the samples studied all exhibited strong affinity for Ca^{2+} and that the affinities for NH_4^+ were in the order yellow-brown soil >> red soil > latosol, which is closely related to their mineral composition (Yu and Chen 1982). This order is the same as that reported by Xie *et al.* (1988) and reflected in Table 5.2.

5.3. Forms of adsorbed ammonium

The variation in the slope of an adsorption isotherm implies that the strength with which the ammonium is adsorbed varies with the degree of ammonium saturation of the soil. This is also confirmed by thermal decomposition studies. Results showed that adsorbed ammonium was released upon heating to 50°C, and was driven off nearly completely when heated to 400–600°C. Upon heating to 50°C, the ammonium adsorbed on a podzolic soil began to release. When heated to 100°C, the amount released from the podzolic soil, kaolinite, and bentonite, was 17%, 15% to 20%, and 32% of the total, respectively. However, for bentonite from other sources, little ammonium was released when the temperature was below 200°C, but the release rate could be increased if the samples were heated above 200°C (Barshad 1948; Darrah *et al.* 1986; Gorbunov 1939). Thus, we can conclude that adsorbed ammonium probably exists in, at least, two forms. These are defined as weakly-adsorbed ammonium and strongly-adsorbed ammonium. Studies on the release of adsorbed ammonium from red soils showed that the weakly-adsorbed

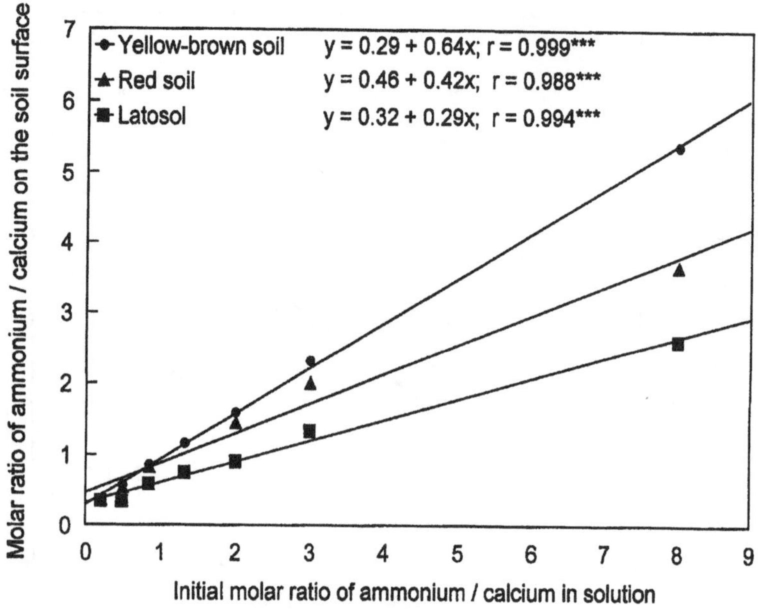

Figure 5.5. Ammonium selectivity of soils (Yu and Chen 1982).

ammonium could be released by steam distillation with water vapour at 100°C, while steam distillation with 0.1 M NaOH was required for the release of the strongly-adsorbed ammonium (Barshad 1951; Chen 1983; Chen and Gao 1959).

5.3.1. Strongly-adsorbed ammonium

The data available (Chen 1957; Chen and Gao 1959) show that the strongly-adsorbed ammonium is thermally stable and that it undergoes cation exchange reactions. In addition, the amount of the strongly-adsorbed ammonium is slightly affected by pH of the system, electrolyte concentration and soil-water ratio (as indicated by the low coefficients of variation given in Table 5.3). Therefore, it is concluded that the strongly-adsorbed ammonium is exchangeable ammonium adsorbed through electrical attraction arising from the permanently negatively charged surface (Chen and Yuan 1990).

5.3.2. Weakly-adsorbed ammonium

Unlike strongly-adsorbed ammonium, weakly-adsorbed ammonium has low thermal stability and high sensitivity to changes in system pH , electrolyte concentration and soil to water ratio (Table 5.3). This suggests that weakly-adsorbed ammonium is associated with variably charged surfaces (Chen and Yuan 1990). Similar results were obtained with volcanic ash-derived soils abundant in allophane. Ammonium adsorbed by these soils was readily leached or hydrolyzed

Table 5.3. Amount of adsorbed ammonium (cmol kg^{-1}) as influenced by solution concentration, pH and water:soil ratio.[1] (Expressed as $X \pm S_x$).

Soil type	Form of adsorbed ammonium	NH_4Cl solution	
		0.005–1.00 mol/L	pH 5.4–8.8
Latosolic red	Strong	2.46 ± 0.26 (11%)[2]	4.38 ± 0.56 (13%)
	Weak	2.43 ± 1.25 (52%)	2.27 ± 1.70 (75%)
Red	Strong	4.60 ± 0.75[3] (16%)	5.14 ± 0.03 (<1%)
	Weak	3.17 ± 1.74 (55%)	2.91 ± 1.65 (57%)

[1] Chen (1957).
[2] Numbers in parentheses are coefficients of variation (S_x/X %).
[3] These results refer to the variation in amount of adsorbed ammonium as a function of soil:water ratio which was divided into 10 intervals ranging from 0.6 to 100; 0.2 M NH_4Cl (pH 8.3) solution was used.

(Birrell and Gradwell 1956), presumably owing to non-coulombic attraction (Wada and Ataka 1958).

The rate of release of weakly-adsorbed ammonium at 100°C can be described by first order kinetics

$$-\frac{da}{dt} = Ka \tag{5}$$

or, in integrated form,

$$\ln a = -Kt + C \tag{6}$$

where a is the amount of weakly-adsorbed ammonium at time t (min) and K is a rate constant. When t = 0, a = a_o, and C = ln a_o.

The rate of release of weakly-adsorbed ammonium from latosolic red soils, red soils and podzolized yellow soils were described by Equation (6) and decreased in the order latosolic red soils > red soils > podzolized yellow soils, and were related to their iron oxides contents (Chen 1957).

In another experiment, the effects of two chemical treatments on ammonium adsorption by a latosolic red soil were studied. The soils were treated overnight with a cold 1:1 HCl:water solution to remove partially active iron oxides, or soaked in 0.5 M $(NH_4)_2SO_4$ to mask part of the hydroxylated surface (M–OH) by selective adsorption of sulfate (Zhao and Chen 1990). By comparison with the control, it was found that the treatments reduced the amounts of weakly-adsorbed ammonium by 38% and 45%, respectively. Strongly-adsorbed ammonium changed slightly, accounting for 105% and 104% of the control, respectively (Chen 1957). These results indicate that weakly-adsorbed ammonium is linked with the hydrous oxide type surfaces which can provide variable charge for adsorption of ammonium. Ammonium thus adsorbed can be hydrolyzed with the formation of NH_4OH which in turn is physically adsorbed (molecular adsorption) and can be released thermally. On the basis of these results, the ammonium-saturated soil system may be described as

$$\underset{\text{strongly-adsorbed form}}{[(\text{soil})\,NH_4^+} \qquad + \qquad \underset{\text{weakly-adsorbed form}}{NH_4OH]} \qquad (7)$$

This assumption is supported to a certain extent by the following:

1. The pH values (5.5–8.7) of latosolic red and red soils, which were determined after saturation with ammonium, by leaching with NH_4Cl solutions of various pHs, were strongly related to the amount of weakly-adsorbed ammonium ($r = 0.975^{**}$, $n = 8$; Chen 1957). This can only be explained if the relationship depicted in (7) is correct; that is, ammonium-saturated soil can be regarded as a system consisting of strongly-adsorbed ammonium and NH_4OH arising from hydrolysis of weakly-adsorbed ammonium.

2. When soil is extracted by a neutral solution of 1 M KCl, strongly-adsorbed ammonium will be displaced into the bulk solution in the form of NH_4Cl, while the hydrolyzed product of the weakly-combined ammonium will be NH_4OH. We can determine the total amount of ammonium ($NH_4Cl + NH_4OH$) using alkaline distillation and the amount of NH_4Cl by the formaldehyde method; the difference between them is the physically adsorbed ammonium. For the latosolic red soils, red soils, podzolized yellow soils and montmorillonite, the amounts of physically adsorbed ammonium were between 0.2 and 4.2 cmol kg^{-1} (Chen and Gao 1959), and were closely related to the weakly-adsorbed ammonium (Chen 1957; Figure 5.6).

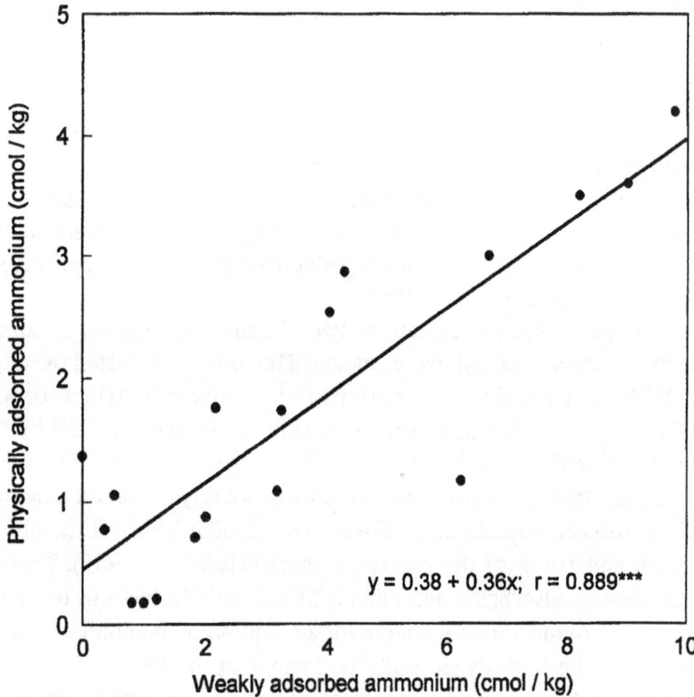

Figure 5.6. Relationship between weakly-adsorbed ammonium and physically-adsorbed ammonium (Chen 1957; Chen and Gao 1959).

In addition, the effective cation exchange capacity (ECEC) of tropical and subtropical soils is usually less than the CEC (determined by NH_4OAc), and the ECEC/CEC ratio correlates significantly with the amount of iron oxide (Chen 1988). This indicates that the difference between ECEC and CEC is related to the variably charged surface, and adds further support to the theory expressed by (7).

In brief, the weakly-adsorbed ammonium is associated with the variably charged surface though the mechanism of its adsorption is complicated. Considering the instability and hydrolysability of adsorbed ammonium resulting from the amount of variable charge with pH and electrolyte concentration, we conclude that weakly-adsorbed ammonium is not desorbed by cation exchange.

5.3.3. *Factors controlling ammonium adsorption*

The amount and the distribution of the forms of adsorbed ammonium are controlled by numerous factors. As mentioned previously, weakly-adsorbed ammonium is associated with hydrous oxide type surfaces, and thus the amount adsorbed can be influenced by almost every soil constituent such as iron oxide, organic matter and the exposed edges of layer silicates (Chen and Yuan 1990).

Particle size is another controlling factor. Table 5.4 shows the amount of ammonium adsorbed by different particle-size fractions of paddy soils derived from five different parent materials. It can be seen that the total amount of adsorbed ammonium decreases sharply with the increase in particle size, and the amount adsorbed by the <2 μm fraction, accounts for 48–96% of the total amount adsorbed (mean 79 ± 17%, n = 13). For soils with hydrous mica as the main mineral, the percentage decreases, while for soils with kaolinite and oxides the value increases (ISSAS 1961). The proportion of weakly-adsorbed ammonium in each particle size fraction depends on the particle size and mineralogical characteristics. Similar results with brown soils were reported by Wu (1982).

Organic matter also plays an important role in the adsorption of ammonium; the amount of weakly-adsorbed ammonium decreases markedly following the removal

Table 5.4. Total and weakly-adsorbed ammonium in different particle-size fractions.[1]

Fraction	Size (μm)	Total adsorbed ammonium (cmol kg^{-1})			Weakly-adsorbed ammonium (%)		
		n	Range	$X \pm S_x$	n	Range	$X \pm S_x$
1	<2	13	15.8–26.6	20.1 ± 4.0	13	23.3–56.9	33.3 ± 9.7
2	2–10	13	0.1–9.3	3.7 ± 2.7	11	38.2–85.7	60.8 ± 13.3
3	10–20	13	0.1–6.9	2.6 ± 2.2	9	48.6–80.6	64.1 ± 12.5
4	20–100	13	0.1–5.5	2.1 ± 1.9	9	45.2–90.9	67.6 ± 15.6

[1] ISSAS (1961).
Samples representing five types of paddy soils, derived from different parent material, were collected from the southern part of China. Trace amounts of adsorbed ammonium were found in two of the No. 2 and four of the No. 3 and No. 4 fractions. They were assumed to contain 0.1 cmol kg^{-1} for statistical analysis of total ammonium, and omitted from the calculations for weakly-adsorbed ammonium.

of organic matter. The reduction was positively related to the amount of weakly-adsorbed ammonium present before the organic matter was removed (Figure 5.7).

The type of clay mineral may influence the total amount of adsorbed ammonium because of their direct effect on CEC. Results obtained with two sets of red-yellow soils collected from Guangdong province (Chen 1988) showed that, after removal of organic matter and iron oxides, there was a significant negative correlation between CEC and the ratio of 1:1 type clay minerals to 2:1 type clay minerals. Also, from Figure 5.8, we can see that clay minerals can affect the distribution of forms. In another study, isothermal adsorption of ammonium was carried out using soil clay fractions separated from latosols, red soils and their mottled horizons, with molar Al_2O_3/Fe_2O_3 ratios in the range of 2.7 to 15.9 and crystal water contents ranging from 8.0 to 17.4%. Results showed that, when the samples were saturated with ammonium, adsorbed ammonium was between 5.7 and 18.1 cmol kg^{-1}, and the ratio of weakly-adsorbed ammonium to total adsorbed ammonium was between 0.11 and 0.64 (Chen 1983). This ratio was in turn dependent on the crystal water content and molar Al_2O_3/Fe_2O_3 ratio (Figure 5.8).

It should be pointed out that the crystal water content of clay depends mainly on the area of hydroxylated surface, or rather, on the quantity of oxides (mainly iron

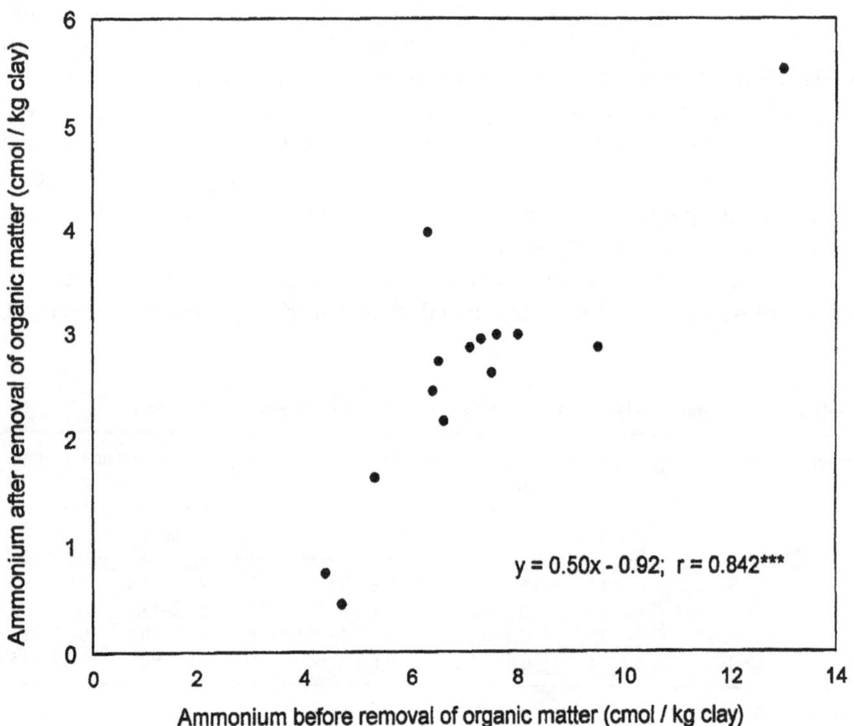

Figure 5.7. Effect of organic matter on weakly-adsorbed ammonium (ISSAS 1961).

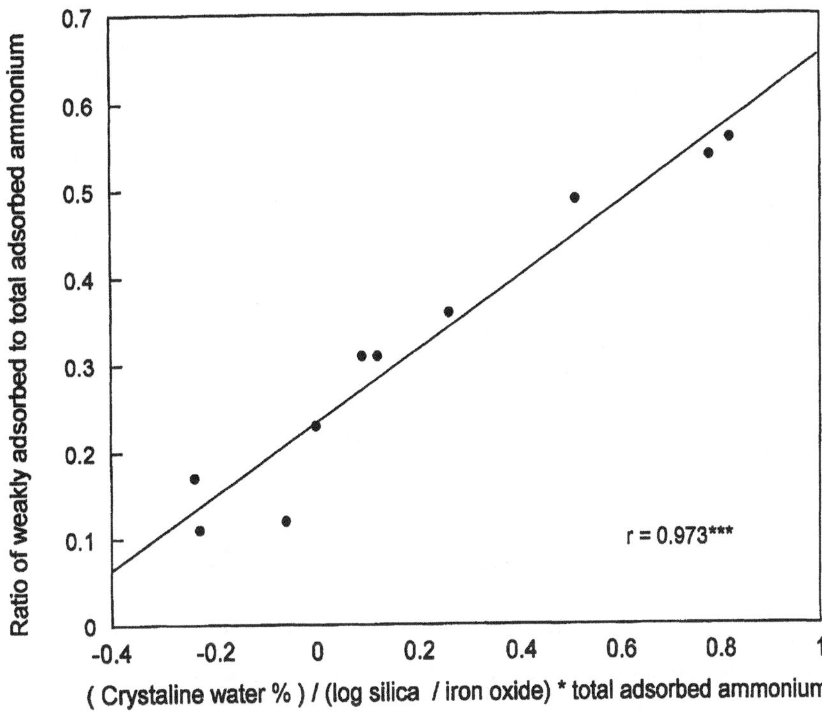

Figure 5.8. Relationship between the chemical composition of the clay fraction and distribution of adsorbed ammonium (Chen 1983).

oxides) and the broken bonds of the layer silicates. Therefore, the hydroxylated surface is also an important factor affecting ammonium adsorption and the distribution of the various forms.

5.4. Desorption of exchangeable ammonium

As indicated above, strongly-adsorbed ammonium is essentially exchangeable ammonium. In general, ammonium only appears in exchangeable form or in strongly-adsorbed form, since a large amount of adsorbed ammonium is required for the occurrence of weakly-adsorbed ammonium. It can be seen from Figure 5.9 that no weakly-adsorbed ammonium is detectable when the total amount of adsorbed ammonium falls to 2.16 cmol kg^{-1} (the intercept with the X axis in Figure 5.9). Under field conditions, weakly-adsorbed ammonium may occur in the sites adjacent to unevenly distributed fertilizer. However, it is temporary, due to ammonium diffusion and migration, and its loss on drying (Chen 1957). Thus, under normal conditions, the amount of weakly-adsorbed ammonium is maintained at a very low level. For these reasons, exchangeable ammonium substantially pertains to strongly-adsorbed ammonium.

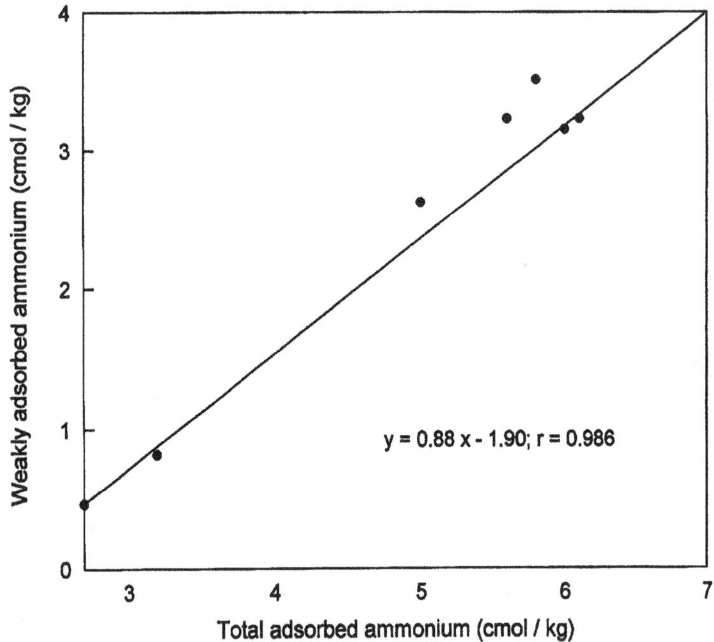

Figure 5.9. Relationship between weakly-adsorbed and total adsorbed ammonium for Latosolic red soil (Chen 1957).

5.4.1. Amounts of exchangeable ammonium

In virgin or unfertilized soils, the amount of exchangeable ammonium may vary from less than 1 mg N kg^{-1} to 30 mg N kg^{-1}. Table 5.5 shows that large variations exist between soils, and during different seasons.

In general the percentage of exchangeable ammonium in inorganic N is below 10%, except in flooded rice soils (ISSAS 1961). Statistical evaluation of the data in Table 5.6 shows that the coefficients of variation for the proportion in the same soil may vary from 19% to 78%, averaging 46%, while it may amount to 57% (mean value) among various soils and samples of different treatments. These results indicate that chemical and biological processes responsible for the fluctuations in exchangeable ammonium may be different in different soils.

Despite the large variation, kinetic relationships can still be found between exchangeable ammonium content and time, in experiments with uniformly mixed soils. Results of two experiments (ISSAS 1978; Guo *et al.* 1986), an incubation experiment under unsaturated conditions and a pot experiment with flooded rice, are given here to provide further evidence of this relationship. It can be seen that the amounts of exchangeable ammonium in the soil decrease with time, whether fertilized or not, and that a significant negative relationship exists. From the point of view of reaction kinetics, the results from the incubation experiment can be described by the Elovich equation, while those of the pot experiment by a zero order reaction (Table 5.7). These suggest that under unsaturated conditions (even if

Table 5.5. Exchangeable ammonium in unfertilized soils.

Soil	Number of determinations	Exchangeable ammonium (mg N kg^{-1})		Reference
		Range	$X \pm S_x$	
Brown	5	2.2–8.6	5.3 ± 2.5	Zhu (1986)
Meadow	5	3.5–5.7	4.9 ± 1.1	Zhu (1986)
Cinnamon	3	17.5–25.4	20.6 ± 4.2	Shi *et al.* (1987)
Albic	4	4.6–13.7	9.6 ± 4.3	Shi *et al.* (1987)
Swampy phaeozem	4	11.4–29.6	16.2 ± 8.9	Shi *et al.* (1987)
Loess	7	0.9–7.4	2.5 ± 2.2	Guo *et al.* (1986)
Alluvial	7	0.6–15.4	4.0 ± 5.5	Guo *et al.* (1986)

Table 5.6. Exchangeable ammonium in uncultivated soils.[1]

Soil		Total inorganic N (mg N kg^{-1})	Percentage of total inorganic N[2]		
			Exch. ammonium	Non exch. ammonium	Nitrate
Brown	Unfertilized	180 ± 30	2.6 ± 1.1	84.6 ± 4.6	12.9 ± 5.4
	Fertilized[3]	278 ± 21	7.7 ± 6.0	62.0 ± 4.0	30.7 ± 10.0
Meadow	Unfertilized	175 ± 25	2.7 ± 0.5	82.3 ± 3.3	15.0 ± 3.2
	Fertilized	263 ± 26	3.8 ± 1.7	60.5 ± 5.2	35.7 ± 7.0

[1] Zhu (1986).
[2] Samples were incubated at 70% WHC and sampled after 28, 48, 58 and 63 days.
[3] Application rate was 100 mg NH_4 N kg $soil^{-1}$.

Table 5.7. Relationship between exchangeable ammonium and incubation time.[1, 2]

Experiment	Soil	Treatment	Number	Range (mg N kg^{-1})	r	a	b	P
Incubation					y = a + b log t			
	Loess	Unfertilized	7	7.4–0.9	–0.970	7.0	–3.5	<0.001
		Fertilized	7	56.4–4.4	–0.928	60.1	–31.7	<0.01
	Alluvium	Unfertilized	7	15.4–0.6	–0.974	15.0	–8.6	<0.001
		Fertilized	7	62.7–1.5	–0.974	60.9	–36.1	<0.001
					y = a + bt			
Pot	Submergenic paddy	Fertilized	5	95.4–7.5	–0.967	107	–2.6	<0.01
	Gleyed paddy	Fertilized	5	89.4–8.3	–0.993	113	–2.5	<0.001

[1] ISSAS (1961, 1978).
[2] y denotes the amount of exchangeable ammonium (mg N kg^{-1}) and t denotes time (days).

uncultivated), exchangeable ammonium is more variable than under flooded condition (even when rice is grown).

Table 5.8 shows the fluctuations of fixed ammonium, exchangeable ammonium, nitrate and the total inorganic N. It can be seen that throughout the incubation

Table 5.8. Inorganic nitrogen (mg N kg soil^{-1}) in unsaturated soils.[1]

Soil	Treatment	Total inorganic	Fixed ammonium	Exch. ammonium	Nitrate
Loess	Unfertilized	205 ± 13 (6%)[2]	184 ± 7 (4%)	2.5 ± 2.2 (90%)	18.5 ± 8.8 (48%)
	Fertilized	284 ± 8 (3%)	213 ± 12 (6%)	19.5 ± 21.1 (108%)	51.8 ± 29.1 (56%)
Alluvium	Unfertilized	265 ± 14 (5%)	233 ± 11 (5%)	4.0 ± 5.5 (137%)	28.7 ± 12.6 (44%)
	Fertilized	338 ± 13 (4%)	259 ± 17 (6%)	14.7 ± 22.9 (156%)	64.2 ± 32.1 (50%)

[1] Calculated from the data of Guo *et al.* (1986).
[2] Values in parentheses are coefficients of variation (S_x/X%).

experiment (61 days), the total inorganic N and fixed ammonium changed only slightly, while exchangeable ammonium and nitrate varied markedly, between them exhibiting strong negative relationships (Table 5.9). It is interesting to note that the absolute values for the regression coefficient 'b' are all greater than 1. This suggests that part of the nitrate-N is derived from organic N.

5.4.2. *Extent of desorption*

The coexistence of various cations in soil solution enables the cation exchange reaction to take place at the solid phase-liquid phase interface, thereby leading to the desorption of exchangeable ammonium. The extent of desorption is, to a certain extent, dependent on the characteristics of the clay minerals. For example, in the presence of 0.4 M KCl, the desorption of ammonium from a yellow-brown soil (with hydrous mica as the main mineral) at different degrees of ammonium saturation varied from 28.4% to 33.7% (mean 30.2 ± 2.2%). However, under the same conditions, the desorption of ammonium from a latosol and a red soil was close to 100% (Yu and Chen 1982). Table 5.10 presents values for the extent of desorption from the clay fractions of several of the main soils of China, in addition to bentonite and kaolinite. It can be seen that the mean value for the amount of ammonium desorbed decreased in the order latosol > red soil > Lou soil > phaeozem (Jiushan), kaolinite > phaeozem (Keshan) > bentonite > yellow-brown soil, whether organic matter was removed or not.

Apart from kaolinite, bentonite and the yellow-brown soil, the coefficients of variation for the desorption values (Table 5.10) were very low, being in the range

Table 5.9. Linear relationship between exchangeable ammonium and nitrate in unsaturated soils (n = 7).[1]

Soil	Treatment	r	a	b	P
Loess	Unfertilized	–0.951	28	–3.7	<0.001
	Fertilized	–0.994	78	–1.4	<0.001
Alluvium	Unfertilized	–0.922	37	–2.0	<0.01
	Fertilized	–0.970	84	–1.4	<0.001

[1] Calculated from the data in Guo *et al.* (1986).

Table 5.10. Desorption of ammonium from soil clay fractions, kaolinite and bentonite.[1]

Soil	Number of samples	Adsorbed ammonium ($cmol\ kg^{-1}$)	Amount desorbed (%)[2]	
			Untreated	H_2O_2 treated
Latosol	7	1.0–10.0	82.4 ± 6.9	87.7 ± 10.9
Red	8	2.3–18.4	80.2 ± 2.5	77.4 ± 4.9
Lou	8	2.7–30.2	70.6 ± 2.0	68.8 ± 3.0
Phaeozem (Jiushan)	8	2.7–42.6	68.5 ± 4.5	62.6 ± 2.9
Phaeozem (Keshan)	8	3.5–49.6	63.4 ± 3.4	60.2 ± 3.4
Yellow brown	8	3.3–37.5	55.8 ± 8.6	53.0 ± 8.0
Kaolinite	8	0.7–8.3	66.9 ± 14.0	
Bentonite	8	2.3–50.6	58.4 ± 9.8	

[1] Xie *et al.* (1988) and unpublished data.
[2] Following adsorption in NH_4Cl solution, samples were equilibrated at 20 ± 1°C in 0.04 M KCl at a soil:solution ratio of 1:20 to desorb ammonium. Concentrations of NH_4Cl ranged from 0.002–0.16 M with 8 intervals.

of 2% to 7%. This implies that the desorption of exchangeable ammonium is governed by the mineral composition rather than by the degree of ammonium saturation. The difference in desorption between soil samples probably arises from the difference in binding energy. This is reflected in the following regression equation:

$$\text{Ammonium desorption (\%)} = 95.4 - 0.066\ K_1 \qquad (r = -0.926,\ n = 12)$$

where desorption refers to the values given in Table 5.10 (except for bentonite and kaolinite), and K_1 is the Langmuir constant relating to the binding energy in Equation (2) (Xie *et al.* 1988). Regarding kaolinite, Sx/X may amount to 14%, implying that the upper limit of its desorption can approximate that of red soil. The lower value for its desorption can be attributed to experimental error.

Up to the present time, there has been little direct verification of the influence of organic matter on the desorption of exchangeable ammonium from soils. Following the removal of organic matter with H_2O_2, the proportion of ammonium desorbed from 5 soils, except for the latosol, decreased by 1.8 to 5.9% only, and the decrease was significantly related to the organic matter content (21–102 $g\ kg^{-1}$; $r = 0.989^{**}$, n = 5; Table 5.10; Xie *et al.* 1988). However, the reverse was observed with the latosol. These results imply that, following the H_2O_2 treatment, the characteristics of the exposed surfaces of the various soils differ from each other. With the latosol, an iron oxide surface was exposed, while for the other 5 samples, hydrous mica surfaces were exposed and these have a stronger affinity for ammonium than the iron oxide surfaces.

In addition, the steric effect arising from the irregularity of clay mineral surfaces (Marshall 1964; Talibudeen 1981) can produce considerable variation in the extent of desorption. Investigations by Wiegner (see Kelley 1950) on the replacement of Ca^{2+} and NH_4^+ from mixed Ca-NH_4 permutite, kaolinite and bentonite provided good evidence for this. It was found that more ammonium was released by 1M

$CaCl_2$ solution from NH_4-Ca samples (prepared by leaching Ca-saturated samples with a solution of NH_4Cl) than from the corresponding Ca-NH_4 samples (prepared by leaching NH_4-saturated material with $CaCl_2$ solution). Expressing the proportion of ammonium desorbed in NH_4-Ca samples as 100%, then the proportion of ammonium desorbed from the Ca-NH_4 samples of permutite, kaolinite and bentonite were 40%, 61% and 94%, respectively.

5.5. Ammonium buffering capacity

The soil is capable of buffering the soil solution against rapid changes in ammonium concentration through adsorption, thereby not only affecting various soil processes involving ammonium, but also enabling a plant to avoid damage from the high ammonium concentrations in soil solution which result from applications of ammonium or ammonium-producing fertilizer.

Commonly, the slope of the adsorption isotherm is taken as a measure of the soil's buffering capacity. The greater the slope, the greater is the amount of ammonium in the liquid phase that can be adsorbed on the solid surface, and the greater the reduction in ammonium concentration in the bulk solution. A quantitative estimate of the amount adsorbed is possible, if we assume several equilibrium concentrations. Recalling the Freundlich equation (3), the buffering capacity (dy/dc) can be expressed as

$$\frac{dy}{dc} = \frac{K}{n} C^{(1-n)/n} \qquad \text{(Chen } et\ al. \text{ 1985)} \qquad (8)$$

When the equilibrium concentration, C, is 1,

$$\frac{dy}{dc} = \frac{K}{n} \qquad (9)$$

Using Equation (9), the buffering capacities over the low concentration range (C = 1) of some of the main soils in China were calculated from the isothermal ammonium adsorption data of Li (1938) and Xie *et al.* (1988), which could be best described by the Freundlich equation. When the concentration of $(NH_4)_2SO_4$ is 1 cmol 400 ml^{-1}, the buffering capacities of the flooded red soils (denoted as ● in Figure 5.10) are low, corresponding with their low selectivities for ammonium (Table 5.2 and Figure 5.4). Figure 5.10 suggests that CEC plays a crucial role in buffering capacity.

When the equilibrium NH_4Cl concentration is 1 mmol $2L^{-1}$, the buffering capacities, of the clay fractions from a phaeozem from Jiushan, a black soil from Keshan, a Lou soil (stratified old loessial soil), a yellow-brown soil, a red soil and a latosol were 1.54, 1.80, 1.19, 1.62, 0.77 and 0.33, respectively (Xie *et al.* 1988). Significant relationships can be found between the buffering capacities and CEC, with a correlation coefficient of 0.920** (n = 6). Similar positive relationships can

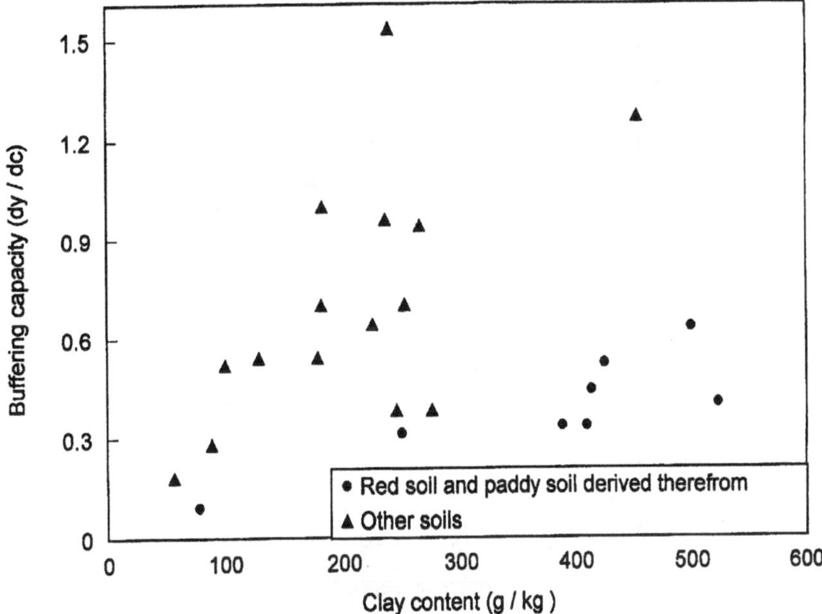

Figure 5.10. Effect of clay content on buffering capacity for ammonium in several soils of China (Li 1938).

also be observed in paddy soils (Figure 5.11). These results also suggest that CEC plays an important role in buffering capacity.

The effect of soil organic matter on buffering capacity is complex because of its intrinsic negative charge, and its bonding with clay minerals. It can be seen from Figure 5.12 that there is a strong relationship between the amount of organic matter and buffering capacity, with a correlation coefficient of 0.863** (n = 10). However, we can obtain a stronger correlation (r = 0.958), if we discard the data relevant to the flooded red soils (denoted as ● in Figure 5.12). This is because the characteristics of the clay minerals and their complexes with organic matter in the red soils are different from those in the other soils. However, this was not the case with the relationship between buffering capacity and CEC (Figure 5.11), since CEC is an overall reflection of the type and amount of clay minerals and organic matter content and complexes formed between them.

5.6. Ammonium diffusion

Diffusion and mass flow are two main processes for the movement of ions and molecules in soil. Diffusion is regarded as the dominant mechanism for the movement of ammonium. Research has been conducted to study the release and diffusion of ammonium after addition of granulated NH_4HCO_3 to paddy soils derived from Xiashu loess under water-saturated and non-percolation conditions (Chi and Wang 1978). The ammonium concentrations at different distances from a granule can be

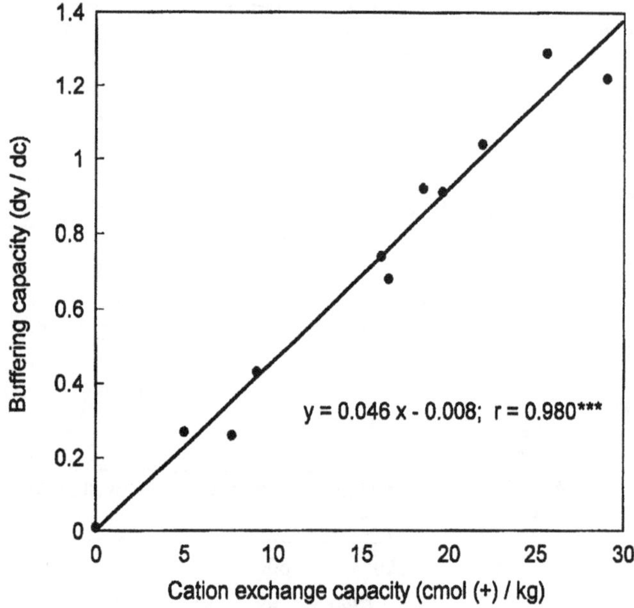

Figure 5.11. Effect of cation exchange capacity of paddy soils on buffering capacity for ammonium (Chen *et al.* 1963).

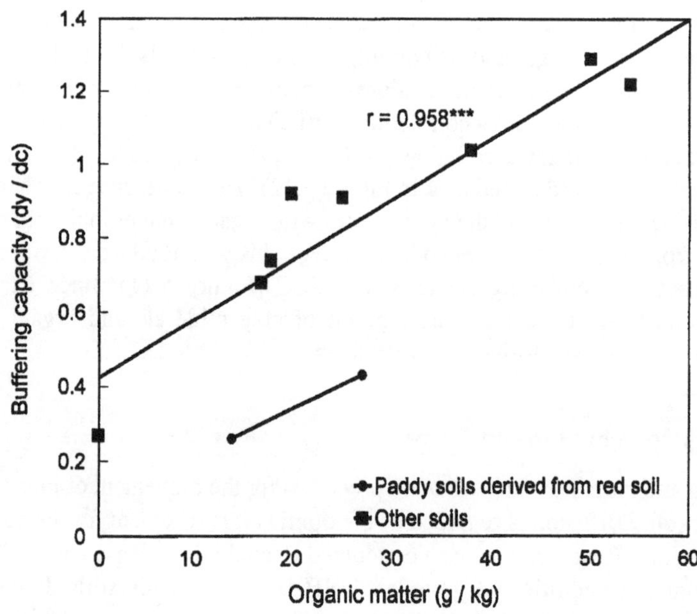

Figure 5.12. Effect of organic matter on buffering capacity for ammonium (Chen *et al.* 1963).

described by the parabolic diffusion equation, $y = a + bt^{1/2}$ (Figure 5.13). The results suggest that, under the experimental conditions, ammonium diffuses in three dimensions and its diffusion volume is proportional to the cube of the distance (d^3), while its concentration is proportional to $1/d^3$ (Table 5.11).

Uptake of nutrients by plant roots results in nutrient depletion around the roots, and a concentration gradient between the non-rhizosphere (source) and rhizosphere soil (sink). This concentration gradient drives the movement of ions from source to sink. A set of experiments on the relationship between 'source' and 'sink' has been carried out by Chen (1962) using a submergenic paddy soil derived from loess (15.9 g organic matter kg^{-1}), and a gleyed paddy soil (25.3 g organic matter kg^{-1}). Samples were weighed into beakers, packed to different bulk densities, and saturated with water (source), and finally covered with a water layer of equal volume (sink). The electrical conductivity (EC) of the sink was determined at different times and the results were found to fit the parabolic diffusion equation (Table 5.12).

These results indicate that the movement of ammonium from source to sink is controlled by a diffusion process. Theoretically, the ion concentration in solution in the low bulk density treatment, should be greater than that in the high bulk density treatment, since the former has a higher water to soil ratio than the latter. However,

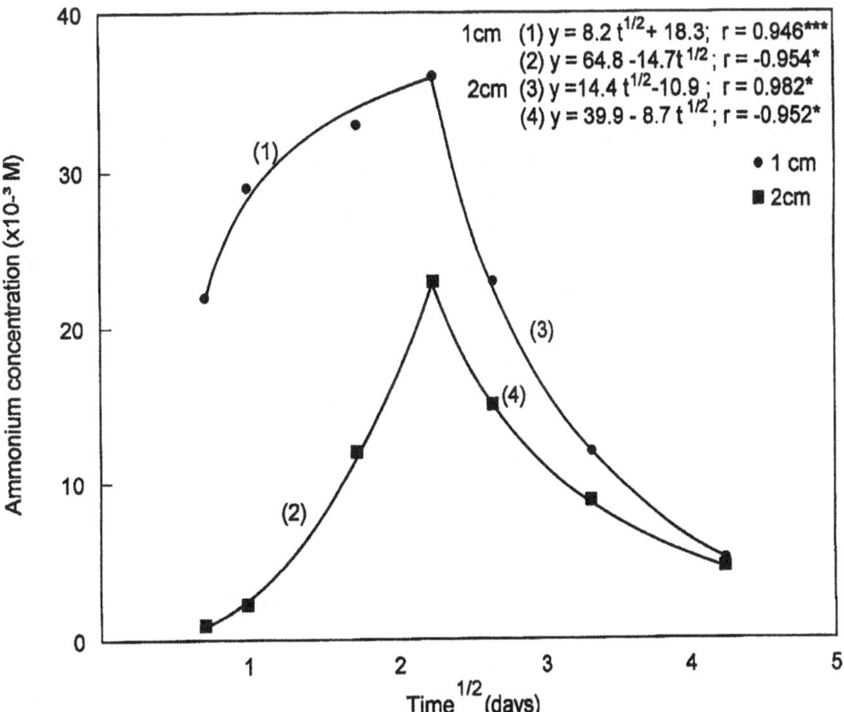

Figure 5.13. Change in ammonium concentration with time at different diffusion distances (Chi and Wang 1978).

Table 5.11. Relationship ($y = a + b/d^3$) between ammonium concentration (y) and diffusion distance (d) at different times.[1]

Diffusion time (days)	Distance for equilibrium (cm)	Number	r	a	b	P
3	7	11	0.974	1.86	31.8	<0.001
5	11	11	0.894	2.95	35.0	<0.001
7	8	11	0.880	2.68	21.4	<0.01
11	8	11	0.818	2.67	10.0	<0.02

[1] Calculated from data of Chi and Wang (1978).

Table 5.12. Influence of bulk density on ion diffusion in soils.[1]

Soil	Treatment (bulk density)	r	a	b	P
Gleyed paddy	Low	0.977	2.14	2.53	<0.001
	High	0.992	−0.15	2.95	<0.001
Submergenic paddy	Low	0.959	2.28	1.83	<0.001
	High	0.986	−0.37	2.32	<0.001

[1] $EC = a+bt^{1/2}$, where EC ($S\ cm^{-1}*10^{-5}$) denotes the electrical conductivity of the soil solution (sink) and t denotes diffusion time (days); (Chen 1962).

experimental results show that this was the case for EC determined in the sink solution 19 hours after fertilization, but was not true at 43 hours. Based on Table 5.12, the turning point is 30 hours for the submergenic paddy soil, and 29 hours for the gleyed paddy soil. It can also be seen from Table 5.12 that 'a' and 'b' values vary in different ways. In the submergenic paddy soil, the 'a' value for the low bulk density treatment was greater than that for the high bulk density treatment, while the 'b' value changed slightly. These results indicate that when the concentration gradient between two treatments differs slightly, soil porosity affects ion diffusion, i.e. ions diffuse faster in soils with greater total porosity.

The accompanying anions affect the ammonium diffusion rate in soil, particularly in neutral or slightly alkaline soil. For example, if $(NH_4)_2SO_4$ is added to these soils, the formation of $CaSO_4$ will lead to a reduction in the Ca^{2+} concentration in the soil solution, and an increase in exchangeable ammonium as well as a reduction in SO_4^{2-}, consequently reducing the ammonium concentration in the soil solution, its diffusion rate and the amount which migrates (Darrah *et al.* 1986).

A model experiment was conducted using a gleyed paddy soil to study the effect of bulk density on the diffusion of Ca^{2+} and NH_4^+. Samples were weighed into tubes to establish soil columns (5 to 7 cm high) of different bulk densities. These were saturated with 0.2 M NH_4Cl solution (source), then each was covered with a water layer of equal volume (sink). One week later, Ca^{2+} and NH_4^+ were determined (Chen 1962). The results (Table 5.13) show that soil porosity affected cation diffusion, especially NH_4^+. From the Ca^{2+}/NH_4^+ ratios, it can be inferred that ammonium has a

Table 5.13. Influence of bulk density on NH_4^+ and Ca^{2+} diffusion in soils.[1]

Bulk density (g/cm^3)	Concentration. in the sink ($mmol\ 100\ ml^{-1}$)		Ca^{2+}/NH_4^+ ratio
	NH_4^+	Ca^{2+}	
1.34	0.38	1.06	2.78
1.13	0.46	1.12	2.48
1.03	0.66	1.26	1.92

[1] Chen (1962).

greater diffusion rate than Ca^{2+}. This is attributed to its high diffusion coefficient, which is nearly 100 times greater than that of Ca^{2+} (Xuan 1985).

Although diffusion is the dominant mechanism for ammonium migration in soil, under certain conditions, mass flow or downward percolation assumes additional significance. When the soil is nearly water-saturated, ammonium in soil solution will be continuously replenished through the release of exchangeable ammonium, application of fertilizer N and mineralization of organic N. A pot experiment with flooded rice showed that ammonium could still be detected in the soil solution 40 days after surface application of $(NH_4)_2SO_4$ (92 mg N kg^{-1}). The averages of five measured values for a submergenic paddy soil and a gleyed paddy soil were 0.35 ± 0.14 and 0.37 ± 0.18 mg N $100g^{-1}$, respectively (ISSAS 1978). Under these conditions, mass flow, downward percolation and diffusion, may all be responsible for the movement of ammonium to depth. If the percolation rate was less than 4mm per day the migration depth of ^{15}N labelled ammonium in a clayey loam 70 days after surface application of $(NH_4)_2SO_4$ was 20 to 30 cm (Chu *et al.* 1977). In a clayey paddy soil growing rice, ammonium was also detectable at a depth of 20 to 30 cm 30 days after application of ^{15}N labelled urea (Sun 1987). A similar study by Cai *et al.* (1985) showed that the migration depth was 15 to 30 cm 2 months after surface application of ammonium-producing fertilizer. In conclusion, the results available, though limited and obtained under various experimental conditions, all suggest that in a flooded rice field, the ammonium in soil solution is likely to be moved mainly through mass flow or percolation. However, the depth of its movement is limited, due to the cation exchange, immobilization, and nitrification taking place during transfer.

5.7. References

Barshad, I 1948. Vermiculite and its relation to biotite as revealed by base exchange reaction, X-ray analysis, differential thermal curves and water content. Amer. Mineral. 33:655–678.

Barshad, I 1951. Cation exchange in soils. I. Ammonium fixation and its relation to potassium fixation and to determination of ammonium exchange capacity. Soil Sci. 72:361–371.

Bhattacharyya, A K 1971. Mechanism of the formation of exchangeable ammonium nitrogen immediately after waterlogging of rice soils. J. Indian Soc. Soil. Sci. 19:209–213.

Birrell, K S and Gradwell, M 1956. Ion-exchange phenomena in some soils containing amorphous mineral constituents. J. Soil Sci. 7:130–147.

Burchill, S and Hayes, M H B 1981. Adsorption. In: Greenland, D J and Hayes, M H B (eds.), The Chemistry of Soil Processes. pp. 221–600. John Wiley & Sons Ltd. Chichester.

Cai, G X, Zhu Z L, Zhu, Z W, Trevitt, A C F, Freney, J R and Simpson, J R 1985. Nitrogen loss from ammonium bicarbonate and urea applied in flooded rice field. (in Chinese). Soils 17:225–229.

Chen, C F 1957. Desorption of adsorbed ammonium in certain red and yellow soils of China. (in Chinese). Acta Pedol. Sin. 5:331–341.

Chen, C F and Gao, C C 1959. Characteristics of adsorbed ammonium in certain red and yellow soils of China. (in Chinese). Acta Pedol. Sin. 7:78–84.

Chen, C F 1962. Preliminary study on the high yield experience of Chen Yong-kang for nutrient regulation of rice by water management. (in Chinese). Chinese J. Soil Sci. 4:1–6.

Chen, C F and Chiang, P S 1963. Ammonium ion adsorption of some paddy soils. (in Chinese). Acta Pedol. Sin. 11:171–183.

Chen, J F 1983. Adsorption characteristics of red soils in China. In Li, Q K (ed.), Red Soils in China. (in Chinese). pp. 91–101. Science Press, Beijing.

Chen, J F, Zhao M Z and Yu S F 1985. Research methods on adsorption of soil. In: Hseung, Y (ed.), Soil Colloids. (in Chinese). Vol. II. pp. 498–546. Science Press, Beijing.

Chen, J F, 1988. Capacity and composition of cation exchange. Chapter 7. In: Yu, T R and Wang, Z Q (eds.), Soil Analytical Chemistry. (in Chinese). pp. 192–223. Science Press, Beijing.

Chen, J F and Yuan, C L 1990. Surface characteristics of soil clays. In Hseung, Y and Chen, J F (eds.), Soil Colloids. (in Chinese). Vol. III. pp. 19–53. Science Press, Beijing.

Chi, K L and Wang, C H 1978. Application of microelectrodes for the study of release of granulated ammonium bicarbonate in paddy soils. (in Chinese). Acta Pedol. Sin. 15:182–186.

Chu, C L, Tsai, K H and Yu, C C 1977. A preliminary investigation on the fate of ^{15}N-labelled ammonium sulfate in rice field. (in Chinese). Science Bulletin of China 22:503.

Darrah, P R, Nye, P H and White, R E 1986. Simultaneous nitrification and diffusion in soils. III. The effects of the addition of ammonium sulfate. J. Soil Sci. 37:53–58.

Fairbridge, R W and Finkl, C W (eds.). 1979. The Encyclopedia of Soil Science. Part I. pp. 7–9, 65. Dowden, Hutchinson & Ross, Inc. Stroudsburg, Pennsylvania.

Gorbunov, N E 1939. Effect of soil drying on desorption of adsorbed cations. (in Russian). Pochvovedenie 8:22–35.

Guo, P C, Zhu, B J and Han, X R 1986. Fixation and release of ammonium by clay minerals in soils. In: Soil Agricultural Chemistry and Soil Biology and Biochemistry Committees, Soil Science Society of China (eds.), Advances and Prospects for Soil Nitrogen Research in China. (in Chinese). pp. 28–33. Science Press, Beijing.

ISSAS 1978. (Institute of Soil Science, Academia Sinica) (ed.), Soils of China. (in Chinese). pp. 285–298. Science Press, Beijing.

ISSAS 1961, (Institute of Soil Science, Academia Sinica) (ed.), Environmental Conditions in Soils for High Yield of Rice. (in Chinese). pp. 134–139. Science Press, Beijing.

Kardos, L T, 1955. Soil fixation of plant nutrients. In: Bear, F E (ed.), Chemistry of the Soil. pp. 177–197. Amer. Chem. Soc., Washington, D C.

Kelley, W P 1950. Cation Exchange in Soils. pp. 48–60. Reinhold Pub.Corp., New York.

Li, Q K 1938. On the rate of fixation of ammonium sulfate by some soils of China. Special Soils Publication. Series B4. pp. 27–40.

Marshall, C E 1964. The Physical Chemistry and Mineralogy of Soils. Vol. I. Soil Materials. pp. 273–282. John Wiley & Sons, Chichester.

Russell, E W 1973. Soil Conditions and Plant Growth. (10th ed.). pp. 92–94. Longmans, London.

Shi, S L, Wen, Q X and Liao, H Q 1987. The contents of nonexchangeable ammonium in the main soils in China. (in Chinese). Soils 19:79–83.

Sun, S T 1987. Fate and efficiency of fertilizer nitrogen(^{15}N labelled urea) applied in wetland rice soil. (in Chinese). Soils 19:177–182.

Sun, Y and Wu S R 1989 Fixed ammonium content in Lou soil (stratified old loessial soil) and its availability for crops. (in Chinese). Chinese J. Soil Sci. 20:205–207.

Talibudeen, O 1981. Cation exchange in soils. In Greenland, D J and Hayes, M H B (eds.) The Chemistry of Soil Processes. pp. 116–177. John Wiley & Sons, Chichester.

Wada, K and Ataka, H 1958. The ion uptake mechanism of allophane. Soil Plant Food. 4:12–18.

Wen, Q X and Zhang X H 1986. Fixed ammonium in soils. In: Soil Agricultural Chemistry and Soil Biology and Biochemistry Committees, Soil Science Society of China (eds.), Advances and Prospects for Soil Nitrogen Research in China. (in Chinese). pp. 34–35. Science Press, Beijing.

Wu, G Y 1982. The nitrogen behaviour and nitrogen supply efficiency of microaggregates of brown soil with different fertility levels. (in Chinese). Chinese J. Soil Sci. 3:6–9.

Xie, P, Jiang J M and Hseung Y 1988. Characteristics of ammonium adsorption by colloids of some main soils in China. (in Chinese). Acta Pedol. Sin. 25:175–183.

Xuan, J X 1985. The movement of ammonium ions to rice roots and their adsorption by rice plants. (in Chinese). Chinese J. Soil Sci. 16:269–271.

Yu, S F and Chen, J F 1982. Preliminary study on ammonium adsorbed by soils from binary solution of NH^{+}_{4}–Ca^{2+} chloride. (in Chinese). Acta Pedol. Sin. 19:248–256.

Zhao, M Z and Chen J F 1990. Adsorption of anions. In: Hseung, Y and Chen, J F (eds.), Soil Colloids. (in Chinese). Vol. III. pp. 376–441. Science Press, Beijing.

Zhu, B J 1986 The release of nonexchangeable ammonium in soil and its relation to plant growth. (in Chinese). Chinese J. Soil Sci. 17:31–33.

6
Nitrification

LI LIANG-MO

6.1. Introduction

Nitrification is the process by which microorganisms oxidize ammonium to nitrate and derive energy from the oxidation for subsistence. Since nitrification is interlinked with the mineralization-immobilization and N loss processes, it is an important N transformation which is related not only to the efficiency of fertilizer N, but also to the contamination of the environment. In this chapter some aspects of nitrification relating to investigations in China will be discussed.

6.2. Nitrifying organisms

While chemoautotrophic nitrifying bacteria are the main microorganisms involved in the nitrification process, the heterotrophic organisms should not be neglected (Adams 1986a, b; Doxtader and Alexander 1966). The sequential nitrification carried out by heterotrophic and autotrophic bacteria is of great significance in N transformations.

6.2.1. Chemoautotrophic nitrifying organisms

It is very difficult to get pure species of these bacteria, so there has been considerable confusion in the taxonomy of nitrifying bacteria (Zheng *et al.* 1983; Belser 1979; Walker 1975). In 'Bergey's Manual of Determinative Bacteriology' (Buchanan and Gibbons 1974), 4 genera of ammonium oxidizing bacteria and 3 genera of nitrite oxidizing bacteria have been described, all of which are chemoautotrophs. *Nitrosomonas* and *Nitrobacter* are the most dominant nitrifying bacteria in soils, and *Nitrosomonas europaea* and *Nitrobacter winogradskyi* are two commonly recognized standard species.

It is generally considered that autotrophic nitrifying microbes are obligate aerobes. However, Chen and Zhou showed that the ammonium oxidizing bacteria in paddy soils were facultative anaerobes (Chen *et al.* 1981; Zhou and Chen 1983), with the maximum amount of nitrite produced under aerobic and anaerobic conditions being equal. Moreover, these autotrophic nitrifying organisms were so closely associated with heterotrophic denitrifying organisms that the nitrite formed by ammonium oxidation was quickly reduced to N_2 and lost. A similar result

Zhu Zhao-liang et al. *(eds.): Nitrogen in Soils of China, 113–134.*

was obtained by Jin (1988), who showed that the ammonium oxidizing bacteria are facultative anaerobes, while the nitrate bacteria are absolutely aerobic (Table 6.1; Jin 1988).

6.2.2. *Heterotrophic nitrifying organisms*

Many species of heterotrophic bacteria can produce traces of nitrite in the ammonium salt-containing media, and some fungi can oxidize nitrite in the media. There are also a few species of bacteria such as *Arthrobacter* and fungi such as *Aspergillus flavus* which can produce nitric acid when an ammonium salt is the only N source; in this process fungi also produce 3-nitropropionic acid. The amounts of nitrite and nitrate formed by heterotrophic bacteria are far lower than those formed by the autotrophic bacteria (Chen *et al.* 1981). In spite of their much lower oxidizing ability compared with the autotrophs, the heterotrophic bacteria are numerous in soil, so the role they play in this respect should not be overlooked (Alexander 1977). Special attention should be given to the active contribution made by the sequential nitrification by heterotrophs and autotrophs in soils (Castignetti 1980).

6.3. Factors controlling nitrification

6.3.1. *pH*

Acidity is one of the major factors controlling nitrification in soil (Dancer *et al.* 1973). In general , there are few or no autotrophic nitrifying bacteria present under acid soil conditions; if any are present, they are adaptive strains. The autotrophic nitrifying organisms grow well in the pH range 6.6–8.0, although each of the individual genera has its optimum range (Walker 1975). The data of Sahrawat (1982) indicated that, within the pH range 3.4–8.6, nitrification activity was positively correlated with pH. Martikainen (1984; 1985a, b) found that in forest soils treated with fertilizer N the accumulated nitrite + nitrate was positively correlated with the

Table 6.1. Growth of nitrifying bacteria in liquid media containing ammonium or nitrite (media inoculated with suspension of upland and paddy soils; Jin 1988).

Soil	Treatment			
	Aerobic culture		Strictly anaerobic culture	
	Non-inoculated	Inoculated	Non-inoculated	Inoculated
	Ammonium solution			
Upland	− − −	+ + −	− − −	+ + +
Paddy	− − −	+ + +	− − −	+ + −
	Nitrite solution			
Upland	− − −	+ + +	− − −	− − −

number of autotrophic nitrifying organisms, while the number of nitrifying organisms was positively correlated with pH. According to Katyal *et al.* (1988), for the 15 soils they tested, little or no nitrification took place in soils with pH 4.6–5.1, nitrification was slow in soils with pH 5.8–6.0, and active nitrification occurred in soils with pH 6.4–8.3. We also found that in the pH range 5.8–8.8 the nitrification rate increased with increasing soil pH, and that there was a significant correlation between them (Figure 6.1).

6.3.2. Texture

Soil texture controls soil aeration and water permeability so that sandy soils are better aerated and more permeable than clay soils. As autotrophic nitrifiers are aerobic organisms their activity is strongly influenced by soil texture, and Li *et al.* (1986) showed there was a marked negative correlation between nitrification rate and clay content (<0.001 mm; Figure 6.2).

6.3.3. Redox potential

It is usually considered that no autotrophic nitrification will occur when the redox potential is below 250 mv. It was suggested that under mildly aerobic conditions the oxidation of ammonium to nitrate was carried out jointly by autotrophic and heterotrophic bacteria, while under anaerobic conditions when the redox potential was below –85 mv, only heterotrophic nitrification took place (Chen *et al.* 1981).

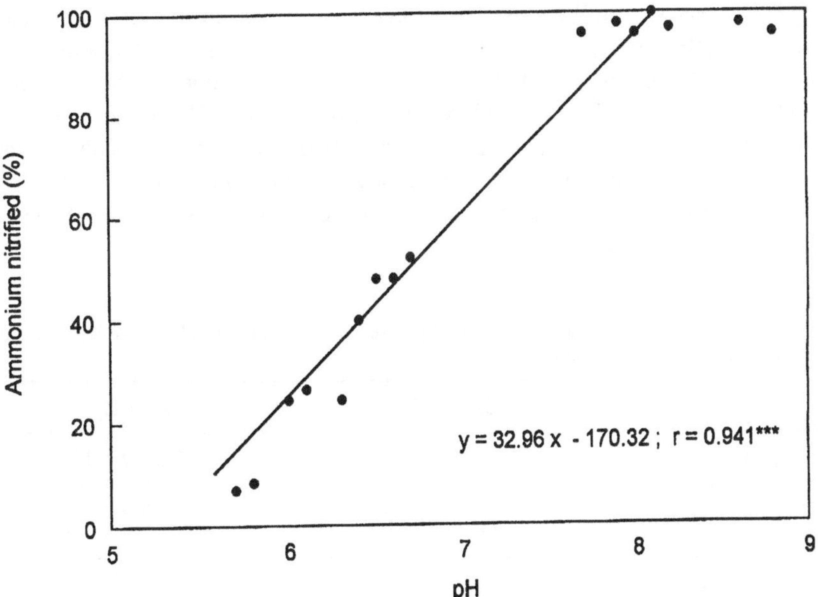

Figure 6.1. Effect of soil pH on nitrification of added ammonium (Li *et al.* 1986).

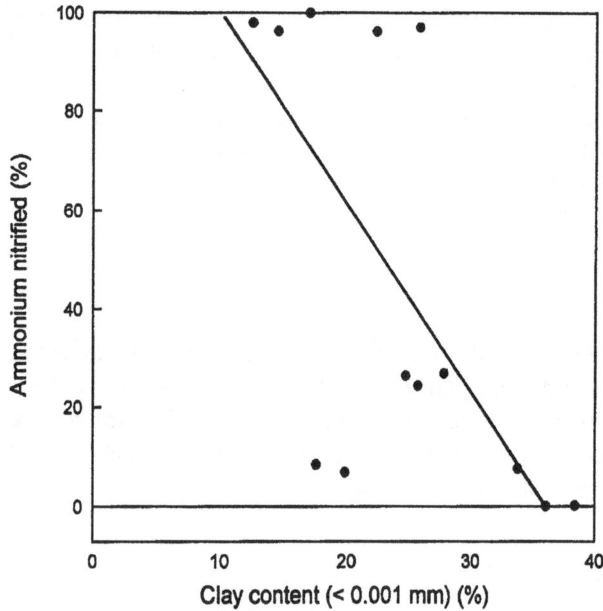

Figure 6.2. Relationship between nitrification of added ammonium and clay content of soil (Li *et al.* 1986).

However, studies by Chen and Zhou (1961, 1963) indicated that ammonium oxidizing bacteria in paddy soils were adapted to the strongly reduced environment and were able to oxidize ammonium. Jin (1988) also showed that ammonium oxidizing bacteria were active in the absence of oxygen, whereas nitrite oxidizing bacteria were absolutely aerobic. After flooding, paddy soils develop oxidized and reduced zones. Between the two soil layers there were large variations in the numbers of ammonium oxidizing bacteria and the nitrifying intensity. The number of ammonium oxidizing bacteria in the oxidized layer varied from 5 to 100,000 times that in the reduced layer (Table 6.2). In addition the amount of added ammonium nitrified in the oxidized layer was greater than that oxidized in the reduced layer (Table 6.3; Li 1986; Chen and Zhou 1963).

Table 6.2. Distribution of ammonium oxidizing bacteria in the oxidized and reduced layers of paddy soils (estimated numbers g dry soil^{-1}).[1]

Soil	Oxidized layer	Reduced layer
Submergenic	$3.92 \pm 4.66 \times 10^5$	$8.47 \pm 7.26 \times 10^3$
Clayey hydromorphic	$8.39 \pm 5.40 \times 10^6$	$11.5 \pm 5.29 \times 10^3$
Slightly acid gleyed	$40.5 \pm 24.9 \times 10^3$	$7.79 \pm 5.20 \times 10^3$
Neutral gleyed	10^9	$1.54 \pm 1.11 \times 10^5$

[1] MPN method (Li *et al.* 1983a).
Ammonium sulfate (120 mg N) was added to 300 g soil, submerged for one month at 28°C and nitrifying bacteria in the oxidized and reduced layers determined by the MPN method.

Table 6.3. Nitrification of added ammonium (% of applied N) in oxidized and reduced layers of paddy soils.[1]

Soil	Oxidized	Reduced
Submergenic	38.3	19.2**
Clayey hydromorphic	45.2	29.8**
Slightly acid gleyed	22.3	13.2**
Neutral gleyed	58.3	21.2***

[1] The incubation was conducted in the same way as in Table 6.2 (Li 1986).
30 ml culture solution was inoculated with 1 ml of 1/10 soil suspension, incubated for 2 weeks then nitrite + nitrate was determined.
** and *** denote that the results were significantly different at P = 0.01 and 0.001, respectively.

6.3.4. *Moisture*

Soil moisture is another factor controlling nitrification in soil. Extremes of moisture are unfavourable for the activity of nitrifying microbes in soil. When the soil moisture content is low, the oxygen content is high. However, the dryness of the soil will affect the survival of the nitrifying bacteria. Studies have shown that, in the pF range 2.7–4.5, the nitrification rate decreases logarithmically ($NR = Ae^{-BpF}$), while it was close to zero in the pF range 4.5–5.0 (Belser 1979).

Malhi and McGill (1982) showed that the maximum rate of nitrification occurred when the water potential was –33 k Pa, and the minimum rate occurred when the water potential was –1500 k Pa. Similar results were obtained by Flowers and O'Callaghan (1983). The maximum and the minimum nitrifying activity were found at –80 kPa (corresponding to 60% of field capacity) and –1.5 M Pa of soil water potential, respectively. It is generally considered that the optimum moisture content for nitrification in soil is 50–70% of the maximum water-holding capacity. According to our data, the nitrification rate was higher when the soil moisture content was 65% of the maximum water-holding capacity than when it was 30% (Figure 6.3), or when the soil was flooded (Table 6.4). Under flooded conditions, the thicker the water layer on the soil surface, the lower the oxygen content of the soil, and the lower the nitrification rate (Yoshida and Padre 1974).

6.3.5. *Organic manure*

All kinds of organic matter in the medium have inhibiting effects on the nitrifying bacteria, but under natural conditions the presence of organic matter does not always retard the oxidation of ammonium. In sludge or farmyard manure, for instance, there is considerable soluble organic matter, yet nitrification will proceed normally. When organic manure is added to soil extra ammonium is produced for nitrification, and the propagation of autotrophic and heterotrophic nitrifying organisms is stimulated, thus increasing the nitrifying activity in soil.

Flowers and O'Callaghan (1983) and Li *et al.* (1987) found that, when equal amounts of N were added, the soil treated with pig manure had a higher nitrification

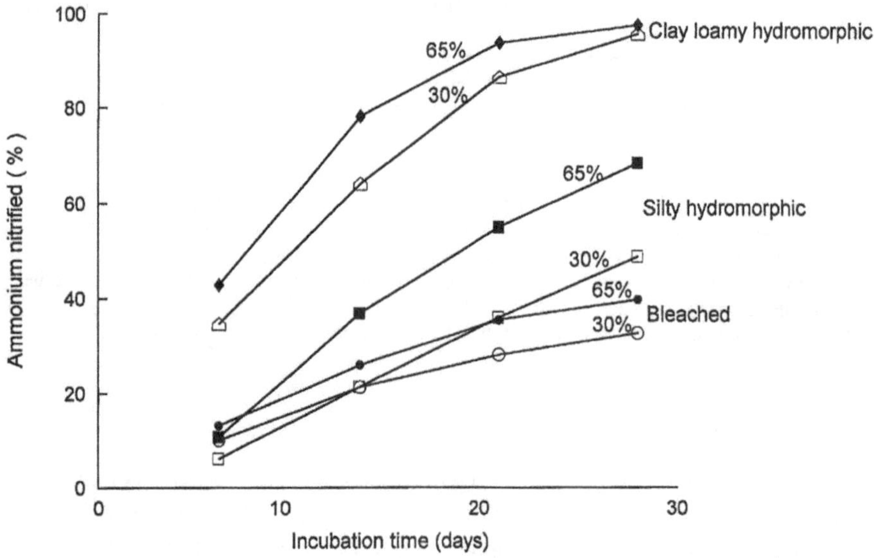

Figure 6.3. Effect of soil moisture (% water holding capacity) on the nitrification of added ammonium in paddy soils.

Table 6.4. Nitrification of added ammonium (% of applied N) in hydromorphic paddy soil under different moisture conditions.[1]

Treatment	Incubation time (days)				
	7	14	21	28	36
60% of the water-holding capacity	5.0	42.7	61.6	93.9	Not determined
Flooded (0.5 cm water layer)	–	Tr	0.4	1.1	3.1

[1] 20 g of fresh soil was placed in 150 ml flask and ammonium sulfate (5 mg N) added. The flasks were covered with plastic film and incubated at 28°C. The samples were analyzed for ammonium, nitrite and nitrate on days indicated (Li *et al.* 1982).

rate than the soil treated with ammonium sulphate. In the treatments with different rates of pig manure, the soil receiving the highest rate of pig manure had the highest rate of nitrification. The increase in nitrifying activity through the application of organic manure is associated with the increased number of nitrifying bacteria. For example, in the Broadbalk plots of Rothamsted Experiment Station the greatest number of nitrifying bacteria were found in plots treated with farmyard manure (Soriano and Walker 1973).

Hao *et al.* (1961) found that a soil which had been cultivated for only 10 years after reclamation but had received large quantities of organic manure had approximately the same number of nitrifiers as a moderately manured farmland with a 70–80 year history of cultivation (Table 6.5). Hao *et al.* (1961) also found that there were more nitrifying bacteria in soils of high fertility than in soils of low fertility

Table 6.5. Effect of manuring farmland on number of nitrifying bacteria in soil ($\times 10^3$ g dry soil^{-1}).[1]

Location	Low to moderate rates	High rates
Poyang, Jiangxi	2.5	70
Leping, Jiangxi	1.1	7.0
Nanjing, Jiangsu	0.033	0.33
Changshu, Jiangsu	1320	1370

[1] Hao (1961).

(Table 6.6). Chen *et al.* (1981) also found that the number of nitrifiers and the nitrification intensity were greater in soils with a high organic matter content than in soils low in organic matter (Table 6.7).

6.4. Nitrifying activity of different soils

The nitrifying activity of soils varies with environmental conditions and soil properties. Thus, acid sulphate soils did not nitrify (Sahrawat 1980), the nitrifying activity in red soils and latosols was low, and in calcareous soils the nitrifying activity

Table 6.6. Nitrifying bacteria in paddy soils with different levels of fertility ($\times 10^4$ g dry soil^{-1}).[1]

Soil	Location	Fertility level	Number
Acid hydromorphic	Jiangxi	High	1.89
		Low	2.35
	Zixi, Jiangxi	High	2.30
		Low	0.09
		High	2.01
		Low	0.28
Neutral hydromorphic	Xiaogan, Hubei	High	18.90
		Low	0.88
	Nanjing, Jiangsu	High	41.30
		High	20.30
		Low	2.32
		Low	19.70
Calcareous hydromorphic	Xinhailian, Jiangsu	High	17.30
		Low	2.57

[1] Hao (1961).

Table 6.7. Nitrifying bacteria and intensity in soils of different humus content.[1]

Soil	Humus (g kg^{-1})	Bacteria ($\times 10^3$ g dry soil^{-1})	Nitrifying intensity (mg NO_3^-N kg soil^{-1})
Eroded cinnamon	–	0.270	1.1
Newly reclaimed eroded cinnamon	5	0.470	3.2
Cultivated cinnamon	11	11.8	9.2
Vegetable	19	16.4	10.3

[1] Incubated at 28°C for two weeks (Chen *et al.* 1981).

was very high. Paddy soils in different areas vary in their ability to nitrify. The paddy soil developed from the Quaternary red clay in Jiangxi Province had a lower nitrifying activity than the paddy soil developed from loess in Jiangsu Province, and the activity of the latter was lower than that of the paddy soil in the Zhujiang River Delta in Guangdong Province (Table 6.8).

Much attention has been given to the nitrifying activity of paddy soils and its relation to denitrification loss (Katyal *et al.* 1988). Chen and Zhou (Chen and Zhou 1961, 1963, 1964; Zhou and Chen 1983) found that not only were there ammonium oxidizing bacteria in paddy fields, but their numbers were larger than in upland soils. In paddy soils these microorganisms survived and multiplied at different depths in the plow layer (Tables 6.9 and 6.10), and nitrification took place either in

Table 6.8. Nitrifying activity in different types of paddy soils.[1]

Location	Parent material	Soil type	pH	Incubation time (days)			
				7	14	21	28
				-- Ammonium oxidized (%) --			
Jiangsu	Loess	Hydromorphic	6.03	5.0	42.7	61.6	93.9
Jiangxi	Quaternary red clay	Hydromorphic	5.32	–	–	19.0	49.3
Guangdong	Alluvium from Zhujiang River	Submergenic	6.56	26.8	73.2	82.0	87.1

[1] Experimental conditions as in Table 6.4; moisture content adjusted to 65% WHC (Li *et al.* 1983b).

Table 6.9. Ammonium oxidizing bacteria ($\times 10^3$ g dry soil^{-1}) in the plow layers of paddy and upland fields.[1]

Test No.	Paddy field		Upland field	
1	11.32	(Fallow)	0.03	(Wheat)
2	13.46	"	0.09	"
3	22.22	"	0.96	"
4	40.35	(Rice)	1.85	(Cotton)
5	23.08	"	0.69	"
6	33.98	"	0.94	"

[1] Chen *et al.* (1981).

Table 6.10. Ammonium oxidizing bacteria ($\times 10^3$ g dry soil^{-1}) in different layers of flooded paddy soils.[1]

Soil layer (cm)	Sampling time		
	Before planting	At seedling revival	At tillering
0–1	112.0	75.8	98.7
1–15	69.6	86.5	109.0
15–25	19.5	61.1	103.5
0–25	90.5	73.6	–

[1] Chen *et al.* (1981).

the oxidized or reduced layers. Our data also indicate that the nitrifying activity in the oxidized layer was much higher than that in the reduced layer (Li *et al.* 1983b).

The nitrifying activity of different soils in the same region varied with the type of soil. For example, in the Taihu Lake region some of the soils have a lag phase while others do not. The nitrifying activity of the soils decreased in the following order, neutral gleyed paddy soil and submergenic paddy soil > clay loamy hydromorphic paddy soil > silty hydromorphic paddy soil > slightly acid gleyed paddy soil > bleached paddy soil; the ammonium nitrified by these soils after one week's incubation was 98%, 96%, 40%, 26%, 8.5%, and 7%, respectively (Li *et al.* 1987).

In general, the nitrifying activity of a soil is related to the number of nitrifying bacteria. Studies by Martikainen (1984, 1985b) showed a positive correlation between the number of nitrifying bacteria and the amount of nitrate formed. According to our investigations in the Taihu Lake region, the largest number of nitrifying bacteria was found in the neutral gleyed paddy soil and the submergenic paddy soil which have a high nitrifying activity, but the smallest number of nitrifying bacteria was not in the slightly acid gleyed paddy soil which has the lowest activity (Table 6.11).

As shown in Table 6.12, most of the soils in the red soil region have a low capacity for nitrification due to their low fertility and high acidity (Pan *et al.* 1988). When ammonium sulphate was used as a N source, the nitrifying activity in upland and citrus orchard soils was fairly high, while in paddy soils and tea plantation soils it was very low. This seems to be related to the low pH values (3.73–4.07), the high

Table 6.11. Distribution of ammonium oxidizing bacteria in paddy soils of the Taihu Lake region.[1]

Soil	Number (per g dry soil)
Slightly acid gleyed	$2.38 \pm 0.91 \times 10^4$
Silty hydromorphic	$3.28 \pm 0.0 \times 10^4$
Bleached	$3.55 \pm 1.78 \times 10^4$
Hydromorphic	$7.40 \pm 3.78 \times 10^4$
Neutral gleyed	$1.2 \pm 2.0 \times 10^5$
Submergenic	$2.64 \pm 2.21 \times 10^5$

[1] MPN method (Li *et al.*, 1983b).

Table 6.12. Nitrifying activity (% ammonium N oxidized) in soils from the red soil region.[1]

Soil	Parent material	Crops grown	Time of incubation (days)			
			7	14	21	28
Red	Quaternary red clay	Buckwheat, peanuts	33.7	67.9	75.7	84.6
Red	" "	Citrus	0.9	20.9	–	44.1
Red	" "	Tea tree	0	0	1.3	2.5
Hydromorphic paddy	" "	Rice	0	0	0	0
Strongly gleyed paddy	" "	Rice	0	Tr	0	0

[1] Experimental conditions as in Table 6.4 (Pan *et al.* 1988).

content of Al ions (Brar and Giddens 1968; Li 1984) and the low nutrient status of the soils. Wickramasinghe *et al.* (1985) pointed out that the nitrifying activity of the strongly acid soils was considerably lower than that of the neutral grassland soil, and that the small number of nitrifiers in the first group was closely related to the low pH of the soils. Application of ammonium or potassium sulphate further lowered the soil pH and inhibited nitrification. Application of urea raised the soil pH and hence did not suppress nitrification (Boer *et al.* 1988; Martikainen 1985a, b). Pan *et al.* (1988) found that the nitrifying activity in soils of the red soil region was markedly increased by application of urea, and the increase became more pronounced with time (Figure 6.4).

In recent years more attention has been devoted to the study of heterotrophic nitrification in soil (Adams 1986 a, b; Doxtader and Alexander 1966; Ishaque and Cornfield 1976; Kreitinger *et al.* 1985; Rho 1986; Tate 1977). Selective inhibitors have been used to estimate the relative importance of autotrophic and heterotrophic nitrification in soil (Shattuck and Alexander 1963; Walter *et al.* 1979). Pan *et al.* (1988) found that addition of nitrapyrin to inhibit autotrophic nitrification, maintained the nitrate content of two paddy soils at a relatively low level, but in the tea plantation red soil the nitrate content was higher, and it tended to increase with time of incubation (Figure 6.5). This suggests that heterotrophic nitrification occurred in the tea plantation red soil. Similar results were reported by Ishaque and Cornfield (1976).

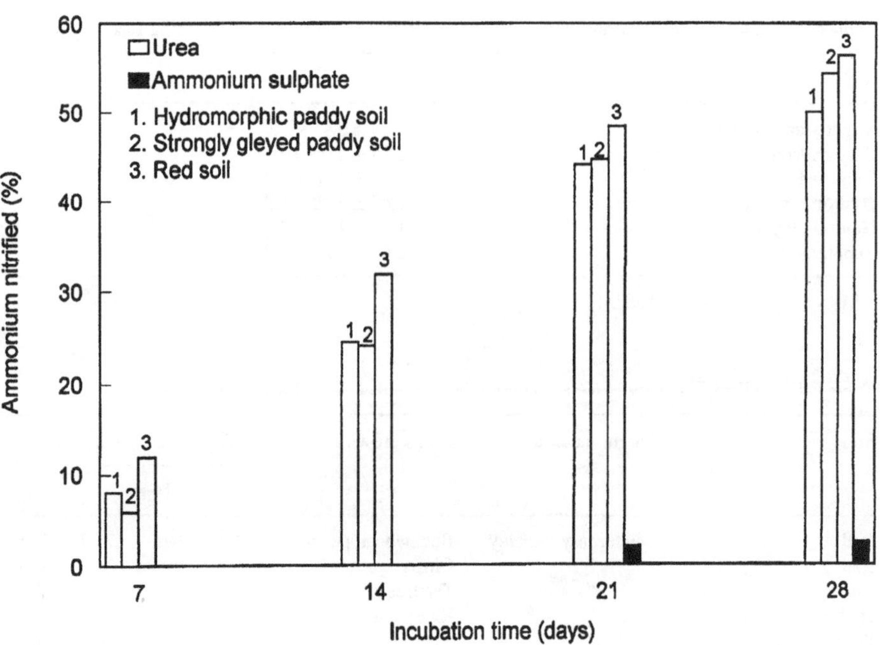

Figure 6.4. Effect of fertilizer type on nitrification in soils of the red soil region (Pan *et al.* 1988).

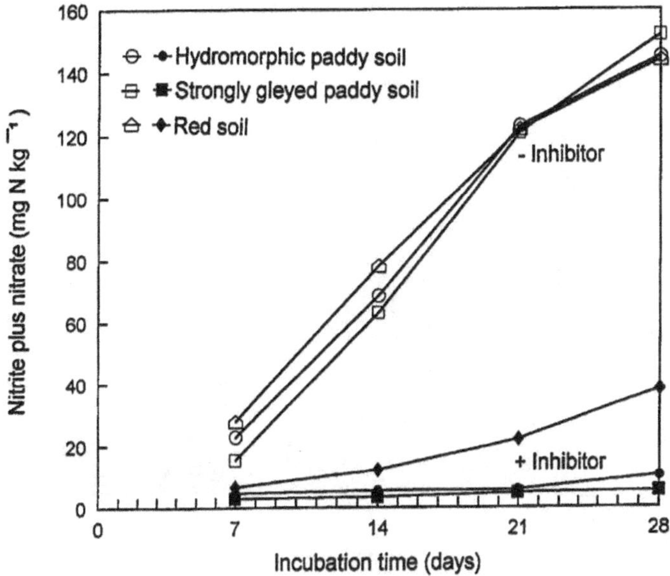

Figure 6.5. Effect of nitrapyrin on the nitrite plus nitrate content of soils derived from Quaternary red clay (Pan *et al.* 1988).

6.5. Formation of nitrous oxide during nitrification

The production of nitrous oxide from a soil was considered to imply that the soil was anaerobic and that denitrification was occurring. Since Yoshida and Alexander (1970) discovered that nitrous oxide was formed in pure cultures of *Nitrosomonas europaea*, investigations have been conducted on the mechanism of nitrous oxide formation during nitrification and the emission of nitrous oxide from soil in the field (Blackmer *et al.* 1980; Robertson and Tiedje 1987; Sahrawat and Keeney 1986). It was found that nitrous oxide emission increased with ammonium concentration and oxygen partial pressure, while the addition of a nitrification inhibitor prevented nitrous oxide formation (Bremner and Blackmer 1978; Goreau 1980). The results presented in Figure 6.6 indicate that nitrous oxide emission was greater at 68.5% of water holding capacity, WHC, than at a lower moisture content (31.7% of WHC); the total amounts of nitrous oxide released over a period of 12 days were 4.33 and 0.634 μg N g soil^{-1}, respectively (Jin 1988).

The effect of temperature on nitrous oxide emission was determined by the moisture content. At the optimum soil moisture content (65% of WHC), nitrous oxide emission at 28°C and 38°C was approximately the same, but when the soil was drier (32% of WHC), the emission at 38°C was greater than that at 28°C (Table 6.13; Jin 1988). There is evidence that nitrous oxide is evolved during nitrification when the soil thaws in spring (Goodroad and Keeney 1984b).

pH is another factor controlling the production of nitrous oxide in soil (Jin 1988; Goodroad and Keeney 1984a). More nitrous oxide was emitted from a calcareous

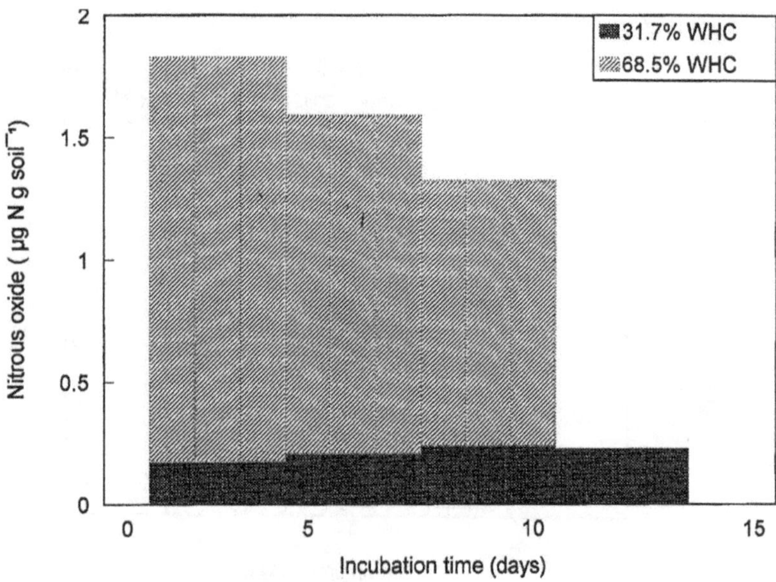

Figure 6.6. Effect of soil moisture on formation of nitrous oxide during nitrification (vegetable soil in Beijing suburbs, pH 8.4, incubated at 38°C; Jin 1988).

Table 6.13. Effect of temperature on production of N_2O (μg N g soil^{-1}) during nitrification.[1]

Time (days)	28°C	38°C
0–3	0.178	0.405
3–6	0.161	0.412
6–9	0	0.302
9–12	0	9.345
0–12	0.339	1.46

[1] Ammonium sulfate added to give 200 mg N kg soil^{-1}, and soil moisture content was adjusted to 32% of the water holding capacity; pH was 8.4 (Jin 1988).

soil than from a slightly acidic soil (Jin 1988). Li *et al.* (1991) found that in addition to the amount of nitrous oxide released, the pattern of emission varied with soil reaction (Figure 6.7). Nitrous oxide was emitted rapidly during nitrification in a calcareous paddy soil and this was associated with the high nitrifying activity of this soil (Li *et al.* 1981b). Emission of nitrous oxide ceased 7–11 days after incubation commenced. However, for the neutral paddy soil there was a distinct lag phase in nitrous oxide emission until day 9, then emission increased greatly. In the acidic paddy soil derived from red soil no nitrous oxide was emitted and this seems to be due to the extremely low nitrifying activity of this soil (Pan *et al.* 1988).

The rate of emission of nitrous oxide during nitrification appears to be related to the kind and rate of N fertilizer applied (Bremner and Blackmer 1978; Bremner *et al.* 1982; Cates and Keeney 1987; Duxbury and McConnaughey 1986). Zhang *et al.*

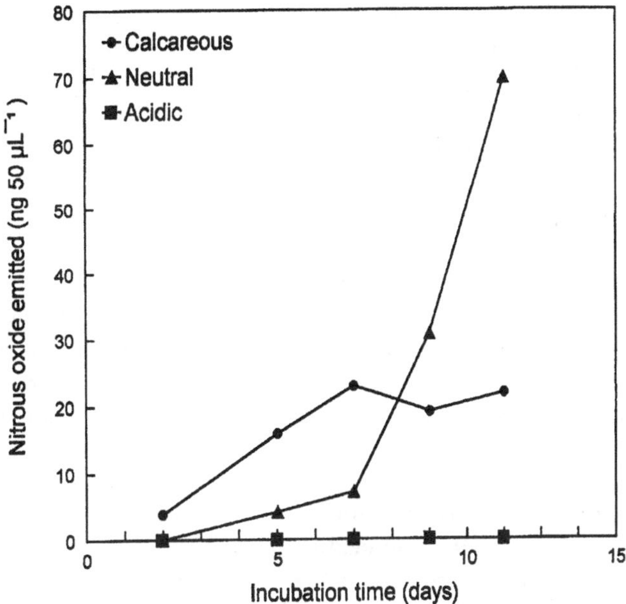

Figure 6.7. Nitrous oxide released from paddy soils during nitrification (5 mg N as ammonium sulphate added to 20 g of fresh soil, water adjusted to 65% of WHC and incubated at 28°C; Li *et al.* 1991).

(1985) found that emission of nitrous oxide during nitrification of ammonium or ammonium-producing fertilizers decreased in the order: urea > ammonium sulfate > ammonium bicarbonate.

It is important to note that the amount of fertilizer N lost as nitrous oxide from some agricultural soils is significant and should not be overlooked. Thus Bremner *et al.* (1982) found that the amount of nitrous oxide emitted from a soil treated with anhydrous ammonia was 7% of the N added. In our study conducted under conditions favourable for nitrification (pot experiment with a fluvo-aquic sandy loam, Chao soil, with a moisture content of 60% of WHC), the total emission of nitrous oxide following addition of ammonium sulfate ranged from 0.3 to 2.7 μg N m^{-2} s^{-1}, and averaged 1.75 ± 0.898 μg N m^{-2} s^{-1} (Li *et al.* 1991).

6.6. Nitrification inhibitors

Nitrification inhibitors have been used to inhibit the transformation of ammonium to nitrate and thus reduce N losses through denitrification and leaching (Hauck 1971; Huber *et al.* 1969; Lewis and Stefanson 1975). In addition, the effect of urease inhibitors (Zhou 1984) and insecticides and herbicides (Aulakh and Rennie 1984; Hauck 1971) on nitrification have been investigated.

Investigations on nitrification inhibitors were commenced in China in the early 1960s, and a National Cooperation Network under the direction of the Petrochemical Ministry was established in the early 1970s, to make a more extensive study of a few selected nitrification inhibitors. The chemicals tested included 2-chloro-6-trichloromethyl pyridine (nitrapyrin), 2-amino-4-chloro-6-methyl pyrimidine (AM), 4-amino-1,2,4-triazole hydrochloride (ATC), thiourea (TU), guanylthiourea, 2,5-dichloronitrobenzene, and dicyandiamide, of which nitrapyrin and guanylthiourea were identified as promising.

6.6.1. Selective inhibition

Many studies have shown that nitrapyrin has significant inhibiting effects on the growth and activity of ammonium oxidizing bacteria. In calcareous and non-calcareous paddy soils amended with nitrapyrin and equal amounts of applied N, the nitrification rate was 3–33% of that in the control treatment, and in extreme case was only 0.01%. The number of ammonium oxidizing bacteria amounted to only 0.15% (or 0.04% in the extreme case) of those in the control (Li *et al.* 1981 a, b; Figure 6.8). Similar effects were found in a pot experiment with wheat and a field experiment with rice (Guo and Liu 1978; Li *et al.* 1981a). Because nitrapyrin is easily degraded under flooded conditions, its inhibiting effect in the paddy field lasted for a shorter time (~one month) than in an upland field where the effect persisted until the wheat plants matured (Li *et al.* 1981a). AM and ATC were also fairly good inhibitors, while thiourea was less effective (Tables 6.14 and 6.15).

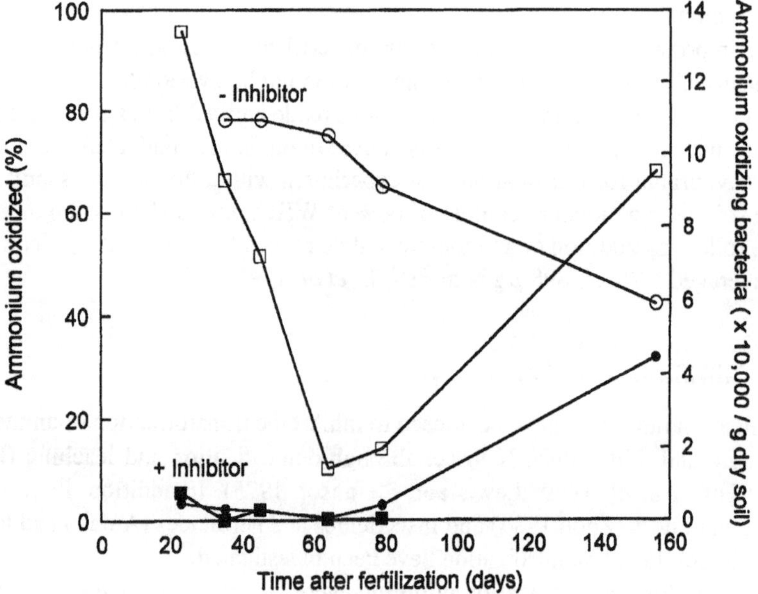

Figure 6.8. Effect of nitrapyrin on nitrification in soil (pot experiment with wheat, hydromorphic paddy soil; Li *et al.* 1981b). ○, ●, ammonium; □, ■, bacteria.

Table 6.14. Effect of nitrification inhibitors on ammonium oxidation in soil (% of added N).[1]

	Sampling time			
	Days after winter fertilization		Days after spring fertilization	
	11 (Jan. 17)	23 (Jan. 29)	10 (Apr. 1)	20 (Apr. 11)
Treatment				
Control	15.1	17.0	25.5	3.1
Nitrapyrin	7.0	7.8	14.8	1.5
AM	10.3	6.1	22.3	1.5
ATC	4.6	15.4	3.6	2.5

[1] Field experiment with wheat grown on hydromorphic paddy soil (Guo and Liu 1978).

Table 6.15. Effect of nitrification inhibitors on the number of ammonium oxidizing bacteria in soil (10^4 g dry soil^{-1}).[1]

Treatment	Days after sowing					
	22	33	43	61	76	166
Control	13.42	933.33	72.46	14.36	19.77	95.37
Thiourea	27.47	522.88	33.16	30.91	9.55	83.63
Nitrapyrin	7.66	1.44	3.21	0.30	0.47	Trace
AM	1.51	0.46	2.67	0.36	0.47	Trace

[1] Pot experiment with wheat grown on hydromorphic paddy soil (unpublished data of Li Liang-mo *et al.*).

6.6.2. *Effects on microorganisms*

Nitrapyrin at its commonly used concentration (2–3% of fertilizer-N) had no adverse effects on the respiratory activity of microorganisms or other activity associated with N transformations such as ammonification and non-symbiotic N fixation, although it is detrimental to symbiotic N fixation (Li *et al.* 1981b). Nitrapyrin damages the roots of leguminous crops, so it is necessary to control its concentration when used with legumes. Sand culture experiments indicate that nitrapyrin did not adversely affect nitrogenase activity in vetch nodules when the concentration was 1–3 μg ml^{-1}, it even had a stimulating effect at a concentration of 1–2 μg ml^{-1}. However, when the concentration was increased to 5 μg ml^{-1} nitrapyrin lowered the nitrogenase activity considerably (Li *et al.* 1981b). Likewise, nodulation and the fresh weight of vetch shoots varied with the concentration of nitrapyrin; no effects were observed when the concentration was below 2 μg ml^{-1}, but at concentrations >3 μg ml^{-1} the number of nodules and the fresh weight of the shoots were markedly reduced (Li *et al.* 1981b). It is expected that, in soil culture, the damage of nitrapyrin to legumes may be small. However, under field conditions, it is considered unwise to use nitrapyrin for a current legume crop. When nitrapyrin is added to a crop, attention must be given to possible residual effects on a succeeding leguminous crop.

In water culture, nitrapyrin markedly inhibited the growth of rice roots and shoots when the concentration was 5 $\mu g\ ml^{-1}$, but at a concentration of 1 $\mu g\ ml^{-1}$ it stimulated the development of roots, and at a concentration of 0.1 $\mu g\ ml^{-1}$ it increased N uptake by rice seedlings (Li *et al.* 1981b).

6.6.3. Effect on fertilizer nitrogen loss

It would be expected that amendment of N fertilizer with a nitrification inhibitor would allow the N to persist in soil as ammonium for longer, and for losses by denitrification and leaching to be reduced. Incubation studies showed that addition of nitrapyrin to flooded soils increased the recovery of labelled ammonium by 4–11% ($p<0.01$) (Li *et al.* 1981b); similar results were obtained in field trials with flooded rice.

The efficiency of nitrification inhibitors in reducing fertilizer-N loss varied with soil property, fertilizers type, kind of inhibitor, and method of fertilization. As shown in Table 6.16, addition of nitrapyrin in general reduced N loss from the slightly acid and acid soils by 7–8%, but it was ineffective in the calcareous soil. Although nitrapyrin markedly inhibited nitrification in the calcareous soil, N was presumably lost by ammonia volatilization under conditions of high pH (Bundy and Bremner 1974).

Pot culture and field experiments also suggested that, irrespective of whether the test plant was rice or wheat, addition of a nitrification inhibitor was more effective in reducing fertilizer-N loss from non-calcareous soils than from calcareous soils (Table 6.17). As shown in Table 6.17 for flooded rice, the addition of nitrapyrin reduced N loss from urea, ammonium bicarbonate and ammonium sulphate by 7.7–10.2%, 7.1–9.0, and 1.2%, respectively. Some inhibitors are more effective for reducing N loss than others. Thus He *et al.* (1981) showed that N loss from an application of urea to flooded rice growing on a non-calcareous soil was reduced by 7.0, 7.4, and 10.1%, respectively by the addition of nitrapyrin, ASU or ATC. In general, nitrification inhibitors were more effective for reducing N loss when fertilizer N was top-dressed onto flooded rice than when it was deep-placed. It is suggested that this occurs because the nitrifying activity in the deep layers of a flooded soil is lower than that in the surface layer.

Table 6.16. Effect of nitrapyrin on nitrogen loss from soil (% of nitrogen added).[1]

Soil	Treatment	Loss
Strongly calcareous	Control	54.5
	+ Nitrapyrin	52.5
Slightly acid	Control	16.5
	+ Nitrapyrin	9.0**
Acid	Control	30.8
	+ Nitrapyrin	23.5**

[1] Ammonium sulfate added to give 400 mg N kg soil^{-1} and incubated under flooded conditions for one month (Li *et al.* 1981b).
** Significantly different at P = 0.01.

Table 6.17. Effect of nitrification inhibitors on N loss (% of applied N).

Location	Soil	Crop	Experiment	Fertilizer	Treatment[2]	Loss	Reference
Hubei	Noncalcareous	Rice	Microplot	Urea	SA–SB	20.7	Guo *et al.* (1980)
				+ nitrapyrin	SA–SB	10.5	
				ABC	SA–SB	48.5	
				+ nitrapyrin	SA–SB	41.4	
Jilin	Noncalcareous	Rice	Microplot	Urea	Bs–SB	33.3	He *et al.* (1981)
				+ nitrapyrin	Bs–SB	26.3	
				+ ASU	Bs–SB	25.9	
				+ ATC	Bs–SB	23.2	
				Urea	DP–S (15 cm)	4.1	
				+ nitrapyrin	DP–S (15 cm)	4.1	
				+ ASU	DP–S (15 cm)	6.0	
				+ ATC	DP–S (15 cm)	3.2	
Jiangsu	Calcareous	Rice	Microplot	AS	Bs–SB	47.5	Li *et al.* (1975)
				+ nitrapyrin	Bs–SB	45.6	
				AS	DP–S	30.2	
				+ nitrapyrin	DP–S	29.8	
				AS	TD–MS	26.2	
				+ nitrapyrin	TD–MS	21.5	
Hebei	Calcareous	Rice	Pot	AS		59.3	Wen *et al.* (1979)
				+ nitrapyrin		58.1	
				AN		84.4	
				+ nitrapyrin		84.5	
				ABC		66.7	
				+ nitrapyrin		57.7	
				Urea		53.0	
				+ nitrapyrin		45.3	
Shaanxi	Calcareous	Wheat	Microplot	Urea	Bs–Inc	58.3	Zhang and Li, (1981)
				+ nitrapyrin, ASU, CMB		56.9–52.8	
				ABC	Bs–Inc	53.4	
				+ nitrapyrin, ASU, MAST		55.5–59	
Beijing	Calcareous	Wheat	Pot	AS	SA	11.1	Wang *et al.*, (1981)
				+ nitrapyrin, ASU		9.2–9.3	

[1] Unpublished data of Li Liangmo, Zhang Shuang, Zhou Xiuru, and Pan Yinghua.

[2] SA–SB = split application, surface broadcast; Bs–SB = basal fertilizer, surface broadcast; DP–S = deep placed at 15 cm; TD–MS = top-dressed at mid stage; DP–S = deep placed; Bs–Inc = basal fertilizer incorporated; CMB = chloromethylbenzene; MAST = 2-amino-4-methyl-6-trichloromethyl triazine.

6.6.4. *Improving efficiency of fertilizer nitrogen*

Although the yield of some crops, such as wheat, rice, maize and sugarcane has been increased by the application of nitrification inhibitors (Lewis and Stefanson 1975; Li *et al.* 1981a; McCormick *et al.* 1984; Patrick *et al.* 1968), yield increases were not always obtained (Guo and Liu 1978; Zang and Zhou 1983; Zang *et al.* 1980; and unpublished data of Li Liang-mo *et al.*)

A number of factors may influence the effectiveness of nitrification inhibitors, viz. soil properties, N management techniques, kinds of nitrification inhibitors and type of crop. As shown in Table 6.18, the effect of nitrapyrin on yield was greater on soils with low fertility. Soil texture affects the degradation of inhibitors and thus their effectiveness. Redemann *et al.* (1964) found that nitrapyrin was lost from a sandy loam more quickly than from a soil high in organic matter and clay. Table 6.19 shows that greater yield increases were obtained on coarse-textured soils than on light- textured ones.

The efficiency of nitrification inhibitors also varied with the type of N fertilizer. For example, thiourea was more effective in increasing the yield of wheat when used with ammonium bicarbonate than with ammonium chloride or ammonium sulphate (Guo and Liu 1978). Also nitrapyrin was more effective than thiourea for increasing the yield of flooded rice, whereas the reverse was true for wheat (Guo and Liu 1978). It should be noted that it is easy to demonstrate the effectiveness of nitrification inhibitors for reducing N loss or increasing crop yields in incubation and pot experiments, but not in field experiments. Factors controlling the effectiveness of inhibitors appear to be more complex in field experiments than in incubation and pot experiments, and the reduction in N loss from the application of nitrification inhibitors is generally not large enough to be reflected in crop yields (Hauck 1971).

Table 6.18. Effect of nitrapyrin on yield of rice (t ha^{-1}) in paddy fields as affected by soil fertility.[1]

Soil	Fertility level	Treatment	Yield
Sandy loam submergenic	Low	Control	4.58
		Nitrapyrin	5.36**
	High	Control	7.89
		Nitrapyrin	8.28
Sandy loam submergenic	Low	Control	5.76
		Nitrapyrin	6.48**
	High	Control	6.38
		Nitrapyrin	6.35
Clay loam gleyed	Low	Control	4.92
		Nitrapyrin	5.48**
	High	Control	6.17
		Nitrapyrin	6.30
Clay loamy hydromorphic	High	Control	7.05
		Nitrapyrin	7.02

[1] Unpublished data of Li Liangmo, Zang Shuang, Zhou Xiuru and Pan Yinghua.
** significantly different at P = 0.01.

Table 6.19. Effect of nitrification inhibitors on yields of rice and wheat (t ha^{-1}) in Jiangsu Province).

Texture	Crop	Fertilizer	Treatment	Yield	Reference
Sandy loam	Rice	Urea	Control	6.77	Li *et al.* unpublished
			Nitrapyrin	7.18	
			Tu	6.80	
Sandy loam	Rice	Urea	Control	4.58	Li *et al.* unpublished
			Tu	5.00	
Loam	Rice	Urea	Control	6.29	Zang *et al.* (1980)
			Nitrapyrin	7.58	
			Tu	7.46	
Clay loam	Rice	Urea	Control	4.33	Zang *et al.* (1980)
			Nitrapyrin	4.70	
Clay loam	Rice	Urea	Control	4.40	Zang and Zhou (1983)
			MDCT[1]	4.94	
Clay loam	Rice	Urea	Control	6.78	Zang *et al.* (1980)
			Nitrapyrin	7.52	
Clay loam	Rice	Urea	Control	4.35	
			Nitrapyrin	4.97	Zang *et al.* (1980)
			Tu	4.84	
Clay loam	Rice	Urea	Control	5.60	Li *et al.* unpublished
			Nitrapyrin	6.17	
			Tu	6.26	
Clay loam	Rice	Urea	Control	5.21	Li *et al.* unpublished
			Nitrapyrin	6.23	
			Tu	6.14	
Clay loam	Wheat	ABC	Control	2.51	Guo and Liu, (1978)
			Tu (3%)	3.06	
			Control	2.82	
			Tu (6%)	3.49	
		AC	Control	2.98	
			Tu	3.23	
		AS	Control	4.05	
			Tu	4.27	
			Nitrapyrin	4.05	
			AM	4.16	
			ATC	4.00	

[1] 2-methyl-4, 6-di (trichlormethyl) triazine.

6.7. References

Adams, J A 1986a. Nitrification and ammonification in acid forest litter and humus as affected by peptone and ammonium–N amendment. Soil Biol. Biochem. 18:45–51.

Adams, J A 1986b. Identification of heterotrophic nitrification in strongly acid larch humus. Soil Biol. Biochem. 18:339–341.

Alexander, M 1977. Introduction to Soil Microbiology (2nd Edition). John Wiley and Sons, Inc., New York. pp. 251–271.

Aulakh, M S and Rennie, D A 1984. Azide effects upon nitrous oxide emission and transformations of nitrogen in soils. Can. J. Soil Sci. 65:205–212.

Belser, L W 1979. Population ecology of nitrifying bacteria. Ann. Rev. Microbiol. 33:309–333.

Blackmer, A M, Bremner, J M and Schmidt, E L 1980. Production of nitrous oxide by ammonia-oxidizing chemoautotrophic microorganisms in soil. Appl. Environ. Microbiol. 40:1060–1066.

Boer,W De, Duyts, H and Laanbrock, H J 1988. Autotrophic nitrification in a fertilized acid heath soil. Soil Biol. Biochem. 20:845–850.

Brar, S S and Giddens, J 1968. Inhibition of nitrification in Bladen Grassland soil. Soil Sci. Soc. Am. Proc. 32:821–823.

Bremner, J M and Blackmer, A M 1978. Nitrous oxide emission from soils during nitrification of fertilizer nitrogen. Science 199:295–296.

Bremner, J M, Breitenbeck, G A and Blackmer, A M 1982. Effect of anhydrous ammonia fertilization on emission of nitrous oxide from soil. J. Environ. Qual. 10:77–80.

Buchanan, R E and Gibbons, N E 1974. Bergey's Manual of Determinative Bacteriology (Eighth Edition). pp. 450–456. The Williams & Wilkins Company, Baltimore.

Bundy, L G and Bremner, J M 1974. Effects of nitrification inhibitors on transformation of urea nitrogen in soil. Soil Biol. Biochem. 6:369–376.

Castignetti, D 1980. Sequential nitrification by an Alcaligenes sp, and Nitrobacter agilis. Can. J. Microbiol. 26:1114–1119.

Cates, R C and Keeney, D R 1987. Nitrous oxide production throughout the year from fertilized and manured maize fields. J. Environ. Qual. 16:443–447.

Chen, H K and Zhou, Q 1961. Investigation on nitrification and nitrifying microorganisms in rice-field soils: I. Nitrification in rice soils. (in Chinese). Acta Pedol. Sin. 9:56–64.

Chen, H K and Zhou, Q 1963. Investigation on nitrification and nitrifying microorganisms in rice soils: II. Propagation of enriched culture of nitrifying bacteria under incubation and the nitrosification intensity in rice soils. In: Abstracts of Papers of the 1963 Annual Meeting of Soil Science Society of China (Part One). (in Chinese). pp. 46–47.

Chen H K and Zhou, Q 1964. Investigation on nitrification and nitrifying microorganisms in rice soils: III. Isolation of ammonium oxidizing bacteria from pure culture. In: Wuhan Microbiology Institute and Huazhong Agricultural College (eds.), Abstracts of Special Topic Reports and Research Reports of the Symposium on Soil Microbiology in 1963. Research Series. (in Chinese). pp. 45–46.

Chen, H K, Li, F D, Chen, W X and Cao, Y Z 1981. Soil Microbiology. (in Chinese). pp. 190–300. Shanghai Science and Technology Publishing House, Shanghai.

Dancer, W S, Peterson, L A and Chesters, G 1973. Ammonification and nitrification of nitrogen as influenced by soil pH and previous treatments. Soil Sci. Soc. Am. Proc. 37:67–69.

Duxbury, J M and McConnaughey, P K 1986. Effect of fertilizer source on denitrification and nitrous oxide emission in a maize field. Soil Sci. Soc. Am. J. 50:644–648.

Doxtader, K G and Alexander, M 1966. Nitrification by heterotrophic soil microorganisms. Soil Sci. Soc. Am. Proc. 30:351–355.

Flowers, T H and O' Callaghan, J R 1983. Nitrification in soils incubated with pig slurry or ammonium sulphate. Soil Biol. Biochem. 15:337–342.

Goodroad, L L and Keeney, D R 1984a. Nitrous oxide production in aerobic soils under varying pH, temperature and water content. Soil Biol. Biochem. 16:39–43.

Goodroad, L L and Keeney, D R 1984b. Nitrous oxide emissions from soils during thawing. Can. J. Soil Sci. 64:187–194.

Goreau, T J 1980. Production of NO_2^- and nitrous oxide by nitrifying bacteria at reduced concentrations of oxygen. Appl. Environ. Microbiol. 40:526–532.

Guo, W J and Liu, M L 1978. Effect of nitrification inhibitor, thiourea, on increasing the yield of wheat. (in Chinese). Soils (10):217–219.

Guo, Z F, Chen, S S, Wang, Y S and Tang, N X 1980. Studies on the absorption, fixation and loss of fertilizer-N and on increasing the efficiency of fertilizer-N use in rice fields. (in Chinese). Application of Atomic Energy in Agriculture (1):24–29.

Hao, W Y 1961. Characteristics of microbiology in paddy soil. In: Editorial Board for Series on Studying Bumper Crops in Agriculture, Chinese Academy of Sciences (ed.), Soil Conditions for High Yields of Rice. (in Chinese). pp. 188–207. Science Press, Beijing.

Hauck, R D 1971. Nitrification Inhibitors. In Goring C S I, and Hamaker, J W (eds.), Organic Chemicals in the Soil Environment. Vol. 2, pp. 655–666. Marcel Dekker, New York.

He, C Y, Zhao, Y Z, Wang, Q, Li, Z G, Kui, X L and Gao, J F 1981. The efficacy of nitrification inhibitors in increasing the efficiency of nitrogenous fertilizer in paddy soils. (in Chinese). Chinese J. Soil Sci. (3):1–3.

Huber, D M, Murray, G A and Crane, J M 1969. Inhibition of nitrification as a deterrent to nitrogen loss. Soil Sci. Soc. Am. Proc. 33:975–976.

Ishaque, M and Cornfield, A H 1976. Evidence for heterotrophic nitrification in an acid Bangladesh soil lacking autotrophic nitrifying organisms. Trop. Agric. 53:157–160.

Jin, F 1988. Study on nitrous oxide produced in soil by nitrification under different conditions. (in Chinese). MS Thesis, Beijing Agricultural University. p. 44.

Katyal, J C, Cater, M F and Vlek, P L G 1988. Nitrification activity in submerged soil and its relation to denitrification loss. Biol. Fertil. Soils. 7:16–22.

Kreitinger, J P, Klein, T M, Novick, N J and Alexander, M 1985. Nitrification and characteristics of nitrifying microorganisms in an acid forest soil. Soil Sci. Soc. Am. J. 49:1407–1410.

Lewis, D C and Stefanson, R L 1975. Effect of N-serve on nitrogen transformations and wheat yields in some Australian soils. Soil Sci. 119:273–279.

Li, L M 1984. An outline of research on nitrification in soils. (in Chinese). Progress in Soil Sci. 12:1–10

Li, L M 1986. Outline and prospects of the studies on nitrification-denitrification in soils of China. In: Soil Agricultural Chemistry and Soil Biology and Biochemistry Committees, Soil Science Society of China (eds.), Advances and Prospects for Soil Nitrogen Research in China. (in Chinese). pp. 25–47. Science Press, Beijing.

Li, L M, Wu, Q T, Li, Z G and Pan, Y H 1991. Fluxes of nitrous oxide from different soils. (in Chinese). Soils 23:24–27.

Li, L M, Zang, S, Zhou, X R and Pan, Y H 1981a. Effect of nitrapyrin on the inhibition of nitrification in some paddy soils of China. In: Institute of Soil Science, Academia Sinica (ed.), Proc. Symposium on Paddy Soil. pp. 837–844. Science Press, Beijing.

Li, L M, Zang, S, Zhou, X R and Pan, Y H 1981b. Effect of nitrapyrin on nitrification and microbial activity. (in Chinese). Acta Pedol. Sin. 18:58–70.

Li, L M, Zang, S, Zhou, X R and Pan, Y H 1982. Preliminary study on nitrification-denitrification in soil. Collections of Scientific Conference on Taihu Region. (in Chinese). Institute of Soil Science, Nanjing.

Li, L M, Zhou, X R, Pan, Y H and Zang, S 1983a. Study on N loss in paddy soils. In: Institute of Soil Science, Academia Sinica (ed.), Proceedings of Fifth Congress of the 1983 Annual Meeting of Soil Science Society of China, Part 2. (in Chinese). pp. 58–59.

Li, L M, Zang, S, Zhou, X R and Pan, Y H 1983b. Studies on nitrification-denitrification in soils. In: Editorial Board of Environmental Science Information Network, Chinese Academy of Sciences (ed.), Proceedings of Nitrogen Pollution in Environment and Nitrogen Cycling. (in Chinese). pp. 160–168. The Institute of Environmental Chemistry, Chinese Academy of Sciences Beijing.

Li, L M, Pan, Y H, Zhou, X R, Wu, Q T and Li, Z G 1986. Nitrification and nitrogen loss in different soils. In: Soil Science Society of China (ed.), Current Progress in Soil Research in the People's Republic of China. pp. 135–143. Jiangsu Science and Technology Publishing House, Nanjing.

Li, L M, Pan, Y H, Zhou, X R, Wu, Q T and Li, Z G 1987. Nitrification in the main soils of the Taihu-Lake region and the factors affecting it. (in Chinese). Soils 19:289–293.

Malhi, S S and McGill, W B 1982. Nitrification in three Alberta soils: Effect of temperature, moisture and substrate concentration. Soil Biol. Biochem. 15:397–299.

Martikainen, P J 1984. Nitrification in two coniferous forest soils after different fertilization treatments. Soil Biol. Biochem. 6:577–582.

Martikainen, P J 1985a. Number of autotrophic nitrifiers and nitrification in fertilized forest soil. Soil Biol. Biochem. 17:245–248.

Martikainen, P J 1985b. Nitrification in forest soil of different pH as affected by urea, ammonium sulphate and potassium sulphate. Soil Biol. Biochem. 17:363–367.

McCormick, R A, Nelson, D W, Sutton, A L and Huber, D M, 1984. Increased nitrogen efficiency from nitrapyrin for corn. Agron. J. 76:1010–1014.

Patrick, W H Jr, Peterson, F J and Turner, F T 1968. Nitrification inhibitors for lowland rice. Soil Sci. 105:103–105.

Pan, Y H, Li, L M, Wu, Q T and Li, Z G 1988. Study on activities of nitrification and denitrification in red soils under different utilization. (in Chinese). Soils 20:184–187.

Redemann, C T, Meikle, R W and Widofsky, J G 1964. The loss of 2-chloro-6-(trichloromethyl)-pyridine from soil. J. Agric. Food Chem. 12:207–209.

Rho, J 1986. Microbial interactions in heterotrophic nitrification. Can. J. Microbiol. 32:243–247.

Robertson, G P and Tiedje, J M 1987. Nitrous oxide sources in aerobic soils: nitrification, denitrification and other biological processes. Soil Biol. Biochem. 19:187–193.

Sahrawat, K L 1980. Nitrogen mineralization in acid sulphate soils. Plant Soil 57:143–146.

Sahrawat, K L 1982. Nitrification in some tropical soils. Plant Soil 65:281–286.

Sahrawat, K L and Keeney, D R 1986. Nitrous oxide emission from soils. Adv. Soil Sci. 4:103–148.

Shattuck, G E Jr and Alexander, M 1963. A differential inhibitor of nitrifying microorganisms. Soil Sci. Soc. Am. Proc. 27:600–601.

Soriano, S and Walker, N 1973. The nitrifying bacteria in soils from Rothamsted classical field and elsewhere. J. Appl. Bact. 36:523–529.

Tate, R L 1977. Nitrification in Histosols: a potential role for the heterotrophic nitrifier. Appl. Environ. Microbiol. 33:911–914.

Walker, N 1975. Nitrification and nitrifying bacteria. In: Walker, N. (ed.). Soil Microbiology. A Critical Review. pp. 133–146. Butterworths, London.

Walter, H M, Keeney, D R and Fillery, I R 1979. Inhibition of nitrification by acetylene. Soil Sci. Soc. Am. J. 43:195–196.

Wang, F J, Liu, X S, Peng, G Y, Wen, Y F and Wang, B Z 1981. Effect of nitrification inhibitors on N uptake by crops. (in Chinese). Application of Atomic Energy in Agriculture (2):34–39.

Wen, X F, Wang, B Z, Wang, F J, Peng, G Y and Mai, H K 1979. Study of the effect of nitrification inhibitors on rice yield with isotope ^{15}N. (in Chinese). Acta Pedol. Sin. 16:380–386.

Wickramasinghe, K N, Rodgers, G A and Jenkinson, D S 1985. Nitrification in acid tea soils and a neutral grassland soil: Effect of nitrification inhibitors and inorganic salts. Soil Biol. Biochem. 17:249–252.

Yoshida, T and Alexander, M 1970. Nitrous oxide formation by *Nitrosomonas europaea* and heterotrophic microorganisms. Soil Sci. Soc. Am. Proc. 34:880–882.

Yoshida, T and Padre, B C Jr 1974. Nitrification and denitrification in submerged Maahas clay soils. Soil Sci. Plant Nutr. 20:241–247.

Zang, S, Zhou, X R, Pan, Y H and Li, L M 1980. Effect of N-serve on the yield of rice. (in Chinese). Soils (4):139–142.

Zang, S and Zhou, X R 1983. Effect of 2,4-dichloro-6-trichloromethyl pyridine on the efficiency of N use and on rice yield. (in Chinese). Anhui Chemical Industry (1):38–40.

Zhang, F Z, Gao, Z M, Han, S H and Li, P J 1985. Studies on the content and behaviour of nitrous oxide in soils. (in Chinese). Journal of Ecology (1):1–5.

Zhang, W and Li, Y C 1981. The use of nitrogen fertilizers by winter wheat and the effect of nitrification inhibitor. (in Chinese). Application of Atomic Energy in Agriculture (1):41–45.

Zheng, S M, Yan, W M and Qian, X M 1983. Autotrophic Microorganisms. (in Chinese). pp. 25–47. Science Press, Beijing.

Zhou, L K 1984. Soil urease activity and transformation of urea fertilizer in soil. (in Chinese). Progress in Soil Science (1):1–9.

Zhou, Q and Chen, H K 1983. The activity of nitrifying and denitrifying bacteria in paddy soil. Soil Sci. 135:31–34.

7
Biological nitrogen fixation

YAO HUI-QIN

7.1. Introduction

Biological nitrogen fixation (BNF), the microbial conversion of atmospheric N to a plant-useable form, helps to replenish the soil N lost by plant removal, ammonia volatilization, denitrification and leaching. Microorganisms capable of fixing atmospheric nitrogen are widely distributed on the surface of the earth, and various BNF systems composed of microorganisms and plants exist (Table 7.1). It has been estimated that the total amount of N fixed biologically each year is of the order of 175 million tons (Dixon and Wheeler 1986), and it is apparent that BNF has contributed greatly to agricultural production throughout the world. Dou (1989a) estimates that N input through BNF was greater than fertilizer N input in Australia, New Zealand and USA. In China, N input through BNF in 1987, though much less than fertilizer input, still amounted to 14.5% of the total input in agriculture (Table 14.9).

Table 7.1. Diazotrophic microorganisms (Dou 1989a).

NON-SYMBIOTIC NITROGEN FIXATION SYSTEMS
Organotrophs
Aerobes: *Azotobacter vinelandii, A. beijerinckii, A. chroococcum, A. agile*
Anaerobes: *Clostridium pasteurianum*
Facultative anaerobes: *Klebsiella pneumoniae, Bacillus polymyxa*
Photosynthetic Bacteria
Autotrophs: *Cyanobacteria nostoc, Anabaena*
Heterotrophs: *Rhodospirillum, Chromatium, Rhodopseudomonas, Chlorobium*
Chemolithotrophs: *Thiobacillus ferroxidans*
SYMBIOTIC NITROGEN FIXATION SYSTEMS
Legume-*Rhizobium*
Parasponia-Rhizobium
Dicotyledon-Actinomycete Frankia
Cyanobacterium
Cyanobacterium-Pteridophyte – Azolla
Cyanobacterium-Fungus-lichens
Cyanobacterium-Cycas
Cyanobacterium-Haloragaceae
Cyanobacterium-Bryophyta
Associative nitrogen-fixing bacteria – Spirillum lipoferum

Zhu Zhao-liang et al. *(eds.): Nitrogen in Soils of China, 135–158.*

The consumption of fertilizer N is increasing with the global demand for food. By the year 2000, the global consumption of fertilizer N is estimated to increase to 95 million tons, of which 21 million tons will be consumed in China (Li and Cao 1985). Nevertheless, the contribution of BNF to soil N will still be considerable.

In addition, more and more attention is being given to the problems of energy use, food production and environment pollution, which are directly or indirectly related to the wide spread use of fertilizer N. Consequently, people are re-evaluating BNF and much work has been done on the exploitation and utilization of N fixation in the last two decades.

During the last 40 years, a lot of research has been carried out in China in the field of biology, ecology, biochemistry, cytology and genetics of nitrogen-fixing microorganisms, and the exploitation of microbial resources. This chapter presents a brief introduction to the relevant research in the field of BNF.

7.2. Nitrogen fixing systems

7.2.1. Legume Rhizobium symbioses

It is estimated that there are 19,700 species of legumes and that most can nodulate. This symbiotic system not only contains a huge variety of legumes, but it fixes N at a fast rate. The N fixed by this symbiotic system is estimated to account for 50% of the N fixed globally, and 70% of the N fixed in the agro-ecosystem (Dixon and Wheeler 1986). So, symbiotic N fixation by legumes plays an extremely important role in agricultural production. For this reason, the study of *Rhizobium* has been active for over a century.

The rate of fixation depends on crop variety, and soil and climatic conditions. Thus, for perennial leguminous pastures it varies up to 300 kg N ha^{-1} yr^{-1} and for annual leguminous crops it ranges up to 90 kg N ha^{-1} yr^{-1} (Burns and Hardy 1975).

Early in the study on *Rhizobium*, it was found that fairly successful cross-inoculation of *Rhizobium* strains with the formation of nodules may be achieved between certain crops. Such a 'cross-inoculation group' is based on the specificity. Plants of different groups cannot be inoculated by rhizobia from a different group with the formation of effective nodules. Fred (1932) listed 22 genera of plants and divided them into 7 cross-inoculation groups. The rhizobia of each group were named after the predominant host plant. Since then it has been found that the boundaries between the groups overlap and some strains of rhizobia form nodules on plants occurring in several different groups. For instance, *Phaseolus vulgaris,* which nodulates with fast growing strains of *R. leguminosarum* bv *phaseoli*, can sometimes nodulate with slow growing strains. Cross infections occur between lupin, soybean and cowpea groups (Lange 1961). Slow growing strains of rhizobia are capable of forming nodules on *P. lunatus* and *P. acutifolius*, but *R. leguminosarum* bv *phaseoli* cannot induce nodule formation on these two plants of the kidney bean group (Carroll

1934). From plant of the family Ulmaceae, Trinick (1973) isolated rhizobia that nodulated cowpea. Recently a fast growing strain of *Bradyrhizobium japonicum* was isolated from Chinese *R. japonicum* by Keyser *et al.* (1982). This strain can nodulate a soybean plant that is also a host for slow growing strains of rhizobia. These results suggest that classification based solely on symbiotic properties is not satisfactory.

In the 1960s, it was suggested that *Rhizobium* may be classified by numerical analysis and DNA homogeny (Graham 1964). After reviewing a large number of publications, Jordan (1984) put forward a new genus, *Bradyrhizobium*, thus dividing the nodule bacteria into 2 genera, *Rhizobium* and *Bradyrhizobium* (slow growing strains). The former can be further separated into 3 species, *R. leguminosarum phaseoli* and *trifolii*), *R. meliloti*, and *R. loti*. The new classification needs further improvement because of the wide existence of intermediary strains.

Chen *et al.* (1988) determined physiological, biochemical and biological characters of 56 strains of rhizobia and 3 strains of Agrobacteria found in China. Based on the Ssm similarity coefficient and minimum distance cluster equation, the rhizobia of were classified numerically at the genus level using a computer, and separated into fast growing rhizobia, slow growing rhizobia and agrobacteria. The rhizobia of *Astragalus* was found to be an independent group previously included in the *Rhizobium* genus and *Rhizobium* of *Stylosanthes* was identified as another independent group in the slow growing *Rhizobium* genus. They also studied new strains of rhizobia isolated from host plants found in some regions of China and determined their location in the classification system.

Current *Rhizobium* classification systems are mostly associated with important annual leguminous crops, while those associated with perennial leguminous trees are scarce. Basak and Gayul (1980) reported that *Rhizobium* strains from 15 species of leguminous trees under 4 genera, such as Acacia, could cross-infect and nodulate plants of the cowpea group. They suggested that the rhizobia from these leguminous trees be allocated to the cowpea genus. Zhou and Han (1987) found that pure cultures of nodule bacteria isolated from leguminous trees could cross-infect each other to form nodules in most cases. On the basis of physiological and biochemical characteristics, they grouped the *Rhizobium* strains in *Robinia pseudoacacia*, *Sophora viciifolia*, *Wisteria sinensis* and *Amorpha fructicosa* as fast growing, and those in *Albizia chinensis*, *Acacia auriculiformis* and *Albizia julibrissin* as slow growing. In accordance with the new classification system proposed by Jordan (1984), Zhou and Han (1987) suggested that nodule bacteria from leguminous trees be allocated to the genera of *Rhizobium* and *Bradyrhizobium*.

There are over 700 genera and more than 19,700 species in the legume family, but only 3,000 species have been examined for nodulation. Nodule bacteria have been isolated and studied from a much smaller number of species. Consequently, in addition to collecting and preserving the known resources, investigations should be made to find new resources. *Parasponia* was isolated from *Parasponia* in Papua New Guinea (Trinick 1973) and fast growing *Bradyrhizobium japonicum* was discovered in China by Keyser *et al.* (1982).

Wang *et al.* (1987) reported that fast growing *Bradyrhizobium japonicum* was widely distributed in the various soil types over the Northeast China Plain, and the Changbai Mountain below 1100 m, but not in soils above that elevation. Using a serological method, Ge (1986) found that the fast growing strain was distributed widely in all of the major soil types in China.

Marked differences exists between fast and slow growing strains of rhizobia in morphology, serological character, growth rate, physiological and biochemical reactions (Zhang and Wang 1987; Xu 1984; Ge 1986; Keyser *et al.* 1982). For instance, fast growing strains have a generation time of 3.6 hr, while for slow growing strains the generation time is 6.8 hr. After incubation, the pH value of the medium for fast growing strains was 5.6 while for slow growing strains it was 7.1. In addition fast growing strains use a wider spectrum of carbon sources and have a higher tolerance to salt than slow growing strains. Qi *et al.* (1987) reported that out of the 40 fast growing strains investigated 78% could tolerate salt at a concentration of 0.3 M, and 50% could tolerate 1.0 M. Eighty per cent were resistant to chloromycetin, 33% had resistance to 2 or more antibiotics and some individual strains had resistance to all 5 antibiotics tested. This character of drug resistance is of some practical value in genetic engineering, permitting the fast growing strains to be used as recipients with selected markers in DNA transformation.

Research in genetics shows that many fast growing strains of *Bradyrhizobium japonicum* have mega plasmids, large plasmids and small plasmids (Ning *et al.* 1986) with Nod and Nif genes located in the large plasmids. In this way they differ from the slow growing strains which have Nod and Nif genes positioned within the nucleoplasm. Therefore, it may be feasible to transfer the Nod and Nif genes on plasmids to non-leguminous plants (Chen *et al.* 1988).

With respect to symbiotic nitrogen fixation, Duteau (1984) reported that the fast-growing strain USDA 191 could form effective symbiosis with most of the soybean cultivars of North America. Xu (1984) found that some fast growing strains of *Bradyrhizobium japonicum* displayed effective symbiosis with the soybean cultivar Tiefeng 18 and Kaiyu 8. Using the local soybean cultivar as the test crop, Zhang (1985) found that strain QB 113 gave consistent yield increases in 4 years of experiments. Similar results were found by Qu (1988). Thus there are excellent prospects for utilizing the fast growing strains of *Bradyrhizobium japonicum* in agricultural production.

The fast growing *Bradyrhizobium japonicum* is similar to other fast growing rhizobia in growth rate and other physiological and biochemical characteristics. However, it is also similar to the slow growing *Bradyrhizobium japonicum* in differentiation and survival within cells of the host plant, and in symbiosis with soybean (Cao and Ai 1988). Thus, Keyser *et al.* (1982) suggested that in taxonomy, it should be put in a category between the fast and slow growing strains.

From the above discussion it is obvious that the fast growing *Bradyrhizobium japonicum* is superior to the slow growing strains in physiological, biochemical and genetic characters, and in nodulation and biological nitrogen fixation. Moreover, it is possible to screen out from fast growing *Bradyrhizobium japonicum* the good strains, which are strong competitors, have high nitrogen-fixing activity and are stress resistant. Consequently, further basic and applied research on *Bradyrhizobium japonicum* is warranted.

The rapid development of artificial pasture in China requires the selection and provision of *Rhizobium* inoculant for various legumes. Ning *et al.* (1989) started the selection of rhizobia from 22 leguminous plants in 1981. These included *Astragalus adsurgens* Pall, *Coronilla varia* L., *Hedysarum fruiticosum Pall, Trigonella ruthenica* L., *Stylosanthes guianensis.*, Macroptilium atropurpureum (D.C) Urban *Macrotyloma axillare* (E. Meyer) Verde, *Desmodium* spp., *Desmodium intortum, Astragalus dahuricus* (Pall) DC., *Astragalus cicer* L., *Astragalus melilotoides pall, Astragalus membranaceus* Fisch., *Acacia auriculiformis, Acacia mearnsiidewild, Acacia spp.,* and *Caragana korshinskii* riom. rhizobia of the first 7 host plants have already been used over a wide area and rhizobia of the remainder over a smaller area. This was the first time that many of these rhizobia had been isolated in China, and the first time that some of them, such as *Astragalus dahuricus, A. melilotoides, and Trigonella ruthenica* L. had been found anywhere in the world. Coating seeds with selected stains of rhizobia markedly increased the nodulation and yield of pasture legumes (Table 7.2).

7.2.2. *Azolla*

China has a long history of using *Azolla,* an aquatic fern, which harbours the symbiotic nitrogen-fixing blue-green algae *Anabaena azollae.* In the 17th century, *Azolla* was extensively cultivated as green manure or a fodder crop in Wenzhou region of Zhejiang Province. In the 1940s the *Azolla* cultivation area in Wenzhou, Pingyang and Ruian counties reached around 30,000 ha (Shen *et al.* 1963). In the 1970s, the total area of *Azolla* cultivation in the country reached 1.3 million ha spreading from the regions south of Qinglin Mountain, 35° N, to north and northeast China (Hebei, Liaoning and Helongjiang provinces; Liu 1979).

Besides the native *Azolla imbricata,* the species currently under cultivation and exploitation are *A. nilotica, A. pinnata, A. rubra, A. mexicana, A. filiculoides, A. microphylla* and *A. caroliniana*, which were introduced from abroad. Numerous

Table 7.2. Effect of *Rhizobium* inoculation on nodulation rate and yield.[1]

Legume	Nodulation rate (%)		Dry weight (t ha^{-1})	
	Control	Inoculated	Control	Inoculated
Astragalus adsurgenes	55.0	84.0	1.76	2.75
Trifolium spp.	23.2	89.0	9.36	13.0
Onobrychis viciifolia Scop.	6.7	34.0	3.60	4.25

[1] Ning *et al.* (1989).

field experiments have shown that cultivating *Azolla* in paddy fields, and using *Azolla* as a green manure for the current rice crop can markedly increase yield (Liu 1979).

The rate of nitrogen fixation by *Azolla* varies greatly with species (Table 7.3; Li 1987). Watanabe (1984) showed that the daily maximum N_2-fixing activity under optimum condition (22°, water culture) is >0.3 g N m^{-2}. During the growth of rice, some good varieties of *Azolla* can fix 10 kg N ha^{-1} day^{-1} (Dixon and Wheeler 1986).

It was found that the rate of N fixation by *Azolla* in flooded fields was comparable to that of some leguminous plants (Bai *et al.* 1979b). However, the rate of fixation may be under-estimated when it is determined by the difference method. A $^{15}N_2$ incorporation study showed that 3–4% of the ^{15}N fixed per day was excreted into the medium, and 1/3 of this was readily available to the adjacent rice plants (Liu *et al.* 1980).

In nature, *Azolla* grows in symbiosis specifically with *Anabaena* and *Azolla* cannot survive without *Anabaena*. Separation of these two partners is an essential step in elucidating their individual role and their interrelationship in symbiotic nitrogen fixation. A great deal of effort was put into the isolation and culture of *Azolla*, free from its *Anabaena* symbiont by freezing, chemical treatment, radiation, alteration of nutritional and light conditions, alternative antibiotic treatments, etc. but all failed. However, Bai *et al.* (1979a) employing the bacterium-free stem apex cultivation method and micro techniques successfully prepared pure cultures of algae-free *Azolla* and *Anabaena*. By following this method, Zheng (1980) also prepared algae-free *Azolla*. The development of such a technique has enabled further studies on *Azolla-Anabaena* symbiosis.

By analyzing and comparing the composition and amount of N-containing compounds in the cavity axil of algae-free *Azolla*, and *Azolla* in symbiosis with *Anabaena*, Xu *et al.* (1983) were able to study the material exchange between the *Azolla* host and the *Anabaena* symbiont. They showed directly for the first time that ammonia formed from nitrogen fixation by *Anabaena* was released into the leaf

Table 7.3. Rate of nitrogen fixation by various species of *Azolla.*[1]

Species	Origin	Nitrogen fixation rate ($g\ N\ m^{-2} yr^{-1}$)
A. nilotica		63.1
A. pinnata		53.9
A. microphylla		98.4
A. caroliniana		109.9
A. rubra		43.5
A. mexicana		82.7
A. filiculoides	USA	127.6
A. filiculoides	Germany	117.2
A. imbricata	Hangzhou, China	102.7

[1] Li (1987).

cavity. Isotope tracer studies showed that *Azolla* not only provided *Anabaena* with nutritional compounds (amino acids, glucose, etc.), but also transferred genetic material (DNA precursor thymine ribonucleoside and the RNA precursor uridine) to *Anabaena*. It was inferred from these transfers that glandular hairs of the frond cavity have the dual function of absorption and excretion in the exchange of compounds between the host and the symbiont (Sun *et al.* 1984a). Studies by diffraction and transmission electron microscopy showed that in the frond cavity of *A. pinnata*, the multicellular ramous structure of the glandular hairs greatly expanded the interface between the hairs and the symbiotic cavity. Apparently this morphological adaptation which facilitates exchange between the two symbiotic partners (Sun *et al.* 1984b).

Concerning carbon sources, sucrose is synthesized only by the host and is supplied to the symbiont. Simultaneous determination of photosynthesis, respiration and C_2H_2 reduction by *A. imbricata* demonstrated that *Azolla* provided nitrogenase with ATP and reductant as well (Shi *et al.* 1981). Consequently, the N-fixing activity of *Azolla* is closely related to photosynthesis. In order to improve the efficiency of nitrogen fixation, attention should be given first to those factors that will improve or affect photosynthesis, such as light intensity, light quality, temperature, pH, bicarbonate and nitrate (Shi 1981).

Recombination of *Azolla-Anabaena* symbiosis is another approach to improving stress resistance in *Azolla*. By placing the *Anabaena* containing indusium cap on the fertilized and decapitated megasporocarp, Lin *et al.* (1987) obtained a recombinant association of *A. imbricata var. putain* and *Anabaena*. Afterwards, they succeeded in acquiring re-establishment of symbiosis of inter-and inner-species from *A. microphylla* and *A. filiculoides* by applying the algae-free *Azolla* sporocarp and indusium cap removing technique, thus making it possible to breed new *Azolla-Anabaena* associations.

Of the 8 available *Azolla* species, except for some individuals with strong tolerance, most are confronted with the problems of oversummering and overwintering. Cheng (1978) tried to culture *Azolla* using air-dried sporocarps preserved over summer or winter. He found that some sporocarps germinated normally which indicated that sporocarps could better withstand drought, heat and cold. This suggested a solution to the over-summering and over-wintering problems. The Soil and Fertilizer Institute of the Hunan Academy of Agricultural Sciences investigated the regularity of *Azolla* spore production in a year under field conditions, and found that under natural climatic conditions, *A. filiculoides* has two spore forming periods, viz. mid April–mid June and early October–early November. The first period is much longer with higher spore production rate and a concentrated sporocarp maturation (SFIHAAS 1981). They also devised an optimal culture solution to produce sporocarps. Later, Chen (1983) introduced a rapid reproduction technique for the germination of *Azolla* sporocarps. This work overcame the problem of survival of the sporophytes through winter and summer, but failed to fully meet practical demands because fewer sporocarps were produced under field conditions.

7.2.3. Free living organisms

Altogether 19 genera of bacteria have been identified as nitrogen fixers. *Azotobacter chroococcum, A. vinelandii, A. agilis, A. fuzhouensis, A. beijerinkia, A. insolita, Pseudomonas azotogensis* and *A. galophilum* are found in some soils of China (Hao and Cao 1978). Of them *A. chroococcum* is the one most frequently found in farmlands in China, with a population varying up to tens of thousands per gram of soil. *Azotobacter* is comparatively high in nitrogen-fixing capacity and fixes 10–20 mg of N for each gram of glucose, while the corresponding value for *Clostridium pasteurianum* is only 1.5–7.0 mg.

The USSR started producing cultures of *A. chroococcum* to inoculate soil in the 1930s. China also has a history of over 2 decades of inoculating soil with cultures of diazotrophs (mainly *A. chroococcum* and *C. pasteurianum.*). However, the effect on crop yield was not consistent (Dou 1989b) as was shown in a large number of field experiments. The ineffectiveness may be attributed to (i) the rate of N fixation by azotobacter is not high and a significant increase in N fixation cannot be achieved unless a large amount of carbon-containing substrate is supplied as an energy source (Hao and Cao 1978), (ii) in the presence of combined N, N fixing bacteria will use that form of N instead of fixing N from the air (Li 1986), and (iii) in soils of high fertility N fixing bacteria are often suppressed by other microorganisms (AMTT 1983). Therefore, the contribution of free living N fixers to soil N is questionable, and further investigations are needed.

7.2.4. Blue green algae (cynobacteria)

De (1933) reported that cultivation of blue-green algae in paddy fields in India helped sustain relatively high yields for years without fertilization. Therefore, the idea of using blue-green algae as a biofertilizer in rice fields is attractive.

Blue-green algae (BGA) are known to consist of 20 families, 140 genera and more than 2000 species, of which 120 species in 24 genera fix N. In China, over 600 species have been documented, but it is still not clear which ones are capable of fixing N. Li *et al.* (1959) succeeded in isolating 4 species of BGA, i.e. *Anabaena azoctica, A. azotica forma a, A. variabilis forma* and *Nostoc linckia forma,* from soils in Hubei Province. Scientists from the Zhejiang Institute of Agricultural Science (ZASI 1959) and Li (1962) also isolated BGA and selected effective species. However, the species they isolated have not been classified.

BGA is a photoautotroph and photosynthesizes while fixing N. Chen and Fang (1983a) studied the relationship between N fixation and photosynthesis with *Anabaena* 7120, and found that the N-fixing activity was very sensitive to combined N, and the inhibitory effect of the latter depended on light intensity. The inhibition was strong under low light intensity, and *vice versa.* This is because photosynthesis is the source of the ATP needed for N fixation, and the source of the carbohydrate, which forms amino acids with ammonium, thus promoting reduction of molecular N_2 and metabolism of N. Nitrogen fixation by BGA is also related to hydrogen metabolism. Catalyzed by hydrogenase, BGA absorbs the hydrogen emitted by nitrogenase, thus improving N fixation efficiency. The hydrogen absorbed by BGA is mainly consumed in oxybiotic reactions (Chen and Fang 1983b). Under certain circumstances, where energy and reductant are not sufficient for N fixation, the absorbed hydrogen might be used as an electron donor for nitrogenase thus intensifying N fixation (Chen and Fang 1983c).

Studies on the application of BGA to rice fields in China began in the 1950s. Ye *et al.* (1959) developed a large scale culturing method to meet the need for inoculation in the field, and the effective species were screened by the Zhejiang Institute of Agricultural Science (ZASI 1959).

The rate of nitrogen fixation by BGA differs with species (Table 7.4). Li (1981) estimated that the amount of N fixed by BGA in Chinese rice fields during the late rice cropping season ranged from 22.5 to 37.5 kg N ha^{-1}. Roger and Watanabe (1986) indicated that the amount fixed ranged from a few kg to 80 kg N ha^{-1} and averaged 27 kg N ha^{-1} per crop.

Although BGA may contribute considerably to soil N under current cultivation conditions, the effect in the field is unreliable because of the influence of factors such as insect damage, environmental conditions, and competition from other algae.

7.2.5. *Associative nitrogen fixation*

In the associative system, N-fixing bacteria occur in the rhizosphere, on root surfaces, and in surface cells of the root, thus forming a close relationship with the plant. The relationship between the two is an associative symbioses, although it is less elaborate than a legume/*Rhizobium* symbiosis. It occurs in various crops, and tropical and subtropical grasses and plays a role in providing N to soils and crops. In recent years this type of biological N fixation has aroused a great deal of interest.

Associative N fixing organisms have been isolated successfully from sugarcane, rice, corn, wheat, and other cereal crops in China, and the amount of N fixed by the various associative systems has been determined. Qiu *et al.* (1980; 1981a, b) isolated two strains of N-fixing bacteria, A-15 and E-26, from the rhizosphere of rice in Guangdong. The N fixing activities of these organisms were confirmed by acetylene reduction (ARA), total N determination and ^{15}N tracer methods, and it was

Table 7.4. Rate of nitrogen fixation by 4 species of blue-green algae cultured for 4 days in 100 ml of N-free solution.[1]

Species, serial no.	Fixed N in algae		Fixed N in solution		Total N fixed
	mg	% of total fixed	mg	% of total fixed	mg
Nostoc linckia, 508	0.68	89.5	0.08	10.5	0.76
A. variabilis forma, 670	0.27	31.5	0.59	68.2	0.86
A. azotica forma, 678	0.54	57.3	0.40	42.7	0.94
A. azotica, 686	0.23	22.0	0.79	78.0	1.01

[1] Li *et al.* (1959).

found that about 1/3 of the N fixed was recovered by the rice plants. Preliminary identification indicated that the A-15 strain was *Alcaligenes faecalis,* a slightly aerobic gram negative azotobacillus with no fatty granule, flagella all round, and relying on lactic acid as carbon source. The ARA of pure A-15 culture ranged from 500 to 700 n mole C_2H_4 ml^{-1} hr^{-1}, the specific activity from 70 to 90 n mole C_2H_4 mg bacterioprotein^{-1} min^{-1}, and the N fixation rate was 20 mg N g malic acid^{-1} (Qiu *et al.* 1980; 1981b). An investigation with the stable isotope ^{10}B α track analysis method, indicated that *Alcaligenes faecalis* was clearly associated with the roots of rice, with some bacteria penetrating the root tissue (You *et al.* 1984). The results from the cultivation of bacteria with extracts and secretions from rice roots showed that the association of the A-15 strain with the roots was not simply to obtain a source of carbon. It is probable that some substances that induce and promote the combination exist in the roots and on the root surfaces of rice (Zhang *et al.* 1984b). Strain E-26 is the facultative anaerobic *Enterobacter cloacae*, which had a high ARA, of 200–300 n mole C_2H_4 ml^{-1} hr^{-1} (You *et al.* 1984; Qiu *et al.* 1980).

A strain of *Azospirillum lipoferum* had been successfully isolated from the roots of corn grown in the Nanning area and other places in Guangxi Province (GBNF 1979). Pot experiments conducted in Hubei Province showed that this strain in association with corn could fix N. The nitrogenase activity of preincubated corn roots ranged up to 1000 n mole C_2H_4 g^{-1} hr^{-1}. This species is, to some extent, 'host' specific (GBNF 1979). Jiang and Lou (1984) isolated 58 strains of *Azospirillum brasilense* with higher nitrogenase activity from cereal crops. Of these strains those isolated from the roots of corn, millet and sugar cane were essentially 'host' specific, while those from the roots of wheat were less so.

Wang *et al.* (1985) studied the relationship between hydrogen metabolism and N fixation by *Azospirillum.* Examination of 98 strains of *Azospirillum brasilense* showed that 57 strains emitted hydrogen, while 53 strains displayed hydrogenase activity and had the capacity to take up hydrogen at different rates. When strains with the capacity to take up and emit hydrogen were cultured together, the nitrogenase activity of the mixture was much higher than that of each of the separate strains. In addition, the nitrogenase activity of the hydrogen absorbing strains was higher in the presence of hydrogen than in its absence. These findings support the suggestion that uptake of hydrogen can increase nitrogenase activity. The results of turning Hup^+Fix^+ mutant strains into Hup^-Fix^- mutants by transposon Tn5 mutation technology shows that a chain reaction exists between the genes for absorption of hydrogen and fixation of N in *Azospirillum brasilense* (Wang and Guo 1988).

The nitrogenase activity of the associative organism varies with the crop (Table 7.5) and may vary with the cultivar of the crop (Table 7.6). In general, the nitrogenase activity of the associative N fixing organisms of rice roots is higher than that of other crops. The associative nitrogenase activity of the japonica rice cultivar Huxuan 19 and the glutinous rice cultivar Jinying 15 was much higher than that of other cultivars of the same variety (Table 7.6).

Table 7.5. Nitrogenase activity (n mole C_2H_4 g dry root^{-1} hr^{-1}) in roots of various plants.[1]

Plant	Stage	Date	Activity
Wheat	Seedling	Feb. 29	22.88
	Flowering	May 5	48.45
	Milking	May 29	73.86
Corn	Flowering	July 16	24.38
	Milking	July 21	25.74
Rice	Elongating	July 27	448.50
	Milking	Aug. 27	305.14
Cotton	Flower budding	July 24	21.51
Soybean	Seedling	July 24	22.07
Mungbean	Initial budding	July 24	10.62
	Flowering	Aug. 4	13.52
Alfalfa	Initial flowering	July 24	16.58
Astragalus adsurgenes pall	Branching	July 24	10.30

[1] Tu (1981).

Table 7.6. Nitrogenase activity (n mole C_2H_4 g fresh root^{-1} hr^{-1}) in the roots of rice, corn and sugarcane.[1]

Crop	Cultivar	Stage	Activity
Japonica rice	Ailin-5	Booting	5.86
	Huxuan-19	Booting	82.86
	Nonghong-73	Booting	1.07
	Aijing-23	Booting	13.98
Glutinous rice	Jingying-15	Booting	67.77
	Aipinuo	Booting	4.40
	Guinuo	Booting	51.96
Summer corn	Zhedan	Tasseling	Tr
	Hudan	Tasseling	Tr
	Zhedan	Filament appearing	0.03
	Hudan	Filament appearing	Tr
Sugarcane	Green peel	Tillering	1.63
	Red peel	Tillering	Tr
	Green peel	Elongating	0.05
	Red peel	Elongating	0.05

[1] He (1981).

The data now available indicate that associative N fixation plays an important role in the N nutrition of nonleguminous plants. However, the lack of a suitable method to quantify associative N fixation obscures its contribution.

7.2.6. *Fixation by non-legumes*

Little progress had been achieved in the study of nonleguminous symbiotic N fixation until the isolation technique for *Frankia* was developed (Callaham *et al.* 1978). This technical breakthrough resulted in the rapid development of research in this field. It was reported that there are 192 species of nonleguminous plants in

21 genera and 8 families involved in symbiotic N fixation with actinomycetes (Huang 1987). Some of the species involved are *Betulaceae alnus*; *Casuarinaceae casuarina*; *Rhamnaceae ceanothus*, *Rhamnaceae colletia*, *Rhamnaceae discaria*, *Rhamnaceae kentrothamnus*, *Rhamnaceae talguenea*, *Rhamnaceae trevoa*, *Rosaceae cercocarpus*, *Rosaceae chamaebatia*, *Rosaceae cowania*, *Rosaceae dryad*, *Rosaceae purshia*, *Rosaceae rubus*, *Myricaceae myrica*, *Myricaceae comptonia*, *Cariariaceae cariaria*, *Datiscaceae datisca*; *Eleagnaceae eleagnus*, *Eleagnaceae hippophae*, and *Eleagnaceae shepherdia* (Huang 1987). Pure cultures of *Frankia* have been isolated from the root nodules of plants in 9 genera, and used successfully to inoculate plants (Lechevalier 1984).

In China there are 48 species of nonleguminous trees in 7 genera and 6 families that bear nodules and have the ability to obtain N through biological N fixation (Huang 1987). Some of them are distributed throughout the country and some occur within distinct climatic zones (Table 7.7).

With high stress resistance, nodulated nonlegumes grow well on poor soils and hence play an important role in ameliorating N deficiency in forest soils. For example, one ha of 10-year-old *Hippophae* forest fixes 180 kg N yr^{-1} (Yin and Wang 1989). *Hippophae* functions as a 'pioneering plant' in erosion control and soil amelioration, and supplies firewood and fodder in arid and semi-arid regions. *Casuarina* is a pioneering tree in sandy soils along tropical coasts. In the coastal areas of South China, large tracts of *Casuarina* forests have become a green 'Great Wall' to act as wind breaks and stabilize the sand.

In China large numbers of *Frankia* strains were isolated from trees of 5 genera under 4 families. Studies of pure cultures of these strains showed that except for a few strains, all of them will form nodules when trees of the same genera are inoculated. Available data show that it is not easy to allocate *Frankia* strains to species based solely on morphological, physiological and biochemical characters. Zhao *et al.* (1988) suggested that cross-infection patterns, physiological and

Table 7.7. Distribution of nodulated nonlegumes in China.[1]

Plant Genus	Family	Distribution	Number of species
Alnus	*Betulaceae*	NEC[2], NC, CC, SC, SWC	8
Casuarina	*Casuarinaceae*	SC, Fujian, Guangdong, Hainan	3
Coriaria	*Coriariaceae*	SWC, Gansu, Guangxi, Hubei, Hunan, Shaanxi, Shanxi	15
Dryad	*Rosaceae*	NEC, NWC	1
Elaeagnus	*Elaeagnaceae*	NEC, NC, SC, CC, WC, EC	40
Hippophae	*Elaeagnaceae*	SWC, NWC, Shanxi, Inner Mongolia, Hebei, Henan	2
Myrica	*Myricaceae*	Provinces south of the Changjiang River	4
Rubus	*Rosaceae*	Throughout China	15

[1] Huang (1987).
[2] NEC = Northeast China; NC = North China; SC = South China; CC = Central China; SWC = Southwest China; NWC = Northwest China; WC = West China; EC = East China.

biochemical characters of the pure culture, and growth conditions be used as the basis for grouping *Frankia*. Based on such a principle, Zhao *et al.* (1988) sorted 12 representative strains into 3 groups and 4 subgroups. The 12 strains were selected from 41 strains of *Frankia*, isolated from nodules of 29 host plants of the 5 genera *Alnus, Eleagnus, Hippophae, Myrica* and *Casuarina*. Based on serological features, Sun *et al.* (1987) sorted 30 *Frankia* strains from various host plants into 3 groups, essentially conforming with the classification of Zhao *et al.* (1988). The first group encompasses endophytes from *Alnus, Myrica* and *Comptonia*, the second from *Hippophae* and the third from *Casuarina*.

Little has been reported concerning the mechanism of N fixation by *Frankia*. Zhou *et al.* (1985) reported that vesicles of *Frankia* were concentrated at the fore-tip of the nodules and that nitrogenase activity was higher at the fore-tip than at the rear end, which suggests that the vesicles are the sites for N fixation in *Frankia*.

Obtaining pure cultures is an essential step for screening strains. The problem then is how to maintain the symbiotic characteristics of *Frankia* when kept under pure culture conditions. Yuan (1987) reported that after 24 generations in pure culture *Frankia* C2I513 of *Casuarina* lost its infection ability. This demonstrates that the change from the symbiotic state to pure culture leads to variation in the genetic character of *Frankia*.

7.3. Factors controlling biological nitrogen fixation

7.3.1. Relationships between host plants and rhizobia

Even for host legumes of the same cross-inoculation group, the N fixation activity of the symbionts formed by inoculation with the same *Rhizobium* strain will be different. This is called the host specificity within a cross-inoculation group. In general, the BNF activity of a strain is much greater when it is associated with its original host (i.e. the host from which the strain was isolated) than with other hosts of the cross-inoculation group. Based on their study on the effective symbiotic relations between clover cultivars and rhizobia, Vincent (1956) first sorted clover into 3 analogous groups, each has its own specific strain of rhizobia.

Similar results were obtained in a pot experiment with clover cultivars conducted in China (Table 7.8). When crimson clover (*Trifolium incarnatum*) was inoculated with strain Wu95 isolated from crimson clover, the dry weight produced and the N fixation rate were 5.8 and 9 times that of the control plants (i.e. no inoculation), respectively. However, when French clover was inoculated with strain T36 isolated from white clover, the dry weight and N fixation rate were similar to that of the control. Similar results were obtained when white clover was inoculated with strain T36 and Wu95. This type of effective combination was also demonstrated in a field experiment. The fresh weight of crimson clover increased by 148% on inoculation with Wu95, and that of the white clover increased by 23.5% on inoculation with T36 (Yao *et al.* 1983).

Table 7.8. Specificity of *Rhizobia* to hosts of the clover cross-inoculation group.[1]

Host plant	Strain	Dry matter (g pot^{-1})	Total N (%)	Total N (mg pot^{-1})	Nodules/plant
White clover	Non inoc.	0.192	1.35	2.6	0
(*Trifolium repens* L)	Wu95	0.477	1.62	7.7	35
	T36	1.045	2.34	24.5	34
Crimson clover	Non inoc.	0.330	1.12	4.2	0
(*Trifolium incarnatum* L)	Wu95	1.980	2.01	39.9	38
	T36	0.390	1.08	4.2	34

[1] Pot experiment (Yao *et al.* 1989).

Similar results were found in the case of *Sesbania*. Strain S-1, isolated from the stem nodules of *S. rostrata*, could nodulate the root and stem of *S. rostrata*, and also the roots of other varieties of *Sesbania*. However, the N fixation rates were, 27.8 mg pot^{-1} on *S. rostrata*, 14.0 mg pot^{-1} on *S. cannabina*, and only 1.7 mg pot^{-1} on *S. sesban*. Similarly, Strain C-1, isolated from root nodules of *S. cannabina* had the highest N fixation rate (26.7 mg pot^{-1}) on *S. cannabina*, but lower rates on other varieties of *Sesbania*. The response of *S. sesban* to the inoculation of either S-1 or C-1 was very low (Table 7.9). The same relationship was found in the pea cross-inoculation group. For instance, inoculation with Strain 7084, an effective strain isolated from *Vicia sativa* L. increased the fresh weight of *Vicia sativa* L. by 112%, but it increased the grain yield of peas by only 9.1% (Cao 1974).

Table 7.9. Specificity of rhizobia to hosts in *Sesbania* cross-inoculation group.

Host plant	Strain	Dry matter (g pot^{-1})	Total N (mg pot^{-1})	Nodules/plant
S. rostrata	Non inoc.	0.10	3.61	0
	S–1[1]	0.60	27.80	25
	C–1[2]	0.43	16.78	16
S. cannabina	Non inoc.	0.10	4.32	0
	S–1	0.30	14.04	16
	C–1	0.59	26.67	15
S. sesban	Non inoc.	0.04	1.92	0
	S–1	0.06	1.70	15
	C–1	0.18	6.40	14
S. bispinosa	Non inoc.	0.06	1.35	0
	S–1	0.18	8.14	7
	C–1	0.37	16.34	14
S. calycotricha	Non inoc.	0.11	4.9	0
	S–1	0.14	6.34	9
	C–1	0.34	13.87	16
S. bispinosa	Non inoc.	0.10	4.98	0
	S–1	0.26	10.68	12
	C–1	0.34	13.57	16

[1] Isolated from stem nodules of *S. rostrata*.
[2] Isolated from root nodules of *S. cannabina*.

7.3.2. Competition between rhizobia

Competition between various rhizobia strains occurs as a result of differences in adapting to invading and nodulating conditions. In fields in which legumes have been cultivated for a number of years, the soils have indigenous rhizobia, which have adapted to the environment and are more effective for nodulation than the non-indigenous strains. Consequently, application of inoculant at sowing does not result in increased nodulation or yield. For example, the results of a pot experiment using an antibiotic resistant strain S-7130 to inoculate *Vicia sativa* show that under the same soil conditions, S-7130 was less effective than the indigenous strains (Table 7.10; Yao *et al.* 1983). Only when the population of the introduced strain reached 10,000 times that of the indigenous strain, did the nodulation rate of S-7130 reach 40%.

In a study on the relationship between the number of bacteria in the inoculum for *Trifolium subterraneum* and competition, Holland (1970) found that the soil where *T. subterraneum* was planted had grown clover of other varieties for many years, and that the competition between the indigenous strains and the inoculated strains lowered the effectiveness of inoculation. Only when the population in the inoculum reached 7.5×10^5 for each seed, 4 times the usual recommendation, was there a marked improvement in the nodulation rate and yield.

The competition occurs not only between the introduced strains and the indigenous strains, but also between the inoculum strains. Johnson and Erdman (1961) used yellow-wilt induced strains 76 and 94 of *R. japonicum* as genetically labelled strains to form a mixed inoculum with some common strains in different ratios to study the effect of the proportion of the strains in the inoculum on nodulation rate. The results indicated that the proportion of strains in the mixed inoculum was not correlated with the nodulation rate. Labelled strain 76 had a nodulation rate of 85%, although its proportion in the inoculum was only 1%, while strain 94 had only 1/4 the nodulation rate of strain 76 when it was present in the same proportion. This can be attributed to the difference in propagation rate in the host rhizosphere of the component strains in the inoculum, which may change the proportion of the strains. Nicol and Thornton (1941) found that strains propagating faster than other strains in the rhizosphere had a higher nodulation rate. Therefore, in selecting strains for

Table 7.10. Comparison of nodulating competence of Strain S7130 and indigenous rhizobia.[1]

Ratio of strains in soil		Population/ g dry soil		Nodules examined	Nodulation rate of S7130 (%)
S 7130	Indigenous	S 7130	Indigenous		
1	1	4000	4000	269	0
10^2	1	4000	40	81	0
10^4	1	4×10^7	4000	206	41
10^6	1	4×10^7	40	250	77

[1] Pot experiment (Yao *et al.* 1983).

inoculation, in addition to the host plant species, attention should be given to the ability of the strains to compete with the indigenous strains and the component strains in the inoculum.

7.3.3. *Effect of combined nitrogen*

Organisms capable of either symbiotic or nonsymbiotic N fixation can use combined N in addition to molecular N. However, the presence of large amounts of combined N in the growth medium inhibits N fixation. Table 7.11 shows that the total N in clover plants increased with the application rate of combined N. When the amount of applied N reached a certain level, symbiotic N fixation was almost completely inhibited. A complementary relationship exists between N fixation and combined N taken up by the plant. A further study indicated that the nodulation rate (Tables 7.12 and 7.13) and the population of rhizobia in soil also declined as the combined N application increased (Nutman *et al.* 1978), indicating that the decrease in N fixation was the result of the toxic effect of combined N on the growth of rhizobia.

Table 7.11. Effect of combined nitrogen on dry matter accumulation and nitrogen fixation by clover.[1]

Treatment		Plant (dry wt)	Nodule (dry wt)	Total N in plant	Dry matter increase	Nitrogen fixed	
N rate (g 1.5 kg soil^{-1})	Inoculation status	(g pot^{-1})	(g pot^{-1})	(mg pot^{-1})	(g pot^{-1})	(mg N pot^{-1})	(%)
0	Yes	3.03	0.46	44.6	1.65	21.6	100
	No	1.38	0	23.0			
0.6	Yes	3.53	0.56	52.3	1.85	28.6	133.0
	No	1.65	0	23.7			
0.12	Yes	4.08	0.45	55.6	1.40	18.3	85.3
	No	2.68	0	37.6			
0.18	Yes	4.00	0.28	52.0	1.16	15.6	73.0
	No	2.84	0	36.4			
0.24	Yes	3.90	0.17	53.0	–0.20	0.4	2.1
	No	4.16	0	52.6			

[1] Pot experiment (unpublished data of Yao Hui-qin).

Table 7.12. Effect of ammonium-N on nodulation and nitrogen fixation by *Sesbania*.[1]

N concentration in the medium (mg L^{-1})	Nodules per plant	Nodule dry weight (mg plant^{-1})	Nitrogenase activity (n mole C_2H_4 plant^{-1} min^{-1})
0	39.1	37.5	45.6
10	4.2	16.2	14.6
30	1.2	2.8	2.5
70	0.5	0.2	0

[1] Pot experiment (Li 1987).

Table 7.13. Effect of ammonium-N on nodulation of *Leucaena*.[1]

Concentration (mg N L^{-1})	Plant number	Plants nodulated	Nodulation rate(%)
0	30	30	100
26	28	26	92.8
52	29	8	27.6
78	30	0	0

[1] Pot experiment (Huang 1987).

The effect of combined N varies greatly with its form and the host species. The high fertility-tolerant soybean cultivar Monqing 6 fixed the maximum amount of N before flowering in the treatment with 150 kg N ha^{-1}, whereas the low fertility tolerant cultivar Edou 2 achieved its maximum rate in the 37.5 kg N ha^{-1} treatment (Xiao *et al.* 1983). When the N application rates were the same, ammonium-N had a lower inhibitory effect on N fixation by soybean than nitrate-N (Zhang *et al.* 1985). At equal concentrations of N, the inhibiting effect of urea was the highest, followed by ammonium and nitrate-N; The effect of ammonium nitrate was in between ammonium and nitrate (You *et al.* 1981). In the case of *Azolla*, when ammonium in the surface water reached 11 mg N L^{-1}, fixation was inhibited significantly; when the concentration reached 13 mg N L^{-1}, the rate of fixation decreased by about 50% (Chen and Li 1980)

In general, N fixed symbiotically accounts for one half to three quarters of the total N in legumes (Zhang and Wang 1989; Zhang *et al.* 1985), indicating that, although it will inhibit symbiotic N fixation significantly, the application of N fertilizer is indispensable for maximizing the yield of legumes. Furthermore, it should be noted that at the seedling stage a small dose of combined N may even greatly stimulate N fixation by legumes, particularly in small grain legumes such as white clover, *Sesbania*, etc. This is because the small seeds of these plants have the capacity to store small amounts of nutrients only. Thus as the nodules of the seedlings cannot fix sufficient N to sustain growth, the seedlings may suffer from N deficiency when there is little available N in the soil. As shown in Table 7.11, application of 0.04 g N kg $soil^{-1}$ resulted in increased total N, dry matter, nodule number, and the plant fixed 33% more N as well. Results of a ^{15}N solution culture experiment indicated that although N fixation by *Sesbania rostrata* decreased when the N concentration in the culture solution increased, it was still 27.0% and 16.2% higher than the control when the N concentration of the culture solution was 2.5 and 5 mg N L^{-1}, respectively. Nitrogen fixation decreased markedly only when the N concentration increased to 10-20 mg N L^{-1} (unpublished data of Yao Hui-qin).

Consequently, for leguminous crops, the rate and timing of N application have to be investigated to improve symbiotic N fixation and increase the yield of the crops.

7.3.4. *Effect of phosphate*

The P content of the grain and stems of legumes ranges from 4.4 to 6.5 g kg^{-1} and 1.5 to 1.7 g kg^{-1}, respectively, and is significantly higher than that of cereal crops (3.1–3.5 and 0.9 g kg^{-1}; Liu 1964). In addition, the P sink strength of rhizobia and bacteroids is high (Jacobson 1985). Therefore, legumes need more P than cereal crops, and on P deficient soils, the response of legumes to P application is greater than that of other crops (Zhang, *et al.* 1984a). Every kg P applied to leguminous crops could increase the total N content by 1.6–3.4 kg and increase grain yield by 18–33 kg (Liu 1964). A comparable increase in N content (3.2 kg N per kg P applied) was obtained with white clover grown on a N and P deficient soil (Table 7.14). Similar effects of P on N fixation by *Azolla* have been found (Table 7.15). As shown in Table 7.15, within a certain range, a linear relationship exists between the biomass yield of *Azolla* and the rate of P application. However, P requirement differs with species. For example, the P required by white clover for maximum fixation was much higher than that of alfalfa (Table 7.16). Thus, a better understanding of the optimal rate of P application for different symbionts is required.

The role P plays in symbiotic N fixation is still unknown. Some authors (Bonetti *et al.* 1984; Gates 1974) believe that P has a peculiar function in N fixation. Based on the fact that P deficiency decreased nodule biomass and N fixation, it was inferred that P has a direct effect on nodule formation. However, Jacobson (1985) argued that the sink strength of nodules for P was very high, and P deficiency

Table 7.14. Effect of P application on nodulation and yield of white clover.[1]

Treatment		Plant dry wt (g pot^{-1})	Total P (%)	Total N (mg pot^{-1})	Nodules/ plant
P applied (g kg $soil^{-1}$)	Inoculation (%)				
0	No	0.190 ± 0.01	0.147	7.5 ± 0.5	0
0	Yes	0.080 ± 0.01	0.149	3.1 ± 0.7	0
0.02	No	1.110 ± 0.08	0.239	13.5 ± 0.3	0
0.02	Yes	3.66 ± 0.11	0.204	119.5 ± 1.6	10

[1] Pot experiment (unpublished data of Yao Hui-qin).

Table 7.15. Effect of P application rate on biomass yield and N fixation by *Azolla.*[1]

P applied (kg ha^{-1})	Inoculation rate (g)	Yield increase		N increase	
		(g)	(%)	(mg)	(%)
0	455	250.8	55.1	4.69	34.8
1.6	490	493.5	100.7	11.08	76.3
3.4	490	801.5	163.5	27.55	189.8

[1] Field experiment with 10 m^2 plots; *Azolla* was inoculated on 19 April and harvested on 27 April (Lu *et al.* 1963).

Table 7.16. Effect of rate of phosphorus application to white clover and alfalfa on nitrogen fixation.

P applied (mg kg soil^{-1})	White clover N fixed (mg mg P^{-1})	Alfalfa N fixed (mg mg P^{-1})
9	2.86 b	5.29 a
13	4.06 ab	4.81 b
22	4.58 a	3.80 c
31	3.92 ab	2.43 d

[1] Pot experiment (unpublished data of Yao Hui-qin).

damaged the metabolic functions of the host plant. The increase in N fixation after P application results mainly from the improved metabolism and hence the improved plant growth.

7.3.5. *Effect of VA mycorhiza*

VA mycorhizae infect the roots of legumes forming hyphae thus increasing the ability of plants to absorbs P and other nutrients. Therefore infection with VA mycorhizae stimulates the growth of legumes and at the same time increases nodulation and N fixation (Cheng 1986; Azcon-Aguilar and Baron 1981; Carling and Brown 1980). Pot experiments with a P deficient soil from Fengqiu showed that addition of the correct rate of superphosphate (21 mg P kg soil^{-1}) and inoculation of the plant with *Rhizobium* and VA mycorhiza resulted in increased total N, total P and nodule number, compared with the plant inoculated with *Rhizobium* only (Table 7.17). Similar results were obtained in field trials. The beneficial effect of the dual inoculation is the integrated result of the functions of VA mycorhizae and *Rhizobia*.

The dependence on VA mycorhiza varies with plants (Lin and Hao 1989). For example, at various P application rates, N fixation by dual inoculated white clover was greater than that of the corresponding single inoculated one in terms of N fixed/unit of P applied. However, in the case of *Sesbania rostrata* N fixation was not increased at low and moderate rates of P application. Nitrogen fixation by the

Table 7.17. Effect of single- and dual-inoculation on yield of white clover.[1]

Treatment	Dry wt (g pot^{-1})	Total P (mg pot^{-1})	Total N (mg pot^{-1})	Number of nodules	(%) Roots mycorrhizal	Fresh wt of clover in field trial (t ha^{-1})
Control	1.11	2.75	13.5	0	26	53.1
Single inoculation[2]	3.16	7.38	119.5	10	30	65.5
Dual inoculation[3]	6.63	14.75	216.5	50	50	80.4

[1] Pot and field experiments (unpublished data of Yao Hui-qin).
[2] Inoculation with *Rhizobium*.
[3] Inoculation with *Rhizobium* and VA mycorhiza.

dual-inoculated *Sesbania* surpassed that of the corresponding singly inoculated plant at high P application rates only (Table 7.18).

7.3.6. Effect of pH

Under natural conditions, soil pH is one of the most important factors controlling BNF. Many diazotrophs will normally fix N in a neutral or slightly alkaline soil environment only. When the soil pH drops below 4.5, nodulation and N fixation stop. In soils with low pH, Al and Mn toxicity, and Ca and P deficiency adversely affect the host plant and rhizobia. The indirect effects are often more important than pH *per se*. In solution with pH 4.0, labile Al is toxic to the white clover Rhizobium system; propagation of rhizobia in the rhizosphere and extension of roots are inhibited, nodule formation is suppressed and root hairs are deformed. These adverse effects decrease and even disappear when the concentration of Ca and P in the solution are increased (Wood *et al.* 1984). Although *Leucaena* will nodulate in slightly acidic soil, it does better in neutral to slightly alkaline soil (Zhou and Han 1987).

Rhizobia are susceptible to soil acid. Two months after a suspension of rhizobia was added to an acid soil (pH 5.0), the population in 0.2 g moist soil decreased from 2.3×10^8 to 3.3×10^3. However, when the suspension was added into a soil of the same type, but with pH 7.5, the population of rhizobia decreased only from 3.3×10^8 to 2.2×10^5 (Mulder 1969). However, the acid-tolerance of rhizobia varies greatly between strains. In an acid tolerance experiment in a red soil (pH 5.2) with 46 strains of rhizobia, 65% of the strains died due to their low acid-tolerance. The survival rates of 6 effective strains were different in red soils with different pH values (pH 4.4 and 6.4). The populations in 2 of the 6 strains declined drastically or even vanished 60 days after addition to a red soil of pH 4.4. The

Table 7.18. Effect of single and dual inoculation on nitrogen fixation by white clover and *Sesbania rostrata* at the seedling stage.[1]

P applied (g kg soil^{-1})	Inoculation	N fixed by white clover (g g P applied^{-1})	N fixed by *S. rostrata* (g g P applied^{-1})
0.021	Control	0.16 g	0.25 d
0.021	Single[2]	3.18 c	0.57 a
0.021	Dual[3]	5.08 a	0.52 b
0.031	Control	0.14 gh	0.36 c
0.031	Single	2.38 d	0.54 a
0.031	Dual	3.72 b	0.55 a
0.041	Control	0.04 i	0.13 f
0.041	Single	1.62 f	0.21 e
0.041	Dual	2.40 e	0.25 d

[1] Pot experiment (unpublished data of Yao Hui-qin).
[2] Inoculation with *Rhizobium*.
[3] Inoculation with *Rhizobium* and VA mycorhiza.

populations of the other 4 strains varied little with soil pH. Yao *et al.* (1989) found that plants inoculated with acid tolerant strains of *Rhizobia* grew taller and accumulated more dry weight than those inoculated with acid-intolerant strains. Therefore, selection of acid tolerant, highly effective strains of *Rhizobia* will be of great practical benefit.

7.4. References

AMTT 1983. (Agricultural Microbiology Teaching Team), Guangxi Agricultural College. Guidance in Soil Microbiology. (in Chinese). pp. 253–256. Science Press, Beijing.

Azcon-Aguilar, C and Baron, J M 1981. Field inoculation of Medicago with VA Mycorrhiza and *Rhizobium* in phosphate fixing agricultural soil. Soil. Biol. Biochem. 13:19–22.

Bai, K Z, Yu, S L, Chen, W L, Yang, S Y and Cui, C 1979a. Isolation and culture of algae-free *Azolla* and *Anabaena azolla.* (in Chinese). Science Bulletin 24:664–666.

Bai, K Z, Yu, S L and Shi, D J 1979b. Determination of the nitrogen fixation ability of *Azolla imbricata* under general conditions. (in Chinese). Acta Bot. Sin. 21:197–198.

Basak, M K and Gayul, S K 1980. Studies on tree legumes. Further addition to the list of nodulating tree legumes. Plant Soil. 56:33–37.

Bonetti, R, Nazareth, M, Montanheiro, S and Saito, S M T 1984. The effects of phosphate and soil moisture on the nodulation and growth of *Phaseolus vulgaris.* J. Agric. Sci. 103:95–102.

Burns, R C and Hardy, R W F 1975. Nitrogen fixation in bacteria and higher plants. Springer-Verlag, Berlin. 145 pp.

Callaham, D, Tredici, P D and Torrey, J G 1978. Isolation and cultivation *in vitro* of the Actinomycetes causing root nodulation in Comptonia. Science. 199:899–920.

Cao, Z B 1974. Screening of effective strains of *Rhizobium leguminosarum.* (in Chinese). Soils (5), 205–209.

Cao, Y Z and Ai, Z C 1988. Differentiation and survivability of fast growing *Rhizobium japonicum* in host cells. (in Chinese). Acta Microbiol. Sin. 15:243–249.

Carling, D E and Brown, M F 1980. Relative effect of vesicular-arbuscular Mycorrhizal fungi on the growth and yield of soybeans. Soil Sci. Soc. Am. J. 44:528–532.

Carroll, W R 1934. A study of *Rhizobium* species in relation to nodule formation on the roots of Florida legumes. II. Soil Sci. 37:227–237.

Chen, D F 1983. An investigation on some techniques for the quick propagation of sporecarps of *Azolla filiculoides.* (in Chinese). Plant Physiol. Commun. (5), 34–36.

Chen, J M and Li, S L 1980. Effect of mineral nutrients on growth of duckweed. (in Chinese). Acta Pedol. Sin. 17:390–394.

Chen, W X, Qi, Y L, Li, J L and Yu, D F 1988. Numerical taxonomy of *Rhizobia.* (in Chinese). Acta Microbiol. Sin. 28:102–108.

Chen, Y and Fang, D W 1983a. Photo-regulation of nitrogen fixation of *Anabaena* 7120. (in Chinese). Acta Phytophysiol. Sin. 9:51–59.

Chen, Y and Fang, D W 1983b. Protective effect of molecular hydrogen on the nitrogen fixation activity of *Anabaena* 7120 being damaged by oxygen. (in Chinese). Acta Phytophysiol. Sin. 9:183–191.

Chen, Y and Fang, D W 1983c. Physiological basis for the beneficial effect of molecular hydrogen on blue-green algae. (in Chinese). Acta Phytophysiol. Sin. 9:415.

Cheng, J F 1978. Sporecarps of *Azolla* and culture of sporelings from sporecarps. (in Chinese). Acta Bot. Sin. 20:54–58.

Cheng G S 1986. Effect of mixed inoculation of *Rhizobia* and endophytic mycorrhiza on clover yield. (in Chinese). Soils 18:132–136.

De, P K 1933. The role of blue-green algae in nitrogen-fixation in rice fields. Proc. R. Soc. Lond. B. Biol. Sci. 127:121–139.

Dixon, R O D and Wheeler, C T 1986. Nitrogen fixation in plants. Chapman and Hall. New York. p. 4.

Dou, X T 1989a. Utilization of biological nitrogen. In: Dou, X T (ed.), Biological Nitrogen Fixation. (in Chinese). pp. 13–15. Agriculture Publishing House.

Dou, X T 1989b. Azotobacteria. In: Dou X T (ed.), Biological Nitrogen Fixation. (in Chinese). pp. 111–114. Agriculture Publishing House.

Duteau, N M 1984. Program and abstracts. World Soybean Research Conference. No. 100.

Fred, E B 1932. Root Nodule Bacteria and Leguminous Plants. University of Wisconsin, Madison. p. 245.

Gates, C T 1974. Nodule and plant development in *Stylosanthes humilis*: Symbiotic response to phosphorus and sulfur. Aust. J. Bot. 22:45–55.

Ge, C 1986. Competition between fast and slow growing *R. japonicum* for nodulation of *Glycine Max* and distribution of that fast-growing *R. japonicum* in 8 provinces of China. (in Chinese). Soybean Science (5), 327–333.

Graham, P H 1964. The application of computer techniques to the taxonomy of the root nodule bacteria of legumes. J. Microbiol. 35:511–517.

GBNF 1979. (Group of Biological Nitrogen Fixation Hubei Institute of Microbiology). Study of nitrogen-fixing bacteria in association with maize. (in Chinese). Acta Microbiol. Sin. 19:160–165.

Hao, W Y and Cao, Z B 1978. Soil Microbiology. In: Institute of Soil Science, Chinese Academy of Sciences (ed.), Soils of China. (in Chinese). pp. 417–437. Science Press, Beijing.

He, F H 1981. Azotase activity in root systems of rice, corn and sugarcane. (in Chinese). Zhejiang Agricultural Sciences (6), 274–280.

Holland, A A 1970. Competition between soil- and seed-born *Rhizobium trifolii* on nodulation of introduced *Trifolium subterraneum*. Plant Soil. 32:293–302.

Huang, W N 1987. Symbiotic nitrogen-fixation of trees and nitrogen-fixing actinomycete. In: You, C S, Jiang, Y M and Song, H Y (eds.), Biological Nitrogen Fixation. (in Chinese). pp. 332–339. Science Press, Beijing.

Jacobson, I 1985. The role of phosphorus in nitrogen fixation by young pea plants (*Pisum sativum*). Plant Physiol. 64:190–196.

Jiang, Y P and Luo, X Y 1984. Serological studies of *Azospirillum brasilense* II. Antigenetic analysis of five serological types of *A. brasilense*. (in Chinese). Microbiology 11:249–250.

Johnson, W and Erdman, W 1961. Competition between bacterial strains effecting nodulation in soybeans. Soil Sci. Soc. Am. Proc. 25:105–108.

Jordan, D C 1984. Family III *Rhizobiaceae*. In: Krieg, N R and Holt, J G (eds.), Berger's Manual of Systematic Bacteriology. 9. ed. pp. 234–254. Williams and Wilkins, Baltimore.

Keyser, H H, Bohlool, B B, Hu, T S and Weher, D F 1982. Fast growing *Rhizobia* isolated from root nodules of soybean. Science. 215:1631–1632.

Lange, R T 1961. Nodule bacteria associated with the indigenous leguminosae of Southwestern Australia. J. Gen. Microbiol. 26:351–359.

Lechevalier, M P 1984. The taxonomy of the genus *Frankia*. Plant Soil. 78:1–6.

Li, Z S 1962. Nitrogen fixation of *Cyanophyceae* and *Azolla* in paddy field. (in Chinese). Scientia Agriculturae Sinica (12), 42–44.

Li, S H 1981. Studies on the nitrogen-fixing blue-green algae as biofertilizer in the late rice crops. In: Acta Hydrobiologica Sinica, Aquatics Institute, Academia Sinica (ed.), Collection on Aquatics. No. 7. pp. 417–423. Science Press, Beijing.

Li, F D 1986. Soil factors and agricultural techniques affecting biological nitrogen fixation. In: Soil Agricultural Chemistry and Soil Biology and Biochemistry Committees, Soil Science Society of China (eds.), Advances and Prospects for Soil Nitrogen Research in China. (in Chinese). pp. 203–211. Science Press, Beijing.

Li, Z S 1987. *Azolla*. In: You, C S, Jiang, Y M and Song, H Y (ed.), Biological Nitrogen Fixation. (in Chinese). p. 298. Science Press, Beijing.

Li, Q K and Cao, Z H 1985. The supply and demand of chemical fertilizers in relation to the food production in the world. (in Chinese). Soils 17:107–109.

Li, S H, Ye, Q Q, Liu, F R, Wang, L M, and Gui, X Q 1959. The nitrogen fixation of some blue-green algae from Chinese rice fields. (in Chinese). Acta Hydrobiologica Sinica. 4:439–444.

Lin, C, Liu, Z Z, Zheng, D Y and Tang, L F 1987. Establishment of artificial symbiont of *Azolla*-algae. In: Soil Science Society of China (ed.), Abstracts of Papers, The 6th National Congress of Soil Science and Annual Meeting of 1987. (in Chinese). Nanchang. pp. 54–55.

Lin, X G and Hao, W Y 1989. Dependence of various plants on VA mycorrhiza. (in Chinese). Acta Bot. Sin. 31:721–725.

Liu, Z X 1964. Preliminary analysis of effectiveness of phosphate application for improving nitrogen fixation. (in Chinese). Chinese J. Soil Science (5), 33–35.

Liu, C C 1979. Use of *Azolla* in rice production in China. In: IRRI (ed.), Nitrogen and Rice. pp. 375–394. IRRI, Los Banos.

Liu, Z Z, Chen, B H, Wei, W X, You, C S, Li, J W and Song, W 1980. Preliminary investigation on nitrogen release of *Azolla*. (in Chinese). Scientia Agriculturae Sinica (4), 39–43.

Lu, S Y, Cheng, K Z, Shen, Z H and Ge, S A 1963. Investigations on biological characters of Azolla, the green manure in rice fields – *Azolla*. (in Chinese). Scientia Agriculturae Sinica (11), 35–40.

Mulder, E G 1969. Natural resources response research. Soil Biol. 9:135–140.

Nicol, H and Thornton, H G 1941. Competition between related strains of nodule bacteria and its influence on infection of the legume host. Proc. R. Soc. Lond. Ser. B. Biol Sci. 130:32–59.

Ning, L F, Gong, H Y, Zhou, L M and Cen, Y H 1986. Study on *Rhizobium* plasmids of some fast growing strains. (in Chinese). Acta Microbiol. Sin. 26:271–276.

Ning, G Z, Li, Y F, Huang, Y L and Hu, J S 1989. *Rhizobia* for forage legumes. Research progress on strains screening and application in China. (in Chinese). Grasslands of China (3), 68–73.

Nutman, P S, Dye, M and Davis, P E 1978. The ecology of *Rhizobium*. In: Loutit, M W and Miles, J A R (eds.), Proc. in Life Science, Microbiol. Ecology. pp. 404–409. Springer Verlag, Berlin.

Qi, B, Li, Z W, Wang, S J and Zhang, X W 1987. Study on some genetic characteristics of fast growing *R. japonicum* from different soil types in Northeast China. Journal of Microbiology, Supplementary Issue. 7:27–31.

Qiu, Y S, Zhou, S P and You, C S 1980. Nitrogen fixing bacteria in rice rhizosphere. (in Chinese). Science Bulletin 25:1008.

Qiu, Y S, Zhou, S P, Mo, X Z, Wang, D S and Hong, J H 1981a. Study of nitrogen-fixing bacteria associated with rice roots. I. Isolation and identification of organisms. (in Chinese). Acta Microbiol. Sin. 21:468–472.

Qiu, Y S, Zhou, S P and Mo, X Z 1981b. Study of nitrogen fixing bacteria associated with rice root II. The characteristics of nitrogen fixation of *Alcaligenes faecalis* strain A15 and *Enterobacter cloacae* strain E-26. (in Chinese). Acta Microbiol. Sin. 468–472.

Qu, Y C 1988. Investigation on techniques for utilizing *Rhizobium japonicum* QB113. (in Chinese). Jilin Agricultural Science (3), 64–67.

Roger, P A and Watanabe, I 1986. Technologies for utilizing biological nitrogen fixation in wetland rice: potentialities, current usage, and limiting factors. Fert. Res. 9:39–77.

SFIHAAS 1981. (Soil and Fertilizer Institute, Hunan Academy of Agricultural Science) Annual regularity of sporogeny of *Azolla*. (in Chinese). Hunan Agricultural Science and Technology (3), 18–20.

Shen, Z H, Lu, S Y, Chen, K Z, Ge, S A 1963. Preliminary investigation on nitrogen-fixation ability of *Azolla*. (in Chinese). J. of Soil Science (4), 46–48.

Shi, D J 1981. Investigation on characteristics of photosynthesis by *Azolla*. (in Chinese). Acta Phytophysiol. Sin. 7:113–120.

Shi, D J, Li, J G, Zhong, Z P, Wang, F Z and Zhu, L P 1981. Studies on nitrogen fixation and photosynthesis of *Azolla imbricata (Roxb) Nakai* and *Azolla filiculoides lam.* (in Chinese). Acta Bot. Sin. 23:306–315.

Sun, J S, Chen, W L, Zhu, Z Q and Zhu, Y M 1984a. Study on *Azolla* and *Anabaena azollae* symbiosis by autoradiography. (in Chinese). Acta Bot. Sin. 26:115–119.

Sun, J S, Zhu, Z Q, Chen, W L and Li, S Q 1984b. Observation of *Azolla* and *Anabaena azollae* symbiosis by electron microscope. (in Chinese). Acta Bot. Sin. 26:343–347.

Sun, H J, Ding, J, Wu, Y, Xu, Q G, Huang, Y L, Su, F Y, Zhang, Z Z and Li, W G 1987. Serological types of *Frankia*. Journal of Microbiology (in Chinese). Supplementary Issue. 7:51–56.

Trinick, M J 1973. Symbiosis between *Rhizobium* and the non-legume *Trema aspera*. Nature. 244:459–460.

Tu, A Q 1981. Study on intensity of associative nitrogen fixation in wheat root-microorganism system. (in Chinese). Shaanxi Agricultural Sciences. (1), 11–13.

Vincent, J M 1956. Strains of *Rhizobia* in relation to clover establishment. Proc. Inter. Grassl. Congress. Palmerston North New Zealand. 1–11.

Wang, Z F, Zeng, K R and Zou, Y Q. 1985. Relation between hydrogen metabolism and nitrogen fixation in *Azospirillum spp*. I. Studies on hydrogen evolution and hydrogen uptaking hydrogenase activity in *A. Brasilense*. Acta Microbiol. Sin. 25:54–59.

Wang, S J, Xue, D L, Qi, B and Zhang, X W 1987. The ecological distribution and identification of the fast growing *R. japonicum* from different types of soil in the Northeastern China. (in Chinese). Acta Microbiol. Sin. Supplementary Issue.7:18–26.

Wang, Z F and Guo, J 1988. Relation between hydrogen metabolism and nitrogen fixation in *Azospirillum spp*. III. A Hup^- mutant of *Azospirillum brasilense* by transposon Tn^5 Mutangenesis. (in Chinese). Acta Microbiol. Sin. 28:24–28.

Watanabe, I. 1984. Limiting factors in increasing N_2-fixation in rice fields. In: Biological Nitrogen Fixation in Africa. Proc. of the 1st Conference of the Africa Association for Biological Nitrogen Fixation, 23–27 July 1984. pp. 436–464.

Wood, M, Cooper, J E and Holding, A J 1984. Aluminium toxicity and nodulation of *Trifolium repens*. Plant Soil. 78:381–391.

Xiao, N H, Li, Z Y and Wu, S T 1983. Variation of symbiotic nitrogen fixation of soybean and effect of nitrogen applied at the seedling stage. (in Chinese). Oil Crops of China (2), 48–52.

Xu, L M 1984. Physiologico-biochemical properties and symbiotic response of the fast growing *Rhizobium japonicum*. (in Chinese). Soybean Science (3), 101–109.

Xu, Y L, Bai, K Z, Yu, S L and Cui, Z 1983. Nitrogen compounds of the leaf cavity liquid of *Azolla* in relation to the symbiosis of *Azolla* and *Anabaena azollae*. (in Chinese). Acta Bot. Sin. 25:82–86.

Yao, H Q, Zhu, Z Y, Cao, J Q, Xu, Y R and Chen, B Y 1983. Influence of some biological factors on competition in nodulation of vetch between inoculant and indigenous strains of *Rhizobia*. (in Chinese). Acta Pedol. Sin. 20:85–91.

Yao, H Q, Cao, J Q and Chen, B Y 1989. *T. rhizobium* used as inoculum for *trifolium*, its efficiency and survival in red earth. (in Chinese). Soils 21:135–138.

Ye, Q Q, Wang, L M, Liu, F R and Li, S H 1959. Mass culture of nitrogen-fixing blue-green algae. (in Chinese). Acta Hydrobiologica Sinica 4:445–451.

Yin, Z D and Wang, Z Y 1989. Investigation on ecological and economic functions of *Hippophae* in Gansu Province. (in Chinese). Hippophae 4:9–11.

You, C B, Li, J W and Song, W 1981. Influence of nitrogen nutrition on the physiological properties of *Azolla*. (in Chinese). Acta Phytophysiol. Sin. 7:97–104.

You, C B, Xiao, J Z, Li, X, Zhou, F Y and Wang, Y W 1984. Association of *Alcaligenes faecalis* with rice root. (in Chinese). Application of Atomic Energy in Agriculture (1), 14–16.

Yuan, C F 1987. Studies on variable ITY of endophyte from root nodules of *Casuarina equisetifolia*. (in Chinese). J Microbiol. 7:18–22.

ZASI 1959. Zhejiang Agricultural Science Institute. Culturing of effective species of blue-green algae. (in Chinese). Zhejiang Agricultural Sciences (1), 5–37.

Zhang, H Y 1985. Effect of the fast growing strains of *R. japonicum* QB 133 on symbiotic nitrogen fixation and yield of soybean. (in Chinese). Soybean Science (4), 267–277.

Zhang, H Y and Wang, S J 1987. Morphological character and physiologico-biochemical properties of 4 fast growing strains of *R. japonicum*. (in Chinese). Acta Microbiol. Sin. 27:92–94.

Zhang, S S and Wang, Z X 1989. Study on absorption of peanut for N fertilizer and N supply from soil and nodule bacteria. (in Chinese). Journal of Leiyang Agricultural College (6), 21–27.

Zhang, G Z, Bao, D J, Dou, S Z, Guo, Q Z and Li, Z Y 1984a. Preliminary study on NPK application to soybean in the Huaibei Plain. (in Chinese). Oil Crops of China (1), 58–61.

Zhang, Y L, Li, X and You, C S 1984b. Association of *Alcaligenes faecalis* A-15 with rice roots. (in Chinese). Plant Physiol.Commun. (6), 32–34.

Zhang, H, Zhang, G Z and Wang, X M 1985. Effect of ammonium sulfate and ammonium nitrate on symbiotic nitrogen fixation by *R. japonicum* and yield of soybean. (in Chinese). Jilin Agricultural Sciences (4), 43–48.

Zhao, Z Y, Huang, J B, Liu, H L, Zhu, B Q and Yang, H F 1988. Comparative study of strains of *Frankia* from various non-leguminosae woody plants. (in Chinese). Acta Microbiol. Sin. 28:279–284.

Zheng, D Y 1980. Culture of algae-free *Azolla* from *Azolla* stem apex. (in Chinese). Fujian Agricultural Science and Technology 4:31–35.

Zhou, H B, Hao, J Q, Yuan, C F and Gong. B 1985. The vesicle formation and nitrogen fixation of *Frankia* symbiosis with some non-leguminous plants. (in Chinese). Acta Phytophysiol. Sin. 11:315–317.

Zhou, X Q and Han, S F 1987. Investigation on rhizobial symbiotic systems of leguminous trees. (in Chinese). Journal of Nanjing Forestry University (2), 28–34.

8
Denitrification

LI LIANG-MO

8.1. Introduction

Denitrification in soil can take place through biological or chemical processes. Biological denitrification is the dissimilatory reduction of nitrate and/or nitrite to gaseous N oxides and N_2 gas under anaerobic conditions. Nearly all denitrifying bacteria are strict aerobes which use nitrate instead of oxygen as the terminal electron acceptor in respiration. The sequence of bacterial denitrification is thought to be,

$$NO_3^- \rightarrow NO_2^- \rightarrow NO \rightarrow N_2O \rightarrow N_2$$

Since denitrification not only leads to loss of N from the soil–plant system, but also to environmental pollution, extensive research has been done on this aspect of N cycling. From the standpoint of environmental protection, pollution caused by nitrate, nitrite and gaseous N oxides is often removed by speeding up denitrification so that the terminal product is N_2 rather than gaseous N oxides (Picard and Faup 1980; Roger *et al.* 1976). In this way, it is possible to prevent the water body from being polluted by nitrate and ensure that the atmosphere is not contaminated with gaseous N oxides.

A great diversity of denitrifying bacteria in soil brings about denitrification. The propagation and activity of these bacteria are affected by environmental conditions, so the denitrifying potential varies in different soils and fields. All human activities, such as cultivation practice and application of fertilizers will influence soil conditions and the activity of denitrifiers, resulting in different losses of fertilizer N from soil. Paddy soils are often more favourable for denitrification and have their own specific mechanism of denitrification loss. In this chapter a brief account will be given of the denitrifying bacteria in soil, denitrifying potential and the factors controlling it, the mechanism of denitrification in paddy soils, and the extent of fertilizer N loss by denitrification.

8.2. Denitrifying bacteria

Microorganisms that carry out denitrification in soil are diverse in species, high in number, and widespread in distribution. The kind and quantity of terminal products

Zhu Zhao-liang et al. *(eds.): Nitrogen in Soils of China, 159–192.*

of nitrate reduction depend on the enzymatic activities of the denitrifiers and the ecological conditions.

8.2.1. Species of bacteria

Payne (1973) grouped the microorganisms capable of denitrification into 15 genera. Focht and Verstraete (1977) included several of the genera identified as denitrifiers by Kessel (1976) viz. *Cryptophaga*, *Flavobacterium*, *Propionobacterium* and *Vibrio*, but did not include bacteria from the genera *Azospirillum* and *Rhizobium* in their list. Most denitrifiers are heterotrophs, but a few autotrophs exist (Payne 1981). Wang *et al.* (1964) made preliminary studies on the isolation of denitrifying bacteria from the rice rhizosphere and their bacteriological characteristics. They concluded that 70–80% of the strains of the dominant denitrifying bacteria belonged to the same community.

The factors influencing the composition of denitrifiers in soil are still not well understood, and there have been few papers concerning this topic. Li *et al.* (1987) investigated the denitrifying bacteria in the rhizoplane and rhizosphere of rice and other cereal crops and found that the bacteria on the rhizoplane were affected considerably by the roots and exhibited greater peculiarities. Thus, on the rhizoplane of rice *Pseudomonas aeruginosa* was dominant, while in the rhizosphere the dominant organisms were *Ps. fluorescens* and *Ps. stutzeri*. Again, on the rhizoplane of the barley plant the dominant species were *Ps. fluorescens* and *Ps. stutzeri*, whereas the main denitrifying organisms in the rhizosphere, were *Ps. fluorescens*, *Ps. mendocina* and *Enterobacter cloacae*. The plant root was shown to have a greater influence than soil type on the distribution of strains. As shown in the study on species and distribution of denitrifiers in the major soil types of the Taihu Lake region (Table 8.1; Li *et al.* 1989b), there was no distinct variation in the composition of denitrifiers between different soils. In all of the soils tested, some species of *Bacillus* and *Pseudomonas* appeared to be dominant; *B. megaterium, Ps. fluorescens* and *Ps. stutzeri* were found in all soils, constituting 40–50% of the total number of denitrifiers. However, some of the species occurred in one of the soils only. For instance, *B. firmus* was found only in the neutral gleyed paddy soil, *B. spaericus* occurred only in the upland soil of Xinzhuang, Changshu, *B. circulans* was found only in the clay loamy hydromorphic paddy soil, while *B. licheniformis* and *Ps. aeruginosa* appeared in paddy soils only. Further work is required to determine whether there is a specific relationship between the physiological characteristics or ecological requirement of denitrifiers and soil properties. Investigations on the distribution of denitrifiers in the red soil region in Jiangxi Province (Table 8.2; Li *et al.* 1992b) showed that some species of *Bacillus* dominated in all soils, and *B. cereus* and *B. megaterium* were widespread. In paddy soils derived from red soil, the proportion of *B. megaterium* was greater than that of *B. cereus*, whereas in impoverished wasteland soils (derived from red sandstone and Quaternary red clay) *B. cereus* accounted for >90% of the total population of denitrifers. The dominance of *Bacillus* in the highly acid red soils deficient in nutrients seems to be associated with the strong tolerance of these microbes to stress.

Table 8.1. Dominant species of denitrifiers (% of total) in the major soil types of Taihu Lake region in Jiangsu Province.[1]

	Location											
	Taicang		Liyang		Wujin		Wuxian		Changshu		Wujiang	
Soil	I[2]	(I)[3]	II	(II)	III	(III)	IV	(IV)	V	(V)	VI	(VI)
Dominant species												
B. megaterium	21.4	16.7	31.6	12.5	20.0	18.2	15.9	12.5	14.3	37.5	36.4	33.3
B. subtilis	–	16.7	–	–	–	–	5.3	12.5	–	25.0	–	–
B. cereus	–	–	–	25.0	6.7	18.2	15.8	25.0	–	–	18.3	16.7
B. licheniformis	–	–	–	–	13.3	–	15.8	–	–	–	–	–
B. mycoides	3.6	8.3			6.7	9.1	–	–	–	–	–	–
B. firmus	–	–	–	–	–	–	–	21.4	–	–	–	
B. circulans	–	–	–	–	–	–	10.5	–	–	–	–	–
B. sphaerieus	–	–	–	–	–	–	–	–	–	12.5	–	–
B. sp.	–	8.3	–	–	–	9.1	–	–	7.1	–	9.1	16.7
Ps. fluorescens	28.6	16.7	47.4	25.0	26.7	–	5.3	25.0	14.3	–	9.1	–
Ps. aeruginosa	–	–	–	–	–	–	10.5	–	–	–	–	–
Ps. stutzeri	17.9	16.7	5.3	–	6.7	–	10.5	12.5	14.3	–	9.1	16.7
Ps. mendocina	21.4	–	–	25.0	–	45.5	–	12.5	28.6	–	–	–
Ps. sp.	–	16.7	10.5	–	–	–	–	–	–	–	–	–
Others	7.1	–	5.3	10.5	6.7	–	10.5	–	–	25.0	18.2	16.7

[1] Li *et al.* (1989b).

[2] I = Submergenic paddy soil
II = Bleached paddy soil
III = Silty hydromorphic paddy soil
IV = Clay loamy hydromorphic paddy soil
V = Neutral gleyed paddy soil
VI = Slightly acid gleyed paddy soil

[3] (I) = Adjacent upland soil
(II) = Adjacent upland soil
(III) = Adjacent upland soil
(IV) = Adjacent upland soil
(V) = Adjacent upland soil
(VI) = Adjacent upland soil

Table 8.2. Denitrifiers in soils of red soil region in Jiangxi province (% of total).[1]

Soil	Parent material	Land use	*B. cereus*	*B. megaterium*	*B. subtilis*	*B. mycoides*	*Micrococcus*	Others
Hydromorphic paddy	Quaternary red clay	Rice	27.3	36.4	9.1	–	9.1	18.2
Cold spring paddy	Quaternary red clay	Rice	33.3	33.3	–	–	–	33.3
Red	Quaternary red clay	Buckwheat	37.5	25.0	12.5	12.5	–	12.5
Red	Quaternary red clay	Citrus	30.8	30.8	7.7	15.4	–	15.4
Red	Quaternary red clay	Tea tree	33.3	22.2	22.2	–	–	22.2
Red	Quaternary red clay	Masson pine	42.9	57.1	–	–	–	–
Purple	Purple sandstone	Wasteland	40.0	40.0	20.0	–	–	–
Red	Red sandstone	Wasteland	90.0	–	9.1	–	–	–
Red	Quaternary red clay	Wasteland	100	–	–	–	–	–

[1] Li *et al.* (1992b).

8.2.2. *Number of bacteria*

Denitrification is closely related to the number of denitrifying bacteria in soil. In a study of the kinetics of denitrification, Focht (1974) found that the reaction followed zero order kinetics, and that the rate of denitrification was proportional to the biomass of the denitrifiers. Considerable effort has been devoted to the enumeration of denitrifiers (Focht and Joseph 1973). Okereke (1984) employed the MPN method to enumerate the denitrifying bacteria capable of reducing N_2O to N_2. This was a new attempt to determine the number of nitrous oxide reducing bacteria. Li *et al.* (1987) developed another method, in which each colony appearing on an ordinary nutrient agar plate was isolated and inoculated on an agar slant. After purification, each strain was again inoculated in a nutrient broth containing nitrate, so as to determine whether these strains can reduce nitrate (including reduction to N_2O and N_2). In this way, they counted the number of denitrifiers. Although tedious, this method proved practical enough to get the number of denitrifiers and the total number of bacteria. Investigations using this method showed that in different soil types in the Taihu Lake region there were, in most cases, more than several million denitrifiers per gram of soil, accounting for 50–80% of the total number of bacteria (Table 8.3; Li *et al.* 1989b).

The number of denitrifying bacteria in soil is strongly affected by soil conditions. Paddy soils usually contain a greater number of denitrifiers than upland soils (Figure 8.1; Li *et al.* 1989b). The surface drainage of paddy fields midway through the season may also lead to variation in the number of denitrifiers (Table 8.4; Chen *et al.* 1981). In the red soil region of Jiangxi Province, red soils derived from different parent materials, except the citrus orchard soil, commonly had several million denitrifiers per g dry soil. In orchard soils the number may be as high as 10 million per g dry soil, while in red sandy soils and purple soils there were only 0.18×10^6

Table 8.3. Number of denitrifiers ($\times 10^6$ g dry soil^{-1}) in the major paddy soils in the Taihu Lake region.[1]

Paddy soil	Fertility	Number	% of total bacteria
Neutral gleyed	High[2]	29.3	62.0
	Low[3]	16.8	50.0
Slightly acid gleyed	High	19.1	50.4
	Low	2.8	55.6
Submergenic	High	42.4	80.0
	Low	6.3	72.7
Clay loamy hydromorphic	High	22.1	72.7
	Low	26.0	99.0
Illuvial hydromorphic	High	25.8	55.6
	Low	3.1	75.0
Silty hydromorphic	High	18.4	58.3
	Low	1.1	50.0

[1] Li *et al.* (1989b).
[2] Soils from plots fertilized for 3 years.
[3] Soils from plots not fertilized in the same experiment.

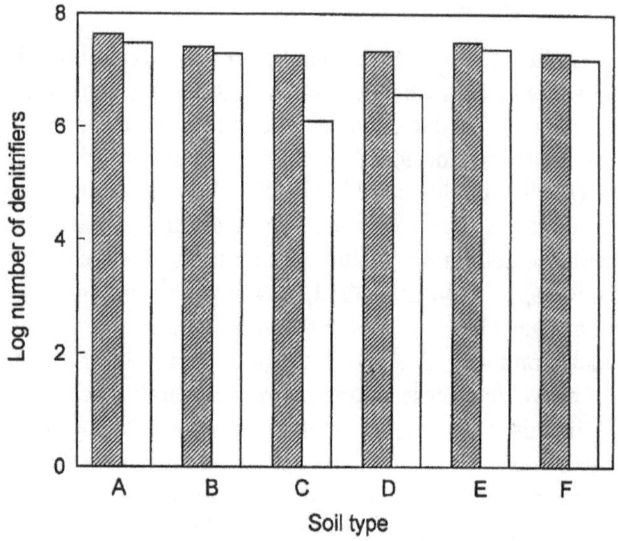

Figure 8.1. Denitrifying organisms in paddy soils and adjacent upland soils. Paddy Adjacent upland. A, Submergenic; B, Bleached; C, Silty hydromorphic; D, Clay loamy hydromorphic; E, Neutral gleyed; F, Slightly acid gleyed (Li *et al.* 1989b).

Table 8.4. Effect of mid-season drainage of paddy field on the number of denitrifiers ($\times 10^6$ g dry soil^{-1}).[1]

Treatment			Number
Surface drained	Flooded	(Transplanting – Jun. 4)	6.0
	Drained	(Jun. 5 – Jun. 13)	5.6
	Reflooded	(Jun. 13)	6.2
Flooded	Flooded	(Transplanting – Jun. 4)	6.1
	Flooded	(Jun. 5 – Jun. 13)	7.1
	Flooded	(Jun. 13)	6.8

[1] Chen *et al.* (1981).

and 0.22 × 10^6 per g dry soil, respectively. The red soil developed on Quaternary red clay in the wasteland had the lowest number of denitrifiers due to its high acidity and deficiency in nutrients (Table 8.5; Li *et al.* 1992b).

8.2.3. Enzymatic activity

Denitrifying bacteria have an extremely intricate enzymatic system. In the progressive reduction of N oxides to N_2, four enzymes are involved, viz. nitrate reductase (nitrate → nitrite), nitrite reductase (nitrite → nitric oxide), nitric oxide reductase (nitric oxide → nitrous oxide), and nitrous oxide reductase (nitrous oxide → dinitrogen) (Payne 1981). Most of the denitrifiers possess the whole reductase system

Table 8.5. Total number of bacteria and denitrifiers ($\times 10^6$ g dry soil^{-1}) in soils of red soil region in Jiangxi Province.

Soil	Parent material	Land use	Total bacteria	Denitrifiers (number)	(%)
Paddy	Quaternary red clay	Rice	8.3	4.8	58
Paddy	Quaternary red clay	Rice	3.1	0.78	25
Red	Quaternary red clay	Buckwheat	7.3	3.6	49
Red	Quaternary red clay	Citrus	18.2	10.9	60
Red	Quaternary red clay	Tea tree	8.4	4.5	53
Red	Quaternary red clay	Masson pine	1.2	0.94	78
Red	Quaternary red clay	Wasteland	0.37	0.05	14
Purple	Purple sandstone	Wasteland	0.88	0.22	25
Red sand	Red sandstone	Wasteland	0.30	0.18	60

[1] Li *et al.* (1992b).

necessary for reducing nitrate to dinitrogen, but some of them lack nitrate reductase and can only use nitrite as the electron acceptor. Others lack nitrous oxide reductase so that nitrous oxide is the terminal product of nitrate reduction, and still others, although they have nitrous oxide reductase, are unable to reduce nitrate or nitrite (Knowles 1982). According to investigations on the enzymatic activity of denitrifying bacteria isolated from soils of the Taihu Lake region, and the red soil region in Jiangxi Province, organisms capable of reducing nitrate to nitrite made up 50–80% and 15–77% of the total number of bacteria, respectively. Organisms capable of reducing nitrate to nitrous oxide constituted 67% and 60% of the total strains tested, and those with nitrous oxide reductase formed 56% and 58%, respectively (Li *et al.* 1989b; 1992b).

Many of the denitrifiers isolated from the soils of the Taihu Lake region had the capacity to reduce nitrate to ammonium. Under strictly anaerobic conditions and in the presence of adequate carbon sources, strains with this capacity may account for 90% of the total strains tested (Li *et al.* 1989b). Among them *B. cereus* #2281 and *B. licheniformis* #934 were two strains capable of the dissimilatory reduction of $^{15}NO_3^-$ to $^{15}NH_4^+$. After incubation for 20 hours, 15.5% of the $^{15}NO_3^-$ was reduced to $^{15}NH_4^+$ by *B. cereus* #2281 (Table 8.6; Li *et al.* 1989b). Caskey and Tiedje (1979, 1980) and McCready *et al.* (1983) also isolated from soil species of *Clostridia* and *Desulfovibrio* which can reduce nitrate to ammonium. Kaspar *et al.* (1981) noted the agricultural significance of this type of dissimilatory reduction.

8.2.4. *Ecological conditions*

Different strains of denitrifiers from the same genus have different N oxide reductases and different enzyme activities. A better understanding of the ecology of denitrifying bacteria will certainly help to exploit and utilize strain resources, on the one hand, and provide a scientific basis for regulating and controlling the growth of strains, on the other.

Table 8.6. Ability of denitrifiers to reduce $^{15}NO_3^-$ to $^{15}NH_4^+$.[1]

Strain	NH_4^+ in culture solution (mg N 10 ml^{-1})	$^{15}N_{ex}$ as % of NH_4^+-N in culture solution	$^{15}NH_4^+$-N formed (% of $^{15}NO_3^-$-N added)
B. cereus (#2718)	1.04	4.92	6.25
B. cereus (#2281)	1.32	9.62	15.5
B. licheniformis (#813)	0.742	1.08	1.00
B. licheniformis (#934)	0.862	4.31	4.60

[1] Li *et al.* (1989b).
Anaerobic incubation for 20 hours. Initial $^{15}NO_3^-$ concentration was 2.77 mg N 10 ml^{-1} culture solution with 29.3% $^{15}N_{ex}$.

Denitrification can occur within the pH range 3.5–11.2, but in general, whether in pure culture or under natural conditions the rate of denitrification is often positively correlated with the pH value (Valera and Alexander 1961). It has also been suggested that the optimum pH range for denitrification is 7–8 (Jiang *et al.* 1989). In soils of the Taihu Lake region and the red soil region of Jiangxi Province, some strains of denitrifying bacteria grew best at pH 7–8. Among them strain #15 (*B. megaterium*) grew better at pH 8.0, whereas strain #22 (*Micrococcus sp.*) grew well at a number of different pH values (Figure 8.2).

Denitrification proceeds under anaerobic conditions. It was reported that the rates of denitrification are negatively correlated with the partial pressure of oxygen

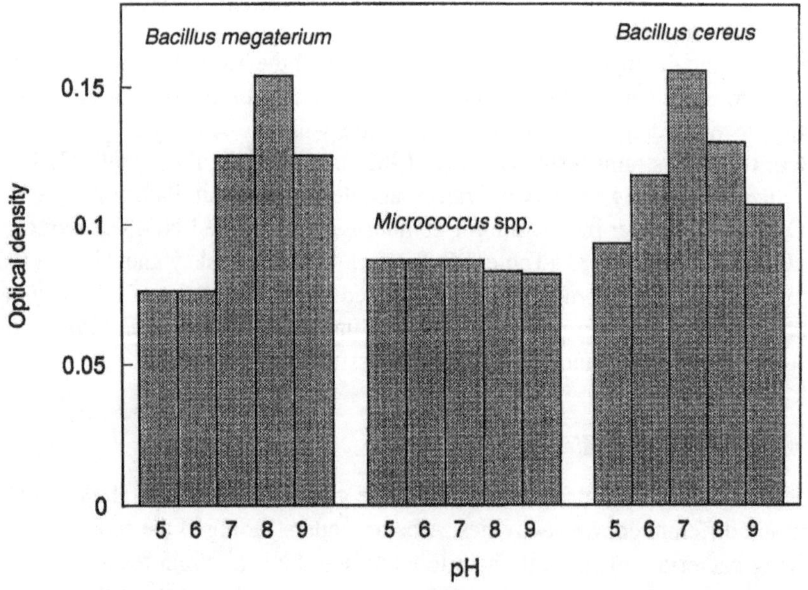

Figure 8.2. Effect of soil pH on the growth of denitrifying organisms (Li *et al.* 1992b).

(Wijler and Delwiche 1954). Li *et al.* (1992b) found that different strains required different oxygen pressures for their growth (Figure 8.3). Thus, strain #26 (*Ps. fluorescens*) grew well when the oxygen concentration was reduced to half that of normal air, and appeared to grow better when the oxygen concentration was reduced to one-tenth of normal air. Strains #22 (*Micrococcus sp.*) and #15 (*B. megaterium*) grew better under anaerobic conditions, while the strain #115 (*B. cereus*) was not so obligate. In the absence of oxygen, different strains varied greatly in their denitrifying activities (Table 8.7; Li *et al.* 1989b).

The essential prerequisite for denitrification is the presence of nitrate, and different species of denitrifying bacteria vary in their tolerance to the concentration of nitrate (Chen *et al.* 1981). When available carbon is held constant, the concentration of nitrate affects the growth of denitrifiers. As shown in Figure 8.4 (Li *et al.* 1992b), the yield of all strains tested was greater when the nitrate concentration was 138 mg N L^{-1} than when it was 276 mg N L^{-1}. The reason for this is unknown.

8.3. Denitrifying potential

The occurrence of denitrification in soil is not due to a single strain's activity, but is an integrated manifestation of the diverse denitrifying bacteria being controlled by many factors in the soil. The acetylene blockage method has been widely applied to the study of denitrifying potential under laboratory conditions. Using this method, a series of factors controlling denitrifying potential in soil have been investigated. These include nitrate concentration (Lalisse-Grundmann *et al.* 1988; Letey *et al.* 1980), moisture and oxygen concentration (Jiang *et al.* 1989; Myrold and Tiedje 1985; Wollersheim *et al.* 1987), soil organic matter, the kind and quantity of applied fertilizers (Jiang *et al.* 1989; Aulakh and Rennie 1987; Lalisse–Grundmann *et al.* 1988; Myrold and Tiedje 1985), tillage (Aulakh *et al.* 1982, 1984; Aulakh and Rennie 1985; Rice and Smith 1982), crop roots (Klemedtsson *et al.* 1987; Mosier and Hutchinson 1981; Reddy and Patrick 1986; Smith and Tiedje 1979; Smith and Delaune 1984), soil conditions (Groffman and Tiedje 1989a, b; Grundmann *et al.* 1988; Letey *et al.* 1980; Murakami *et al.* 1987), herbicides (Yeomans and Bremner 1985a; 1987), and insecticides (Yeomans and Bremner 1985b).

Table 8.7. Activity of denitrifying enzymes in different strains under anaerobic conditions.[1]

Strain	Enzyme	
	$NO_3^- \rightarrow N_2O$ (N_2O produced, $\mu g\ ml^{-1}$)	N_2O reductase (N_2O loss, %)
Ps. fluorescens (33#)	75.6	99.8
Ps. fluorescens (144#)	96.1	5.0
B. megaterium (205#)	50.9	41.4
B. licheniformis (2389#)	29.7	99.7
B. megaterium (123#)	0.0	n.d.

[1] Incubation for 3 days. (Li *et al.* 1992b).

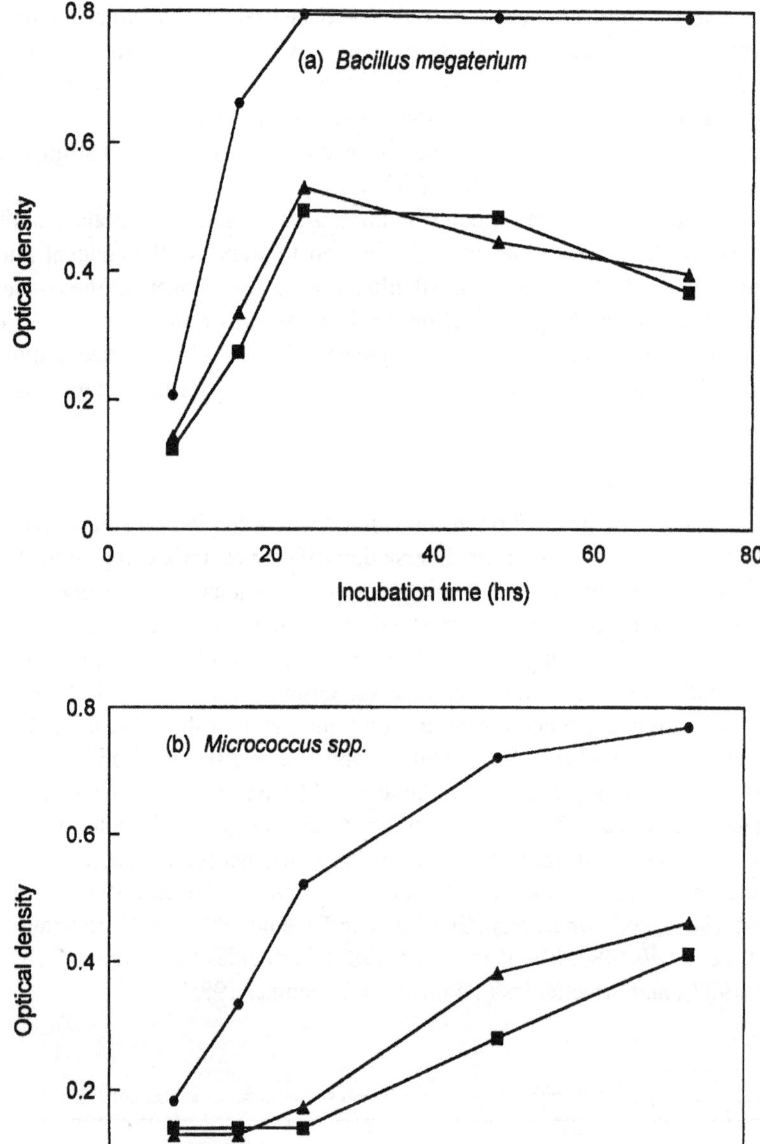

Figure 8.3. Effect of oxygen partial pressure on the growth of denitrifying organisms. Oxygen concentration, normal air ■, 50% normal air ▲, 10% normal air ●. (Li *et al.* 1992b).

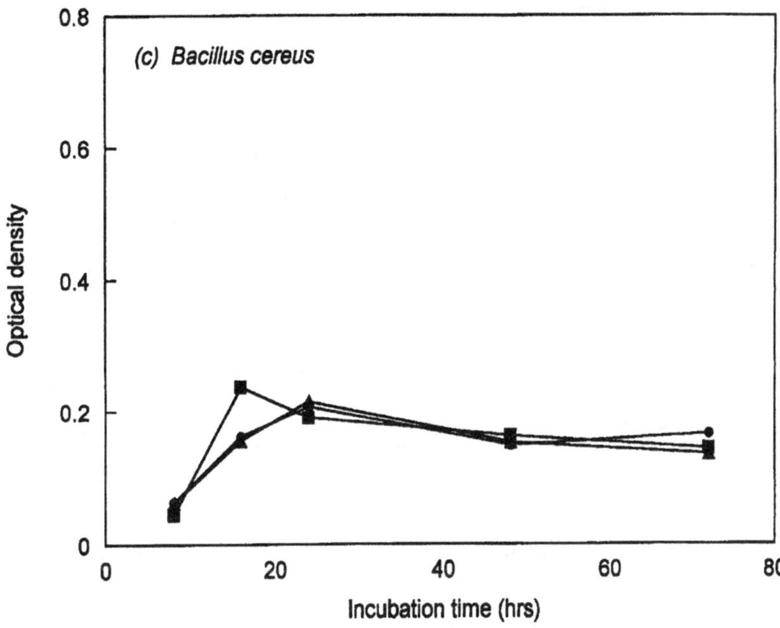

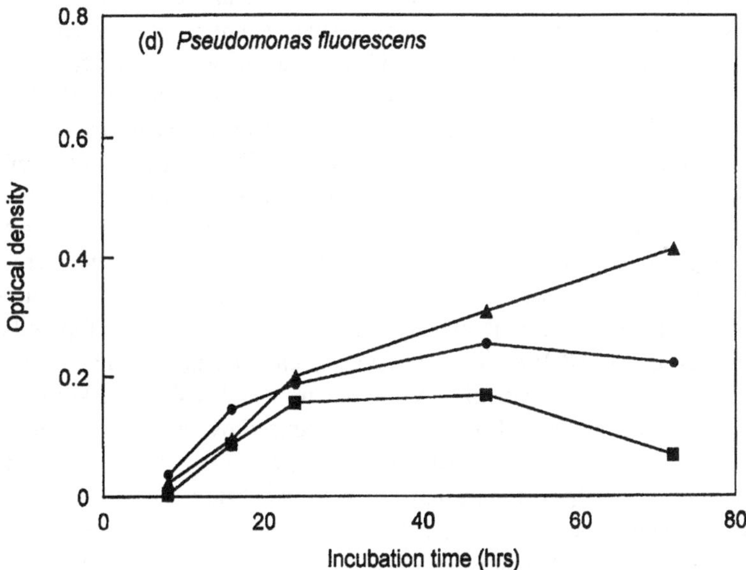

Figure 8.3. (*continued*)

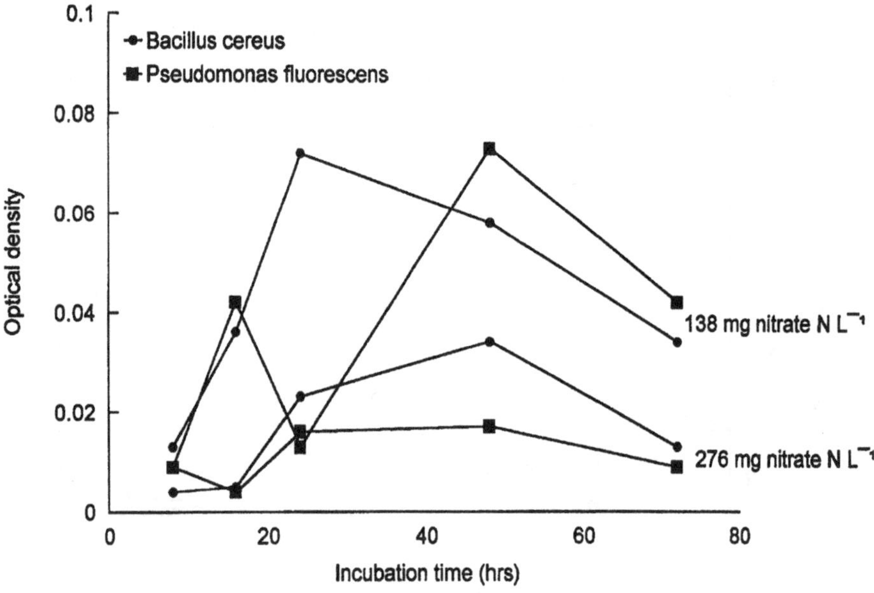

Figure 8.4. Effect of nitrate concentration on the growth of denitrifying organisms (Li *et al.* 1992b).

8.3.1. *Oxygen and moisture content*

Oxygen has long been recognized as an important factor controlling denitrification in soil, since it influences the synthesis and activity of the reductases. Nitric oxide reductase and nitrous oxide reductase are, to some extent, more sensitive to oxygen than others. The time of exposure of soil to the air may affect the formation and reduction of nitrous oxide. As can be seen from the data of Erich *et al.* (1984), exposing freshly sampled soils to air for 24 hours did not influence the formation of nitrous oxide, but completely inhibited its reduction in soil. The moisture content of soil is negatively related to the oxygen content, so the influence of moisture can, to a large extent, be regarded as the influence of oxygen. Many studies have shown that the production of gaseous N by denitrification was highly dependent on soil moisture. When the soil moisture suction was 5 and 10 k Pa, the flux of nitrous oxide N, as measured by the acetylene blockage method, was highest, while at 25 k Pa the flux fell drastically (Ryden and Lund 1980a). With the acetylene blockage method, Jiang *et al.* (1989) investigated the release of nitrous oxide from added nitrate (200 μg g soil^{-1}), and found that the amount of nitrous oxide produced by denitrification in a given period was logarithmically correlated with time (Y = A + B log t; Figures 8.5 and 8.6). It can be seen from Figure 8.5 that soils incubated in argon had a greater denitrification intensity than those incubated in air. Figure 8.6 shows that when incubated in argon, soils with a moisture content corresponding to 60% of the saturated water content released more nitrous oxide than soils whose moisture content was 30% of the saturated water content. The oxygen content affects not only the denitrification intensity, but also the ratio of nitrous oxide to dinitrogen produced (Erich *et al.* 1984; Letey *et al.* 1980).

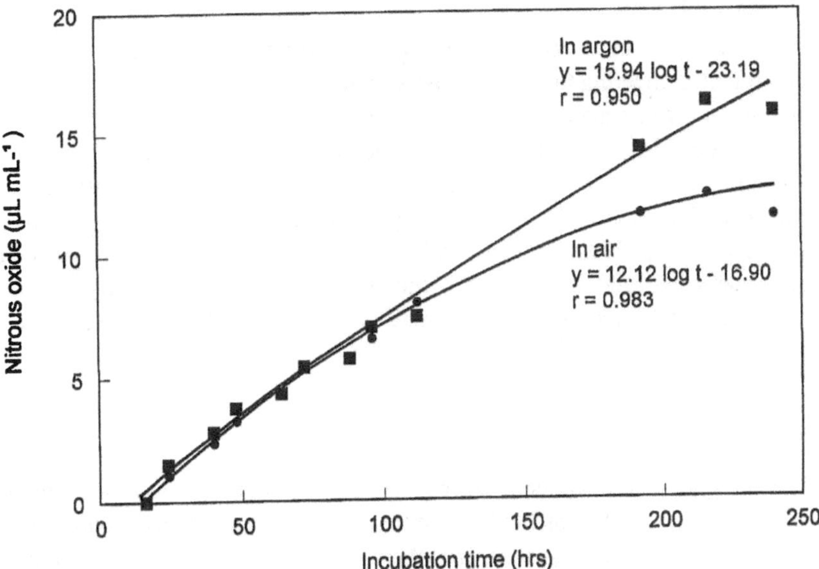

Figure 8.5. Nitrous oxide formed in the atmosphere of different gases (incubated at 100% WHC; Jiang *et al.* 1989).

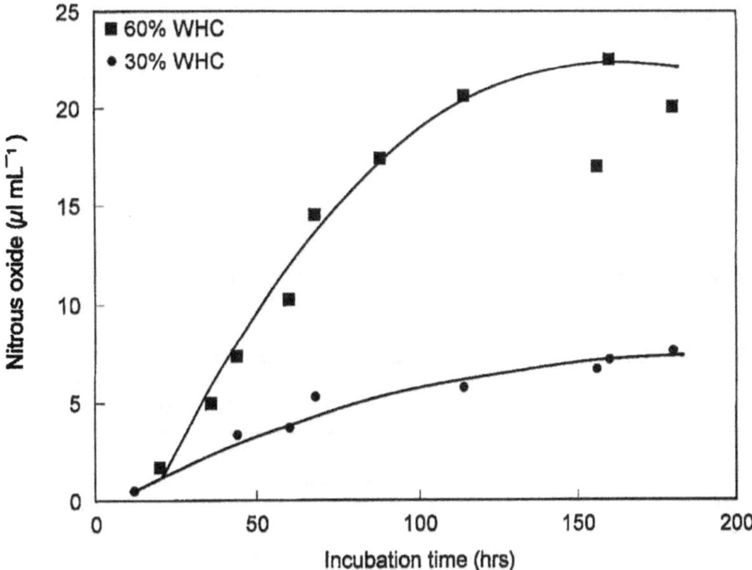

Figure 8.6. Effect of water content on nitrous oxide formation in an argon atmosphere (Jiang *et al.* 1989).

8.3.2. *Organic matter*

The amount of organic matter in soil, especially the available carbon, is another important factor affecting the denitrification potential. Jiang *et al.* (1989) found that in soil high in organic matter, the amount of nitrous oxide formed by denitrification under acetylene blocked conditions was large (Table 8.8). By analyzing total N in soil, Yang and Shao (1981) also found that when the soil organic matter content was between 1% and 2.5%, N loss during submerged incubation was positively correlated with organic matter content. Further, denitrification can be enhanced by adding organic materials to the soil; as shown in Table 8.9, the denitrifying potential of a calcareous paddy soil was increased considerably by adding glucose (Li *et al.* 1991b).

8.3.3. *Plant roots*

A plant, through its metabolism, excretes mineral and organic substances through its roots. These exudates, as well as the material sloughed off roots, provide the nutrients and energy sources for rhizosphere microorganisms. The functioning of plant roots may bring about changes in environmental conditions, such as O_2 tension, moisture, soil pH, etc. within the rhizosphere (Youssef and Chino 1988). Furthermore, as a result of water flow and nutrient uptake by the root, some of the nutrients become enriched, while others are depleted in the rhizosphere (Qin and Liu 1984). All of these influence the growth and activity of microorganisms in the rhizosphere.

Table 8.8. Effect of organic matter content on denitrification in calcareous vegetable soils (μl N_2O–N ml^{-1}).[1]

Organic matter (%)	Time of incubation (hr)					
	16	24	40	48	68	97
3.21	0.98	3.14	8.23	13.4	18.3	24.4
3.86	1.96	6.08	13.1	20.3	31.3	53.0

[1] Incubated in argon at 60% of WHC, using the acetylene blockage method (Jiang *et al.* 1989).

Table 8.9. Effect of glucose on denitrifying potential in calcareous paddy soils (μg N_2O N g soil^{-1}).[1]

Treatment	Time of incubation (day)		
	2	4	6
Soil + KNO_3 + glucose (1%)	0.471 a[2]	1.01 a	1.99 a
Soil + KNO_3	0.295 b	0.48 b	0.68 b

[1] Flooded incubation in atmosphere of nitrogen using acetylene blockage method. 25 mg KNO_3 N added per 100 g soil (Li *et al.* 1991b).
[2] Values in a column with the same character do not differ significantly at P = 0.01.

Rice roots were shown to have noticeable rhizosphere effects on denitrifying bacteria (Li *et al.* 1987). The denitrifying potential of the soil in the rice rhizosphere was greater than that of the non-rhizosphere soil; in an incubation study N loss when $K^{15}NO_3$ was added to a rhizosphere soil was 89.3–92.6%, while it was 70.3–77.8% when added to the corresponding non-rhizosphere soil (Li *et al.* 1984). Raimbault *et al.* (1977) also showed that the activity of nitrous oxide reductase in the rice rhizosphere was 14 times that of the root-free soil.

Barley, wheat and other crops differ from rice in physiological characteristics. As shown by Chen *et al.* (1981) the aerenchyma in rice roots constituted 5–30% of the entire tissue, whereas in barley it was less than 1%. For this reason, the environment in the rhizosphere of barley, wheat, etc. is different from that of rice. In spite of this, the roots of barley, wheat, etc. also stimulate denitrification. According to Smith and Tiedje (1979), the denitrifying activity in the rhizosphere of oats was greater than that of the non-rhizosphere soil, and 1–2 mm away from the root surface the denitrifying activity diminished dramatically. This may be associated with the difference in amount of readily decomposable organic matter present at different distances from the root surface. Smith and Tiedje (1979) also showed that when the nitrate concentration was high, the rate of denitrification in the rhizosphere of oats was faster than that in the non-rhizosphere soil. However, when the nitrate concentration was low, the soil planted with oats had an appreciably lower rate of denitrification than the soil without oats. We investigated the denitrifying activity in the rhizosphere and non-rhizosphere soils of different varieties of wheat seedlings (Li *et al.* 1992a), and the results showed that the rhizosphere soil had a greater denitrifying activity than the non-rhizosphere soil (Table 8.10).

The effect of plant roots on denitrification is governed by many factors. Prade and Trolldenier (1988) investigated the effect of wheat roots on denitrification when nitrate was non-limiting. They found that, when the air-filled porosity was above the critical level (10–12%), little denitrification took place, whether the organic carbon in the soil was high or low (0.12–1.31%) and whether or not wheat

Table 8.10. Denitrifying activity in rhizosphere and non-rhizosphere wheat soils (μg N_2O–N g $soil^{-1}$).[1]

Wheat cultivar	Fertilizer	Soil	Days after fertilization				
			14	21	28	35	42
Zhengyin No. 1	Ammonium sulphate	Rhizosphere	5.91 a[2]	10.0 a	28.5 a	16.9 a	18.6 b
		Non-rhizosphere	1.99 b	5.65 b	5.91 c	4.62 cd	5.76 d
	Urea	Rhizosphere	4.73 a	10.7 a	12.7 b	9.29 b	26.4 a
		Non-rhizosphere	1.83 b	5.23 b	1.73 c	4.05 d	6.87 cd
Baofeng No. 7228	Ammonium sulphate	Rhizosphere	0.85 b	0.60 c	6.74 c	7.44 bc	10.6 c
		Non-rhizosphere	0.60 b	0.38 c	4.31 c	0.73 e	0.72 e
	Urea	Rhizosphere	0.70 b	1.10 c	12.3 c	4.05 d	5.84 d
		Non-rhizosphere	1.46 b	0.25 c	4.69 c	1.55 de	2.66 de

[1] Incubation as in Table 8.9; 167 mg N kg $soil^{-1}$ was applied to the pots (Li *et al.* 1992a).
[2] Values in a column with the same character do not differ significantly at P = 0.05.

was grown on the soil. However, when the air-filled porosity was below the critical level, the denitrification rate increased logarithmically with decreased air-filled porosity. The effect of wheat roots on denitrification was negatively correlated with the air-filled porosity of the soil. Moreover, when the air-filled porosity was below 10–12%, denitrification increased markedly with an increase in soil organic carbon content. These results suggest that the consumption of oxygen around the rhizosphere by respiration of wheat roots stimulates denitrification in soil, and that this stimulation is controlled by the air-filled porosity in the soil.

8.3.4. Tillage

Zero or minimum tillage influences not only the physical and chemical properties and nutrient status of soils (Zhao *et al.* 1981), but their biological characteristics as well. Aulakh *et al.* (1982) found that N loss from zero tilled soil was twice that from conventionally tilled soils, and Aulakh *et al.* (1984) found that the number of denitrifying bacteria in zero tilled soil was 6 times that in conventionally tilled soil. Aulakh and Rennie (1985) found that in a field of wheat fertilized with urea (75 kg N ha^{-1}), N loss through denitrification was 3–7 and 12–16 kg N ha^{-1} yr^{-1} from conventionally tilled and zero tilled plots, respectively; the corresponding figures for the two tillage treatments under fallow with the same rate of N application were 12–14 and 34 kg N ha^{-1} yr^{-1}, respectively. It is argued that in zero tillage the stimulated N loss can be attributed to moisture retention in the soil which increases anaerobiosis, and the straw mulch which provides energy. However, in a flooded incubation experiment it was found that the soil taken from a zero tilled plot in the Taihu Lake region had appreciably lower N oxide and nitrous oxide reductase activity than the soil taken from a conventionally tilled plot (Table 8.11; Wu *et al.* 1988). This effect was particularly noticeable when the soils were incubated for 2–4

Table 8.11. Effect of tillage on denitrification potential in paddy soils.[1]

Soil	Tillage	Time of incubation (day)			
		2	4	6	9
		N-oxide reductase activity[2] (N_2O formed, % of N added)			
Neutral gleyed	Conventional	46.7 a	61.1 a	98.5 a	85.3 a
	Zero	39.3 b	54.9 b	88.6 b	79.8 a
		N_2O reductase activity[3] (N_2O disappearance, %)			
Neutral gleyed	Conventional	64.8 a[4]	84.8 a	93.9 a	
	Zero	5.7 b	60.0 b	87.9 a	
Clay loamy hydromorphic	Conventional	44.4 a	69.9 a	99.5 a	
	Zero	32.8 b	51.1 b	98.9 a	

[1] Wu *et al.* (1988).
[2] Incubation as in Table 8.9.
[3] N_2O injected at rate of 1% (V/V); without C_2H_2.
[4] Values in a column with the same character do not differ significantly at P = 0.05.

days, but as time went on, the difference tended to disappear. Consequently, the effect of zero tillage on denitrification loss from flooded fields may differ from that of upland fields.

8.3.5. Soil type

Different soils in the same region may vary greatly in denitrifying potential due to their different physical and chemical properties. Gould and McCready (1982) determined the denitrifying potential in 5 types of Alberta soil and found that the potential decreased in the following order, black soil > dark brown soil > brown soil = grey luvisol = solodized solonetz soil. They considered that the rate of denitrification in a given soil was dependent on the chemical properties of the soil as well as on the availability of the carbon substrate. We have studied the denitrifying potential of the main types of soil in the Taihu Lake region and concluded that the neutral and slightly acid gleyed paddy soils had the highest denitrifying potential, while the clay loamy and silty hydromorphic paddy soils had the lowest (Table 8.12; Wu *et al.* 1988). These variations may be associated with the moisture regime and organic matter content of the soils. The gleyed paddy soils are of the groundwater type and have a sufficiently high content of organic matter for denitrification to take place. We also determined the denitrifying potential of soils from different regions (Table 8.13; Li *et al.* 1991b), and the results show that the paddy soils derived from red soil in Jiangxi Province had a considerably higher denitrifying potential than the calcareous paddy soils in Henan Province. This may also be related to the organic matter content of the soil, as the Jiangxi Province soils contained 1.15% of organic carbon, whereas those from Henan Province contained only 0.43%.

Soils developed on different parent materials in the same region vary in denitrifying potential because of their different physical and chemical properties. Pan *et al.* (1988) showed that, after 2, 4, 6 and 8 days of incubation, the denitrifying potential (nitrous oxide N emitted as % of applied N) for a purple soil derived from purple

Table 8.12. Denitrifying potential (μg N_2O–N g soil^{-1}) in the main paddy soils of the Taihu-Lake region.[1]

Soil	Organic matter (g kg^{-1})	Incubation time (days)			
		2	4	8	10
Bleached	11.3	108 a[2]	209 a	437 a	383 bc
Neutral gleyed	19.1	67 b	234 a	410 a	617 a
Slightly acid gleyed	24.8	53 b	223 a	330 ab	426 b
Submergenic	11.1	55 b	93 b	270 b	291 bc
Clay loamy hydromorphic	19.3	20 c	102 b	269 b	355 bc
Silty hydromorphic	17.5	23 c	66 b	231 b	242 c

[1] Wu *et al.* (1988).
[2] Values in a column with the same character do not differ significantly at P = 0.05.

Table 8.13. Denitrifying potential (μg N_2O–N g soil^{-1}) in paddy soils from different regions.[1]

Soil derived from	Location	pH	Organic matter (g kg^{-1})	Incubation time (days)		
				2	5	7
Red soil	Yingtan, Jiangxi	5.2	11.5	0.67 a[2]	1.31 a	2.14 a
Calcareous alluvials	Fengqiu, Henan	7.8	4.3	0.31 b	0.50 b	0.75 b

[1] Incubation as in Table 8.9 (Li *et al.* 1991b).
[2] Values in a column with the same character do not differ significantly at P = 0.01.

sandstone was 4.29, 5.09, 4.76 and 4.69%, respectively. However, red soils developed on red sandstone and Quaternary red clay did not denitrify. Analyses showed that the pH and organic matter content of the purple soil were 5.3 and 20 g kg^{-1}, respectively, while the pH values for the red soils were 4.8 and 5.1, and their organic matter contents were 4.8 and 16.4 g kg^{-1}, respectively. It seems that the pH and organic matter in the purple soil were more favourable for denitrification.

8.3.6. *Land use*

The denitrifying potential of a soil can vary with land use because the soil environment changes considerably with irrigation, fertilization and tillage as well as with plant characteristics. For example, with soils from the Taihu Lake region, with the exception of two soils incubated for two days, those planted to flooded rice had a higher denitrifying potential than those planted with wheat (Figure 8.7; Wu *et al.* 1988). In the wheat soil, nitrous oxide emission had usually ceased by the 8th day of incubation, but for flooded rice nitrous oxide was still being emitted after 8 days.

Again, the denitrifying potential of soils derived from Quaternary red clay under different land use decreased in the following order, paddy soil > citrus orchard soil > tea plantation soil > cold spring paddy field and upland soil > coniferous woodland soil (masson pine) (Figure 8.8; Pan *et al.* 1988). It has long been established that paddy soils are conducive to denitrification. The reason that the soils of citrus orchards and tea plantations had a fairly high denitrifying potential may be that these soils received large amounts of organic manure which provides energy for denitrification. The woodland soils of young pines suffered severely from erosion with the result that the pH and organic matter content of the surface soil were too low for denitrification to occur (Pan *et al.* 1988).

8.4. Denitrification in flooded soils

In general, the loss of fertilizer N is greater from paddy soils than from upland soils. This is generally attributed to the existence of an oxidized layer at the soil surface and the underlying reduced layer, and also to the oxidized and reduced layers inside and outside the rice rhizosphere. Consequently, the conditions in

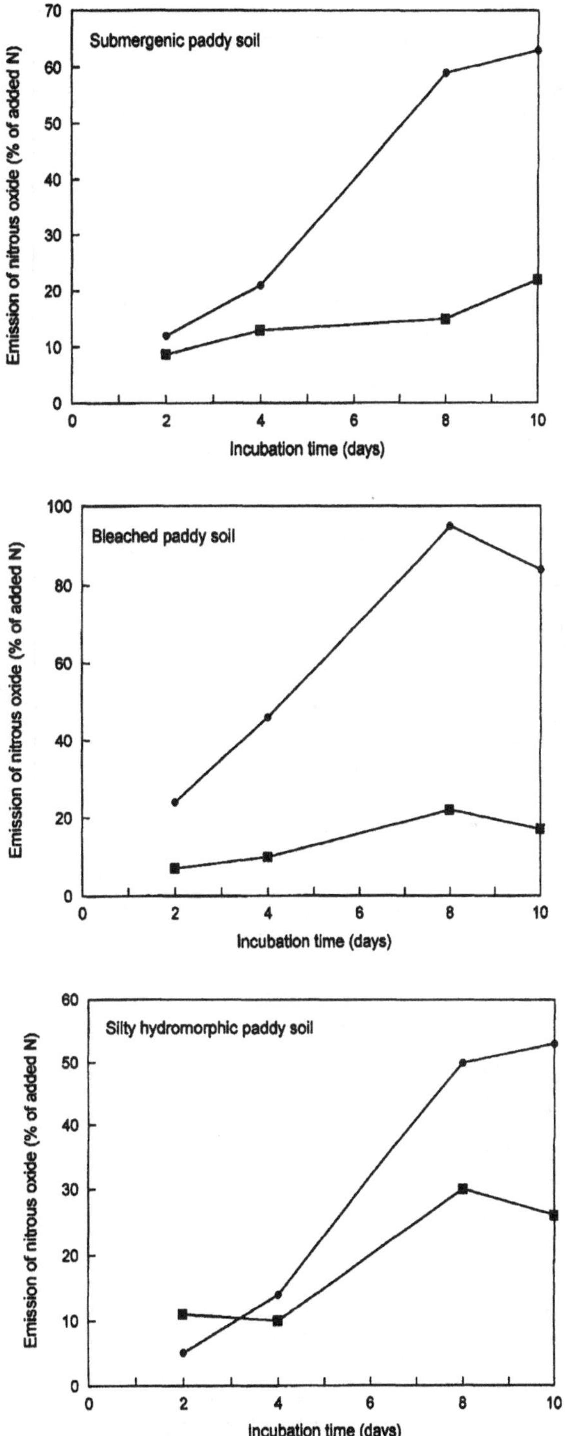

Figure 8.7. Denitrifying potential of flooded rice ● and wheat ■ soils.

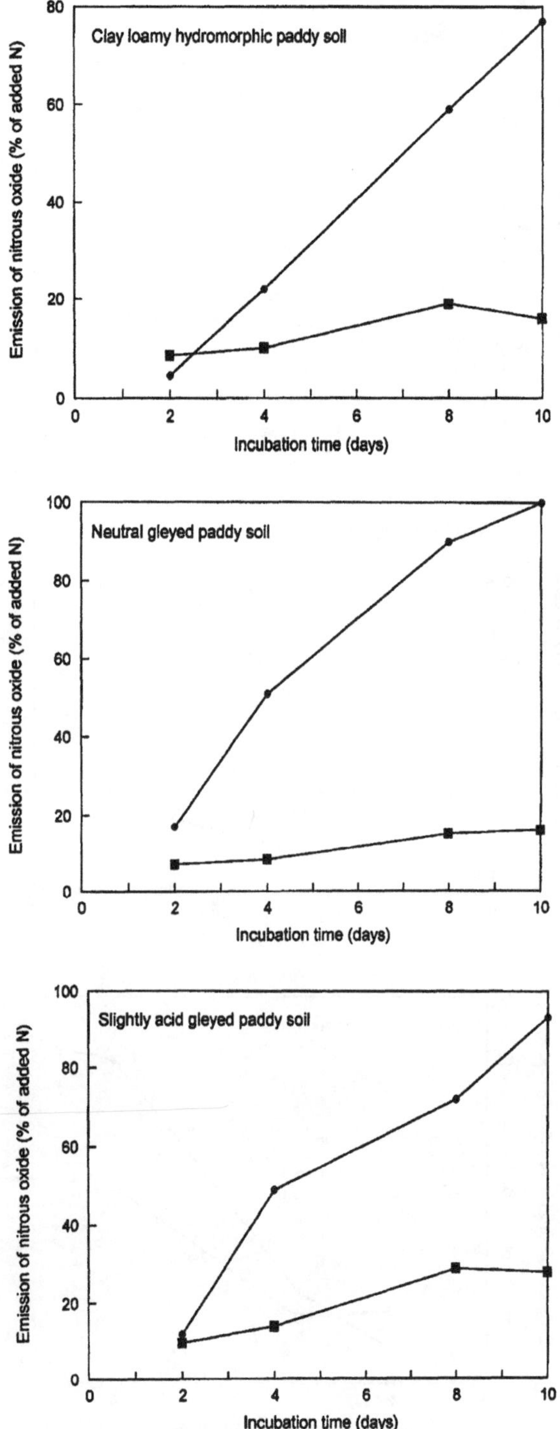

Figure 8.7. (*continued*)

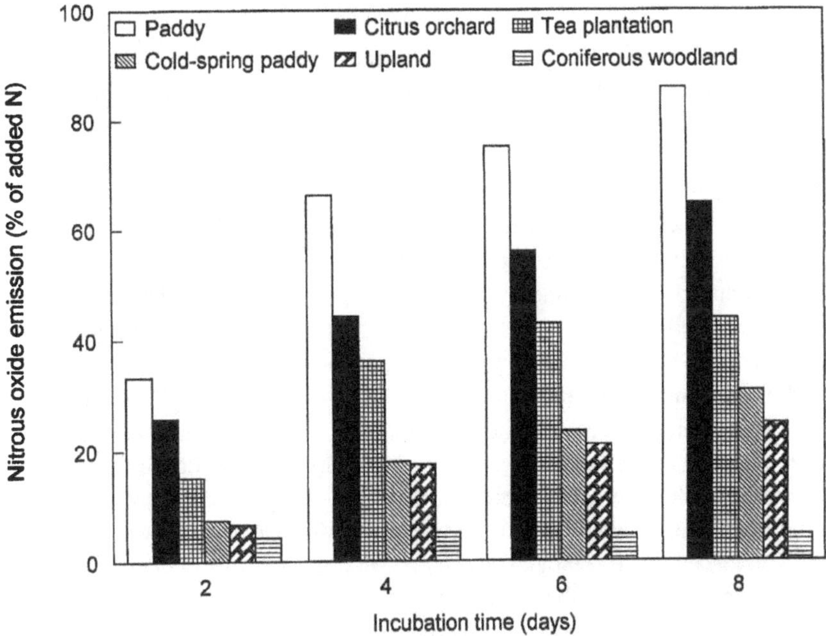

Figure 8.8. Denitrifying potential of soils derived from Quaternary red clay under different land use (Pan *et al.* 1988).

paddy soils are usually more favourable for nitrification-denitrification of ammoniacal fertilizers. The nitrification-denitrification process in paddy soils is discussed below:

8.4.1. *Oxidized reduced layers*

In paddy soils, the nitrification-denitrification which occurs in the oxidized and reduced layers, respectively, has been widely accepted as the mechanism of N loss. Therefore it was proposed that deep placement of ammoniacal fertilizers would decrease N losses (Mitsui 1959). The differentiation of the oxidized and reduced layers in paddy soils is influenced by the properties of the soil and the biological and chemical properties of floodwater. When light textured soils are flooded, differentiation occurs slowly and the oxidized layer appears to be fairly thick but not well defined. By contrast, in heavy textured soils, differentiation proceeds rapidly and the oxidized layer is thin but well defined. Again, for strongly reduced soils, differentiation of the two layers occurs quickly and the oxidized layer is rather thin (Hasebe and Iimura 1982); for weakly reduced soils, the situation is just the opposite. The degree of soil dryness before flooding also affects the differentiation of the layers. According to Hasebe *et al.* (1987), a moist paddy soil, after incubation for 20 days under flooded conditions at 25°C, differentiated into two layers, a 1 cm thick oxidized layer and the underlying reduced layer, while an air dried soil, after flooding for 5 days, divided into two layers, although the oxidized layer was only 1 mm thick. Another factor influencing differentiation is the depth of the floodwater and its dissolved oxygen content (Yoshida and Benjamin 1974). When the floodwater layer is deep, differentiation occurs slowly; when the floodwater layer is shallow and contains a large amount of dissolved oxygen, differentiation occurs quickly.

The oxidized and reduced layers differ greatly in their physical, chemical and biological properties. The oxidized layer often has an Eh value over 350 mv, whereas the reduced layer is about 150 mv, and sometimes is negative. In addition, the reduced layer commonly has a greater pH value and exchangeable acidity than the oxidized layer. These variations result in distinct differences between the activities of the soil organisms in the two layers (Mitsui 1959; Chen *et al.* 1981), including their nitrifying and denitrifying activities. Chen *et al.* (1981) found that the number of nitrifying bacteria in the 0–1 cm oxidized layer was greater than that in the 1–25 cm reduced layer. Li *et al.* (1983) found that both the number of ammonium oxidizing bacteria and its nitrifying activity were greater than those of the reduced layer (Table 8.14). Since the nitrification products, nitrite and nitrate, are not adsorbed by soil, they can easily move downward from the oxidized layer to the reduced layer, and be reduced by denitrifiers to gaseous N oxides and N_2 and lost. According to Ruan *et al.* (see Li 1986), one of the key factors controlling loss of N by denitrification in paddy soils is the rate of nitrification in the oxidized layer. When ammonium sulfate was applied to the surface of the oxidized layer, as much as 40% of the added N was lost, yet when it was incorporated into the soil, only 10% was lost. The loss was negligible when the ammonium sulfate was added to a thin layer under anaerobic conditions (Table 8.15; Li 1986). Katyal *et al.* (1988) also referred to the importance of studying the nitrifying activity at the soil water interface when the floodwater is shallow. It is apparent that the rate of nitrification in the oxidized layer and the rate of diffusion of ammonium from the reduced layer to the oxidized layer are factors limiting the rate of denitrification in flooded soils.

Table 8.14. Ammonium oxidizing bacteria (number g dry soil^{-1})[1] and nitrifying activity ($NO_2 + NO_3$-N formed as % of mineral N) in oxidized and reduced layers of paddy soils.

Soil	Bacteria		Nitrifying activity	
	Oxidized	Reduced	Oxidized	Reduced
Submergenic	$>10^8$	37.3×10^3	–	–
Clay loamy hydromorphic	8.39×10^6	11.5×10^3	45.2	29.3**
Neutral gleyed	$>10^9$	1.54×10^5	58.3	21.2***
Slightly acid gleyed	40.5×10^3	7.79×10^3	22.3	13.2**

[1] MPN method; NH_4^+-containing solution inoculated with soil suspension and incubated for 2 weeks (Li *et al.* 1983).
** Significant at P = 0.01; *** Significant at P = 0.001.

Table 8.15. Effect of oxidized layer on nitrogen loss[1] from flooded soil after application of ammonium sulfate.

Treatment	Method of application	N loss (%)
Oxidized layer–reduced layer[2]	Surface	40.0
Oxidized layer–reduced layer[3]	Incorporation	10.6
Oxidized layer[4]	Incorporation	15.0
Reduced layer[5]	Incorporation	0.0

[1] Calculated as difference between total N before and after incubation (Li 1986).
[2] Ammonium sulphate solution dripped onto 30 g air-dry soil, 5 cm thick.
[3] Ammonium sulphate solution well mixed with 30 g air-dried soil, 5 cm thick.
[4] Ammonium sulphate solution well mixed with 30 g air-dried soil, 0.5 cm thick.
[5] Ammonium sulphate solution well mixed with 30 g air-dried soil, 0.5 cm thick, incubated under anaerobic conditions.

8.4.2. *Rice roots*

Among cereals, rice is known to be the only crop that tolerates waterlogging. Like other hydrophytes, it is capable of transferring oxygen from the leaves through stems to roots. Oxygen then diffuses out of the roots to the surrounding soil, resulting in the formation of a thin oxidized layer around each root. As a consequence the Eh value of the soil in the rice rhizosphere is commonly greater than that of the non-rhizosphere soil (Yu and Liu 1957; Liu and Yu 1963). Although negative Eh values were sometimes obtained due to the measurement techniques, in general, the Eh values and the dissolved oxygen concentrations in the soil of the rice rhizosphere were greater than those of the non-rhizosphere soil (Table 8.16; Li *et al.* 1984). However, it has been reported (Chen *et al.* 1981) that the root has no great influence on the number of nitrifying organisms, and there was not necessarily a high number of nitrifying bacteria around the rice root.

The number of denitrifying bacteria on the rice rhizoplane and in rhizosphere was found to be 187–488 million g dry root^{-1} and 47–459 million g dry soil^{-1}, respectively (Li *et al.* 1984, 1986; Dommergues and Krupa 1978). When rice roots were not sterilized and incubated for 1–2 days in a nitrate containing substrate under anaerobic conditions, 33–90% of the nitrate added disappeared. When the

Table 8.16. Oxidation-reduction potential (mV) and dissolved oxygen (μg L^{-1}) in rhizosphere and non-rhizosphere paddy soils with rice.[1]

Soil		Eh		Dissolved oxygen	
		Tillering	Booting	Tillering	Booting
Clay loamy hydromorphic	Rhizosphere	–91	25	160	860
	Non-rhizosphere	–176	–95	110	430
Silty hydromorphic	Rhizosphere	–28	–50	800	–
	Non-rhizosphere	–117	–201	240	–

[1] Pot experiment (Li *et al.* 1984).

roots were sterilized, however, only 1–10% of the nitrate disappeared (Figure 8.9; Li *et al.* 1984). The difference was not due to the greater uptake of nitrate and nitrite by unsterilized roots. Indeed the uptake was less than that by the sterilized roots (Figure 8.10; Li *et al.* 1984). This suggests that, under favourable conditions, the activity of denitrifiers on the rhizoplane may reach a very high level. In addition, a rhizosphere soil treated with $K^{15}NO_3$ and incubated for two weeks lost more ^{15}N than the corresponding non-rhizosphere soil, but there was no difference in loss between the rhizosphere and non-rhizosphere soils, when ^{15}N urea was added (Table 8.17; Li *et al.*

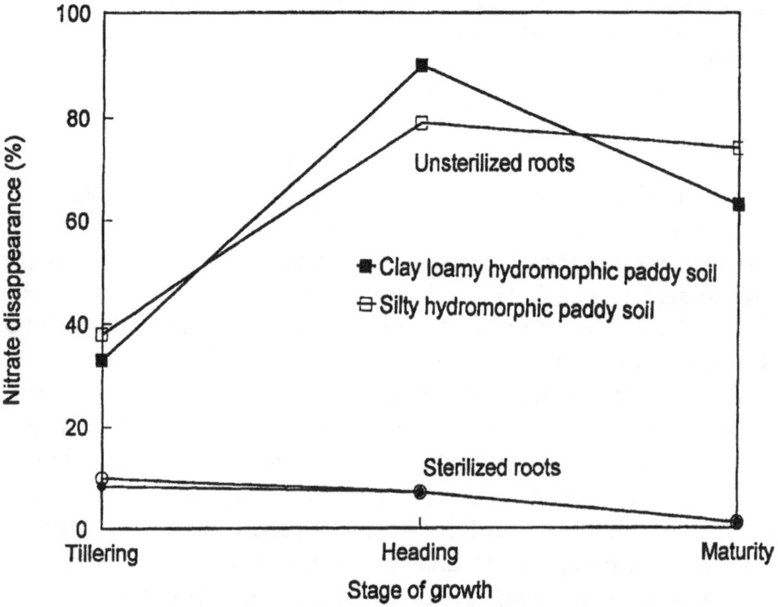

Figure 8.9. Effect of denitrifiers on rice rhizoplane on the disappearance of nitrate (Li *et al.* 1984).

Table 8.17. Nitrogen loss from $K^{15}NO_3$ and ^{15}N-urea added to rice rhizosphere and non-rhizosphere paddy soils.[1]

Soil		$K^{15}NO_3$		^{15}N–urea	
		Residual ^{15}N (mg 100 g^{-1})	Loss (%)	Residual ^{15}N (mg 100 g^{-1})	Loss (%)
Clay loamy hydromorphic	Rhizosphere	4.20	89.5	28.1	29.7
	Non-rhizosphere	11.7	70.7*	27.3	31.6
Silty hydromorphic	Rhizosphere	2.90	92.7	25.6	36.0
	Non-rhizosphere	8.80	77.8*	25.8	35.5

[1] Soils sampled at the tillering stage; 40 mg N added per 100 g soil; incubated under flooded conditions for 2 wk (Li *et al.* 1984).
* Significant at 5% level.

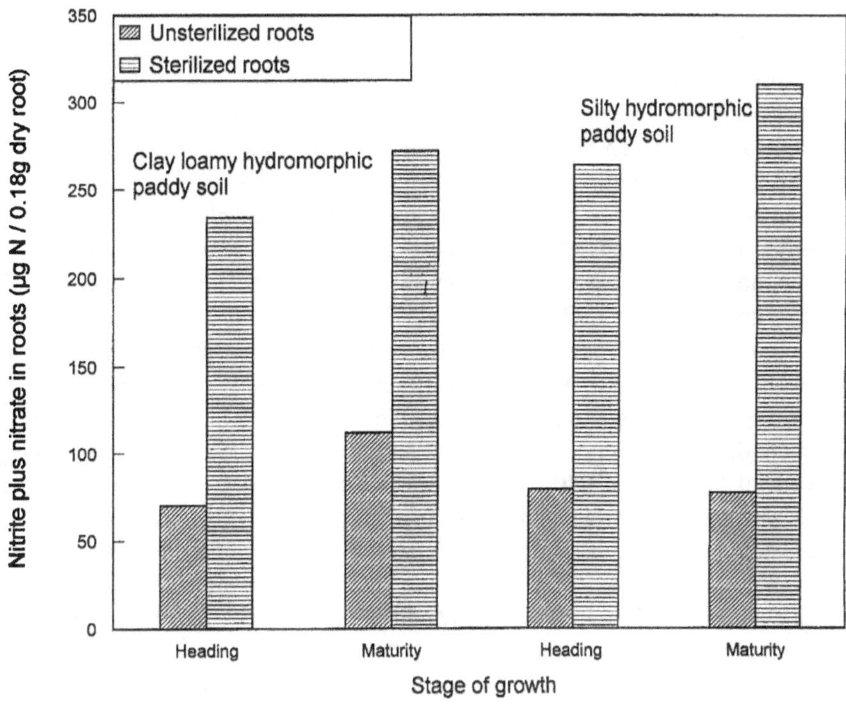

Figure 8.10. Nitrite plus nitrate in sterilized and unsterilized roots (Li *et al.* 1984).

1984). This suggests that the rhizosphere soil does not have a greater nitrifying activity than the non-rhizosphere soil but has a fairly high denitrifying activity.

Smith and Delaune (1984) confirmed that loss of N by nitrification-denitrification occurred in the rice rhizosphere, but they believed that the magnitude of the loss by this mechanism was small. In their experiment, the planted system emitted more nitrous oxide and dinitrogen in the first two days (101 μg N m^{-2} h^{-1}) than the unplanted soil (86 μg N $m^{-2}.h^{-1}$). Four to six days after fertilization no difference was observed between planted and unplanted systems. The initial enhancement of gaseous N emission may occur because of the oxidized rice root rhizosphere. Reddy and Patrick (1986) showed that under rice growing conditions, the rate of N loss after application of $^{15}NH_4Cl$ was 193 mg N m^{-2} day^{-1}, while the N loss due to the rhizosphere effect was 143 mg N m^{-2} day^{-1}. Apparently, the N loss caused by the rhizosphere effect in this study was significant, because it was conducted under high rates of N application.

It should be noted that because there is competition between denitrifying organisms and plants for uptake of inorganic N, plant roots can only enhance loss by denitrification when there is an ample supply of nitrate or nitrite. On the other hand, if the supply of N is limited, or if plant roots strongly assimilate inorganic N, then the growth of rice will decrease loss.

8.4.3. *Amorphous iron and manganese oxides*

Recent studies suggest that other mechanisms for nitrification-denitrification loss from paddy soils exist, in addition to the reactions near the surface oxidized layer and the rhizosphere. For instance, in a pot experiment without the oxidized surface layer and rice plants, 13.5% of the N added as ammonium sulfate was not recovered under flooded conditions (Liao *et al.* 1982). Since there is a large quantity of amorphous metal oxides present in paddy soils (He *et al.* 1981; Savant and McClellan 1987), it is postulated that these oxides may serve as electron acceptors when ammonium is oxidized under anaerobic conditions and lead to N loss through the catalytic action of microbes (Liao *et al.* 1982).

Chao and Kroontje (1966) showed that, in acid or basic solution, nitrate can be reduced by ferrous iron to N_2 through different intermediates, and they suggested that the sequence of gaseous N oxide production was similar to that occurring in the biological denitrification process. Savant and McClellan (1987), when reviewing the work of Sahrawat (1980), Yamane and Okazaki (1982) and Komatsu *et al.* (1978), also suggested that iron oxides take part in the denitrification process in flooded soils. Li *et al.* (1988) found that, after 3–9 days of anaerobic incubation of a sterilized culture solution containing iron oxide and ammonium, the total loss of N reached 16–21% (Table 8.18). Although the pH of the system was raised by ligand exchange during incubation and this possibly resulted in volatilization, loss of ammonia was less than 1% of the added N. Experiments were made under the same conditions using $(^{15}NH_4)_2SO_4$ as the N source, and it was found that 10–15% of the added N was lost, and gaseous N oxides of different mass, such as $^{15}NO_2$, $^{15}N_2O$, $^{15}N^{14}NO$, ^{15}NO, $^{15}N_2$, and $^{15}N^{14}N$, were detected in the culture container (Table 8.19; Li *et al.* 1988).

Similar results were obtained when amorphous manganese oxide was used in place of amorphous iron oxide, and N loss from the system tended to increase with increased addition of amorphous manganese oxide (Table 8.20; Li and Wu 1991). Amorphous iron oxide seemed to be more effective than amorphous manganese oxide in causing N loss. Furthermore, aging of amorphous manganese oxide reduced the effect on N loss (Table 8.21; Li and Wu 1991). Judging from this,

Table 8.18. Effect of amorphous iron oxide[1] on nitrogen loss from ammonium under anaerobic incubation.

Treatment	Incubation time (days)					
	3		6		9	
	Residual N (mg 10 ml^{-1})	Loss (%)	Residual N (mg 10 ml^{-1})	Loss (%)	Residual N (mg 10 ml^{-1})	Loss (%)
No FeOOH, sterilized	4.42		4.15		4.05	
FeOOH (sterilized)	3.47	21.5	3.38	18.6	3.29	19.4
FeOOH + soil suspension	3.71	16.1	3.42	17.6	–	–

[1] Activity of amorphous iron oxide: 98% (Li *et al.* 1988).

Table 8.19. Nitrogen gases formed after incubation of amorphous iron oxide with a solution containing $^{15}NH_4^+$ under anaerobic conditions.

Treatment	Mass[1]						
	47 $^{15}NO_2$	46 $^{15}N_2O$	45 $^{15}N^{14}NO$	44 $^{14}N_2O(CO_2)$	31 ^{15}NO	30 $^{15}N_2$	29 $^{15}N^{14}N$
	Relative output (mm)						
1 Control) (sterilized)	–	–	–	1.0	–	–	–
2 $^{14}NH_4^+$ (pH 6.4) + FeOOH (sterilized)	1.0	23.0	58.0	3.6	90.0	5.0	14.0
3 $^{15}NH_4^+$ (pH 6.4) + FeOOH (sterilized)	2.0	31.0	78.0	4.0	124.0	7.0	17.0
4 $^{15}NH_4^+$ (pH 2.6) + FeOOH (sterilized)	0	2.0	4.0	6.0	5.0	0.6	3.0
5 No. 4 + Soil suspension (not sterilized)	3.0	38.0	98.0	70.0	76.0	8.0	19.2
6 No. 4 + Denitrifiers (not sterilized)	1.0	13.0	30.0	21.0	48.0	3.6	8.6

(Li *et al.* 1988).
[1] Determined with a ZAB–HS Model Organic Mass Spectrometer.

Table 8.20. Nitrogen loss from a solution containing amorphous manganese oxide and ammonium under sterilized anaerobic conditions.[1]

Treatment	Residual N (mg 10 ml^{-1})		Loss (%)
Control (without MnO_2)	4.12	a[2]	
1% Amorphous MnO_2	3.83	b	7.0
2% Amorphous MnO_2	3.75	c	9.0
3% Amorphous MnO_2	3.71	d	10.0
4% Amorphous MnO_2	3.56	e	13.6

[1] Incubated in nitrogen atmosphere for a week. The specific surface area of amorphous manganese oxide was 659 m^2 g^{-1} (Li and Wu, 1991a).
[2] Values in a column with the same character do not differ significantly at P = 0.01.

Table 8.21. Effect of amorphous manganese oxide aged for different times on nitrogen loss from a solution containing ammonium.[1]

No.	Treatment	Residual N (mg/10 ml)	Loss (%)
1	No MnO_2, sterilized	4.33 a[2]	
2	MnO_2 aged for 6 years at room temperature (sterilized)	4.24 a	2.1
3	Newly prepared MnO_2 (sterilized)	3.90 b	9.9
4	No. 2 + soil suspension (not sterilized)	3.65 c	15.7
5	No. 3 + soil suspension (not sterilized)	3.49 d	19.4

[1] Incubation as in Table 8.20. Specific surface areas of aged and new MnO_2 were 433 and 659 m^2 g^{-1}, respectively. (Li and Wu 1991a).
[2] Values in a column with the same character do not differ significantly at P = 0.05.

amorphous manganese oxide newly formed in paddy soils may play an important role in inducing N loss.

It should be noted that the treatments listed in Tables 8.18–8.21, except inoculations with soil suspensions and strains, were carried out under sterile conditions, and that the amorphous iron and manganese oxides were thoroughly sterilized by gamma radiation. Thus, N loss from these treatments results solely from chemical reactions. The results of treatments 4 and 5 in Table 8.21, and treatment 4 in Table 8.22, show that much more N was lost by the joint action of amorphous manganese oxide and microorganisms (Table 8.22). These results suggest that amorphous iron and manganese oxides may act as electron acceptors when ammonium is being oxidized and that N loss may occur during the oxidation or during the reduction of the nitrite or nitrate produced.

8.5. Loss of fertilizer nitrogen

The previous section has dealt only with the denitrifying potential in soils and its associated factors. "How much of the fertilizer N is lost by denitrification?" is a question which still needs to be solved before the efficiency of fertilizer N can be improved. Fertilizer N applied to soil may be lost through ammonia volatilization, nitrification-denitrification, runoff, and leaching. The pathway and amount of fertilizer N lost depends primarily on the type of fertilizer, method of application, cultivation and soil properties. From the large amount of data obtained from ^{15}N balance studies in flooded rice in China, Zhu (1985) concluded that with the prevailing method of fertilization, N loss from ammonium bicarbonate varied from 40 to 70%, that from urea ranged from 30 to 55%, except when deep placed, while the loss from ammonium sulfate varied from 15 to 35% in non-calcareous soils and 30 to 50% in calcareous soils. Obviously, the N loss so estimated is not entirely due to nitrification-denitrification.

Since the late 1970s many workers have devoted themselves to the problem of estimating the amount of N lost by nitrification-denitrification, and many reviews and monographs have been published (Freney and Simpson 1983; Hauck 1986; Hauck and Weaver 1986). The methods in use fall into two groups, (i) indirect and (ii) direct. Group (i) includes measurements of nitrate disappearance and nitrate/

Table 8.22. Losses of nitrogen by the joint action of amorphous manganese oxide and microorganisms.[1]

Treatment	Residual N ($mg\ 10\ ml^{-1}$)	Loss (%)
No MnO_2 (sterilized)	4.28 a[2]	
2% MnO_2 (sterilized)	3.88 b	9.3
Soil suspension (not sterilized)	3.86 b	9.8
2% MnO_2 + soil suspension (not sterilized)	3.55 c	17.1

[1] Incubation as in Table 8.20 (Li and Wu 1991a).
[2] Values in a column with the same character do not differ significantly at P = 0.01.

chloride ratio, while group (ii) includes the acetylene blockage technique and the highly enriched ^{15}N tracer method (Boast *et al.* 1988; Vanden Heuvel *et al.* 1988; Mulvaney 1988; Mulvaney and Vanden Heuvel 1988). Since the principles of the techniques and conditions for the determination by these methods differ, it is difficult to compare the results obtained.

On upland soils, the acetylene blockage method is commonly used to measure the amount of N lost. To overcome the underestimation of loss by nitrification-denitrification due to the inhibition of nitrification by acetylene, a number of chambers were placed in a test field and determinations were made in turn after each of the chambers was treated with acetylene. However, the acetylene blockage method cannot be used to determine loss by denitrification in flooded rice fields because acetylene will not diffuse throughout the whole soil, and also blocks nitrification, thus removing the substrate for denitrification. The ^{15}N tracer method underestimates the loss by denitrification due to the entrapment of gaseous products by the soil.

Cai *et al.* (1985) estimated denitrification loss from a gleyed paddy soil derived from lacustrine deposit by a difference technique. Total N loss was determined in a microplot with ^{15}N labelled fertilizer and ammonia loss was determined with a mass balance micrometeorological technique. As loss by leaching was negligible and runoff was controlled , the difference between total N loss and ammonia loss was taken as a measure of loss by denitrification. Cai *et al.* (1985) observed losses by denitrification of 39% and 37% of the applied N, respectively, when ammonium bicarbonate and urea were applied to flooded rice. The error associated with this method may be significant.

Zhu *et al.* (1989), using the same technique, estimated denitrification loss when ammonium bicarbonate and urea were applied to a calcareous paddy soil in Henan Province; the apparent loss from each of the treatments was 33% of the applied N. At the two sites the variation in apparent denitrification loss was small (range 33–37%). Under conditions conducive to ammonia volatilization, denitrification loss may be less important. Cai *et al.* (1992), using the same method, estimated N loss from urea applied to an acid paddy soil in Zhejiang Province. The apparent denitrification losses were 40.7% and 41% of the applied N, when urea was broadcast into the floodwater or incorporated as a basal dressing, respectively. Ammonia losses were low, the corresponding values being 10.8% and 7.0% of the applied N, respectively. Thus, under the given conditions, denitrification was a much more important pathway for N loss than ammonia volatilization.

Cai (1986) when summing up the results obtained with flooded rice in pot and field experiments showed that when ammonium sulfate or urea were surface broadcast on acid soils and incorporated as basal dressings, N was lost mainly through nitrification-denitrification. However, when ammonium bicarbonate was added to calcareous soils, it was lost primarily by ammonia volatilization. Both pathways of fertilizer N loss were considerable when ammonium bicarbonate or urea was added to acid soils, or when ammonium sulfate or urea was applied to calcareous soils.

Ryden *et al.* (1979) using the acetylene block technique estimated *in situ* the amount of nitrous oxide escaping from a celery field; the plots receiving

335 kg N ha^{-1} (as ammonium sulfate or urea) lost 51.2 kg N ha^{-1} by denitrification over a period of 123 days, or 15.3% of the applied N. Nitrous oxide accounted for 12–18% of the total N lost by denitrification. Ryden and Lund (1980b) also used the acetylene block technique and trapped the nitrous oxide emitted with a 0.5 nm molecular sieve in a continuous flow system, and directly measured the amount of nitrous oxide evolved from vegetable soils. It was shown that the soil applied with 176–528 kg N ha^{-1}, lost 95–233 kg N ha^{-1} yr^{-1} through denitrification; this amounted to 14–52% of the applied N. Nitrous oxide emission comprised 13% to 20% of the denitrification loss at sites with pH above 7, but was about 30% in more acid soils.

Rolston *et al.* (1978) directly determined the fluxes of $^{15}N_2O$ and $^{15}N_2$ emitted from plots treated separately with organic manure and KNO_3 at the rate of 300 kg N ha^{-1}. The results indicated that the magnitude of denitrification loss depended on soil temperature, moisture content, land use patterns, and the kind and rate of fertilizers. The flux of nitrous oxide in all treatments was dependent on the duration and degree of anaerobiosis. The amount of nitrous oxide emitted from different treatments over 20 days was 5–26% of the total N lost by denitrification.

Nitrous oxide is one of the products of denitrification and is also a by product of ammonium oxidation. Since nitrous oxide can deplete the ozone layer and increase the ultraviolet radiation reaching the biosphere, determination of nitrous oxide emission from fertilized fields is important for environmental protection. The maximum nitrous oxide fluxes between agricultural soil and the atmosphere ranged from 8 to 7500 ng N_2O ha^{-1} day^{-1}. The soil treated with organic manure and nitrate emitted most nitrous oxide and the unfertilized soil used for short grass pasture emitted least. Recently, Eichner (1990) summarized the available data on the nitrous oxide emission from fertilized soils.

Using a closed system without acetylene, we investigated the emission of nitrous oxide from soil in a pot experiment with flooded rice. The results showed that in

Table 8.23. Emission of nitrous oxide (μg N m^{-2} s^{-1}) from flooded rice.[1]

Days after fertilization	Duration of sampling		
	06:00–08:00	06:00–09:00	06:00–10:00
2	3.62 ± 2.04	4.84 ± 1.53	4.50 ± 1.56
3	7.78 ± 1.72	8.75 ± 0.92	5.88 ± 1.80
4	6.33 ± 1.85	6.05 ± 4.74	5.78 ± 2.98
5	6.55 ± 2.34	7.22 ± 3.15	5.77 ± 3.01
6	6.23 ± 1.74	7.43 ± 2.62	6.83 ± 2.25
8	6.43 ± 1.14	5.76 ± 1.79	5.43 ± 2.25
9	4.67 ± 0.723	3.86 ± 1.12	4.68 ± 1.57
10	9.25 ± 1.41	7.57 ± 2.25	9.04 ± 2.47
11	4.67 ± 1.14	3.90 ± 1.59	4.40 ± 1.51
12	3.47 ± 1.12	3.63 ± 1.14	4.57 ± 1.14
13	5.27 ± 4.81	2.93 ± 1.05	3.40 ± 0.767

[1] Pot experiment. Top dressed with 2.5 g NH_4NO_3-N 6 kg pot^{-1} (Li *et al.* 1989a).

the gleyed paddy soil applied with NH_4NO_3, the flux of nitrous oxide evolved from the soil in the seedling stage was high (2.93–9.25 μg N_2O–N m^2 s^{-1}, mean 5.71 ± 1.75 μg N_2O–N m^2 s^{-1}; Table 8.23; Li *et al.* 1989a). The large fluxes of nitrous oxide observed may be due to the high rate of ammonium nitrate applied which inhibited the reduction of nitrous oxide to N_2, particularly during the first days of monitoring (Blackmer and Bremner 1978; Rolston *et al.* 1978).

8.6. References

Aulakh, M S, Rennie, D A and Paul, E A 1982. Gaseous nitrogen losses from cropped and summer-fallowed soils. Can. J. Soil Sci. 62:187–195.

Aulakh, M S, Rennie, D A and Paul, E A 1984. Gaseous nitrogen losses from soils under zero-till as compared with conventional-till management systems. J. Environ. Qual. 13:130–136.

Aulakh, M S and Rennie, D A 1985. Gaseous nitrogen losses from conventional and chemical summer-fallow. Can. J. Soil Sci. 65:195–203.

Aulakh, M S and Rennie, D A 1987. Effect of wheat straw incorporation on denitrification of nitrogen under anaerobic and aerobic conditions. Can. J. Soil Sci. 67:825–834.

Blackmer, A M and Bremner, J M 1978. Inhibitory effect of nitrate on reduction of N_2O to N_2 by soil microorganisms. Soil Biol. Biochem. 10:187–191.

Boast, C W, Mulvaney, R L and Baveye, P 1988. Evaluation of nitrogen-15 tracer techniques for direct measurement of denitrification in soil. I. Theory. Soil Sci. Soc. Am. J. 52:1317–1322.

Cai, G X, Zhu, Z L, Zhu, Z W, Trevitt, A C F, Freney, J R and Simpson, J R 1985. Studies on N loss from ammonium bicarbonate and urea applied to flooded rice field. (in Chinese). Soils 17:225–229.

Cai, G X 1986. Ammonia volatilization of nitrogen fertilizers applied to flooded rice. In: Soil Agricultural Chemistry and Soil Biology and Biochemistry Committees, Soil Science Society of China (eds.), Advances and Prospects for Soil Nitrogen Research in China. (in Chinese). pp. 55–67. Science Press, Beijing.

Cai, G X, Yang, N C, Lu, W F, Chen, W, Xia, B Q, Wang, X Z and Zhu, Z L 1992. Gaseous loss of N from fertilizers applied to a paddy soil in southeastern China. Pedosphere. 2:209–217.

Caskey, W H and Tiedje, J M 1979. Evidence for *Clostridia* as agents of dissimilatory reduction of NO_3^- to NH_4^+ in soils. Soil Sci. Soc. Am. J. 43:931–936.

Caskey, W H and Tiedje, J M 1980. The reduction of nitrate to ammonium by *Clostridium sp.* isolated from soil. J. Gen. Microbiol. 119:217–223.

Chao, T T and Kroontje, W 1966. Inorganic nitrogen transformation through the oxidation and reduction of iron. Soil Sci. Soc. Am. J. 30:193-196.

Chen, H K, Li, F D, Chen, W X and Cao, Y Z 1981. Soil Microbiology. (in Chinese). pp. 203–286. Shanghai Science and Technology Publishing House, Shanghai.

Denmead, O T 1979. Chamber systems for measuring nitrous oxide emission from soils in the field. Soil Sci. Soc. Am. J. 43:89–95.

Dommergues, Y R and Krupa, S V 1978. Interactions Between Non-pathogenic Soil Microorganisms and Plants. Elsevier Scientific Publishing Company, Amsterdam. p. 259.

Eichner, M J 1990. N_2O emissions from fertilized soils: Summary of available data. J. Environ. Qual. 19:272–280.

Erich, M S, Bekerie, A and Duxbury, J M 1984. Activities of denitrifying enzymes in freshly sampled soils. Soil Sci. 138:25–32.

Focht, D D and Joseph, H 1973. An improved method for the enumeration of denitrifying bacteria. Soil Sci. Soc. Am. Proc. 37:698–699.

Focht, D D 1974. The effect of temperature, pH and aeration on the production of nitrous oxide and gaseous nitrogen; a zero order kinetic model. Soil Sci. 118:173–179.

Focht, D D and Verstraete, W 1977. Biochemical ecology of nitrification and denitrification. Ann. Rev. Microbiol. Ecol. 1:135–214.

Freney, J R and Simpson, J R 1983. Gaseous loss of nitrogen from plant-soil systems. pp. 33–64. Martinus Nijhoff/Dr. W Junk Publishers, The Hague.

Gould, W D and McCready, R G L 1982. Denitrification in several Alberta soils: inhibition by sulfur anions. Can. J. Soil Sci. 62:333–342.

Groffman, P M and Tiedje, J M 1989a. Denitrification in north temperate forest soils: spatial and temporal patterns at the landscape and seasonal scales. Soil Biol. Biochem. 21:613–620.

Groffman, P M and Tiedje, J M 1989b. Denitrification in north temperate forest soils: relationships between denitrification and environmental factors at the landscape scale. Soil Biol. Biochem. 21:621–626.

Grundmann, G L, Rolston, D E and Kaschanoski, R G 1988. Field soil properties influencing the variability of denitrification gas fluxes. Soil Sci. Soc. Am. J. 52:1351–1355.

Hasebe, A and Iimura, K 1982. Influence of soil moisture contents before flooding on the differentiation of the oxidized and reduced soil layers. Soil Sci. Plant Nutr. 53:497–502.

Hasebe, A, Koike, I, Ohmori, M and Hattori, A 1987. Variation in the process of nitrification and nitrate reduction in submergic paddy soils as measured by 15-N isotope dilution technique. Soil Sci. Plant Nutr. 33:201–211.

Hauck, R D and Weaver, R W 1986. Field Measurement of Dinitrogen Fixation and Denitrification. SSSA Special Publication No. 18. Soil Sci. Soc. Am. Madison, Wisconsin. 115 p.

Hauck, R D 1986. Field measurement of denitrification-an overview. In: Hauck, R D and Weaver, R W (eds.), Field Measurement of Dinitrogen Fixation and Denitrification. SSSA Special Publication No. 18. Soil Sci. Soc. Am. Madison, Wisconsin. pp. 59–72.

He, Q, Chen, J F and Xu, Z Y 1981. Influence of transformation of iron oxide on soil structure. (in Chinese). Acta Pedol. Sin. 18:326–334.

Jiang, D A, Tang, Y D, Ma, Y H, Zhou, L, Gong, Y Y and Chen, Y B 1989. Effect of different conditions on denitrification in soil. (in Chinese). Environmental Science 10:13–19.

Kaspar, H F, Tiedje, J M and Firestone, R B 1981. Denitrification and dissimilatory nitrate reduction to ammonium in digested sludge. Can. J. Microbiol. 27:878–885.

Katyal, J C, Carter, M F and Vlek, P L G 1988. Nitrification activity in submerged soil and its relation to denitrification loss. Biol. Fertil. Soil. 7:16–22.

Kessel, J F van 1976. Influence of denitrification in aquatic sediments on the nitrogen content of natural waters. Agric. Res. Rep. (Wageningen) No 858.

Klemedtsson, L, Berg, P, Clarholm, M, Schnrer, J and Rosswall, T 1987. Microbial nitrogen transformation in the root environment of barley. Soil Biol. Biochem. 19:551–558.

Knowles, R 1982. Denitrification. Microbiol. Rev. 46:43–70.

Komatsu, Y, Takagi, M and Yamaguchi, M 1978. Participation of iron in denitrification in waterlogged soil. Soil Biol. Biochem. 10:21–26.

Lalisse-Grundmann, G, Brunel, B and Chalamet, A 1988. Denitrification in a cultivated soil: optimal glucose and nitrate concentrations. Soil Biol. Biochem. 20:839–844.

Letey, L, Valoras, N, Hadas, A and Focht, D D 1980. Effect of air-filled porosity, nitrate concentration and time on the ratio of N_2O/N_2 evolution during denitrification. J. Environ. Qual. 9:227–231.

Li, L M 1986. Outline and prospects for nitrification-denitrification studies in China. In: Soil Agricultural Chemistry and Soil Biology and Biochemistry Committees, Soil Science Society of China (eds.), Advances and Prospects for Soil Nitrogen Research in China. (in Chinese). pp. 68–81. Science Press, Beijing.

Li, L M and Wu, Q T 1991. Role of amorphous manganese oxide in nitrogen loss. Pedosphere. 1:83–91.

Li, L M, Pan, Y H Li, Z G and Wu, S C 1992a. Transformation of fertilizer-N in rhizosphere soils of wheat. Pedosphere. 2:373–377.

Li, L M, Wu, Q T, Li, Z G and Pan, Y H 1989a. A method of measuring nitrous oxide emission from soils in situ. (in Chinese). Acta Pedol. Sin. 26:305–308.

Li, L M, Wu, Q T, Li, Z G and Pan, Y H 1991b. Fluxes of nitrous oxide from different soils in situ. (in Chinese). Soils 23:24–27.

Li, L M, Zang, S, Zhou, X R and Pan, Y H 1983. Studies on nitrification-denitrification in soil. In: Editorial Board of Environmental Science Information Network, Chinese Academy of Sciences (ed.), Proceedings of Nitrogen Pollution in Environment and Nitrogen Cycling. (in Chinese). pp. 160–168. Institute of Environmental Chemistry, Chinese Academy of Sciences.

Li, L M, Zang, S, Zhou, X R and Pan, Y H 1984. Effect of rice roots on N loss. (in Chinese). Soils 16:5–10.

Li, L M, Pan, Y H, Wu, Q T, Zhou, X R and Li, Z G 1988. Investigation on amorphous ferric oxide acting as an electron acceptor in the oxidation of ammonium under anaerobic conditions. (in Chinese). Acta Pedol. Sin. 25:184–190.

Li, Z G, Li, L M, Pan, Y H and Wu, Q T 1992b. Denitrifiers in soils of red soil regions. Red Soil Ecosystem Experiment Station, Chinese Academy of Sciences (ed.), Studies on Ecosystem in Red Soils. pp. 196–203. Science Press, Beijing.

Li, Z G, Pan, Y H, Wu, Q T and Li, L M 1989b. Numbers, compositions and enzyme activities of denitrifiers of paddy soils in the Taihu-Lake district. (in Chinese). Acta Pedol. Sin. 26:79–86.

Li, Z G, Wan, H M, Wu, L S and Qiao, F Z 1987. Studies on the ecological distribution of denitrifying bacteria in the rhizosphere of rice. (in Chinese). Acta Pedol. Sin. 24:120–125.

Liao, X L, Xu, Y H and Zhu, Z L 1982. Investigation on nitrification-denitrification loss of fertilizer nitrogen in submerged paddy soil. (in Chinese). Acta Pedol. Sin. 19:257–263.

Liu, Z G and Yu, T R 1963. Studies on electrochemical properties of soils. II. Application of micro-electrodes in the study of soils. (in Chinese). Acta Pedol. Sin. 11:160–169.

McCready, R G L, Gould, W D and Brendregt, R W 1983. Nitrogen isotope fractionation during the reduction of NO_3^- to NH_4^+ by *Desulfovibrio sp.* Can. J. Microbiol. 20:231–234.

Mitsui, S 1959. Inorganic Nutrition Fertilization and Soil Amelioration for Lowland Rice. Shanghai Science and Technology Publishing House, Shanghai. 98 p.

Mosier, A R and Hutchinson, G L 1981. Nitrous oxide emission from cropped fields. J. Environ. Qual. 10:169–173.

Mulvaney, R L 1988. Evaluation of nitrogen-15 tracer techniques for direct measurement of denitrification in soil: III. Laboratory studies. Soil Sci. Soc. Am. J. 52:1327–1332.

Mulvaney, R L and Vanden Heuvel, R M 1988. Evaluation of nitrogen-15 tracer techniques for direct measurement of denitrification in soil: IV. Field studies. Soil Sci. Soc. Am. J. 52:1332–1337.

Murakami, T, Owa, N and Kumazawa, K 1987. The effects of soil conditions and nitrogen form on N_2O evolution by denitrification. Soil Sci. Plant Nutr. 33:35–42.

Myrold, D D and Tiedje, J M 1985. Establishment of denitrification capacity in soil: Effects of carbon, nitrate and moisture. Soil Biol. Biochem. 17:819–822.

Okereke, G U 1984. Possible use of N_2O in MPN tubes for enumeration of denitrifiers. Plant Soil. 80:295–296.

Pan, Y H, Li, L M, Wu, Q T and Li, Z G 1988. Study on activities of nitrification and denitrification in red soils under different utilization patterns. (in Chinese). Soils 20:184–187.

Payne, W J 1973. Reduction of nitrogenous oxide by microorganisms. Bacteriol. Rev. 37:409–452.

Payne,W J 1981. Denitrification. pp. 33–53. John Wiley & Sons, Chichester.

Picard, M A and Faup, G M 1980. Removal of nitrogen from industrial wastewater by biological nitrification-denitrification. Water Pollut. Control. 79:213–224.

Prade, K and Trolldenier, G 1988. Effect of wheat roots on denitrification at varying soil air-filled porosity and organic-carbon content. Biol. Fertil. Soil. 7:1–6.

Qin, S W and Liu, Z Y 1984. The nutrient status of soil-root interface. III. Variation of fertilizer nitrogen in rice rhizosphere. (in Chinese). Acta Pedol. Sin. 21:238–246.

Raimbaut, M, Rinaudo, G, Garcia, J L and Boureau, M 1977. A device to study metabolic gases in rice rhizosphere. Soil Biol. Biochem. 2:193–196.

Reddy, K R and Patrick, W H Jr. 1986. Fate of fertilizer nitrogen in the rice root zone. Soil Sci. Soc. Am. J. 50:649–651.

Rice, C W and Smith, M S 1982. Denitrification in no-till and plowed soils. Soil Sci. Soc. Am. J. 46:1168–1173.

Roger, Y S, Adelberg, E A and Ingraham, J L 1976. The Microbial World. pp. 723–724. Prentice-Hall, Inc. Englewood Cliffs, New Jersey.

Rolston, D E, Hoffman, D L and Toy, D W 1978. Field measurement of denitrification. I. Flux of N_2 and N_2O. Soil Sci. Soc. Am. J. 42:863–869.

Ryden, J C and Lund, L J 1980a. Nitrous oxide evolution from irrigated land. J. Environ. Qual. 9:387–393.

Ryden, J C and Lund, L J 1980b. Nature and extent of directly measured denitrification losses from some irrigated vegetable crop production units. Soil Sci. Soc. Am. J. 44:505–511.

Ryden, J C, Lund, L J, Letey, J and Focht, D D 1979. Direct measurement of denitrification loss from soils: II. Development and application of field methods. Soil Sci. Soc. Am. J. 43:110–118.

Sahrawat, K L 1980. Is nitrate reduced to ammonium in waterlogged acid sulfate soil? Plant Soil. 57:147–149.

Savant, N K and McClellan, G H 1987. Do iron oxide systems influence soils under wetland rice based cropping systems? Commun. Soil Sci. Plant Anal. 18:83–113.

Smith, M S and Tiedje, J M 1979. The effect of roots on soil denitrification. Soil Sci. Soc. Am. J. 43:951–955.

Smith, C J and Delaune, R D 1984. Effect of rice plants on nitrification-denitrification loss of nitrogen under greenhouse conditions. Plant Soil. 79:287–290.

Valera, C L and Alexander, M 1961. Nutrition and physiology of denitrifying bacteria. Plant Soil. 15:268–280.

Vanden Heuvel, R A, Mulvaney, R L and Hoeft, R G 1988. Evaluation of N-15 tracer techniques for direct measurement of denitrification in soil. II. Simulation studies. Soil Sci. Soc. Am. J. 52:1322–1326.
Wang, M Q, Luo, Q X and Lin, J M 1964. Studies on denitrifying bacteria in flooded soils. I. Isolation of denitrifying bacteria in rice rhizosphere and their bacteriological properties. (In Chinese). In: Wuhan Microbiology Institute and Huazhong Agricultural College (eds.) pp. 39–43. Abstracts of papers presented at the Symposium on Soil Microbiology
Wijler, J and Delwiche, C C 1954. Investigation on the denitrifying process in soil. Plant Soil. 5:155–169.
Wollersheim, M, Trolldenier, G and Beringer, H 1987. Effect of bulk density and soil water tension on denitrification in the rhizosphere of spring wheat. Biol. Fertil. Soil. 5:181–187.
Wu, Q T, Li, L M, Li, Z G and Pan, Y H 1988. Denitrifying potential of main paddy soils in the Taihu-Lake region. (in Chinese). Soils 20:325–327.
Yamane, I and Okazaki, M 1982. Chemical properties of submerged rice soils. Trans. 12th Inter. Congr. Soil Sci., New Delhi (India). pp. 143–157.
Yang, G Z and Shao, Z C 1981. Soil conditions in relation to denitrification of paddy soil in southern Jiangsu Province. (in Chinese). Chinese J. Soil Sci. (3):4–7.
Yeomans, J C and Bremner, J M 1985a. Denitrification in soil: Effects of herbicides. Soil Biol. Biochem. 17:447–452.
Yeomans, J C and Bremner, J M 1985b. Denitrification in soil: Effects of insecticides and fungicides. Soil Biol.Biochem. 17:453–456.
Yeomans, J C and Bremner, J M 1987. Effects of dalapon, atrazine and simazine on denitrification in soil. Soil Biol. Biochem. 19:31–34.
Yoshida, T and Padre, B C Jr 1974. Nitrification and denitrification in submerged Maahas clay soils. Soil Sci. Pant Nutr. 20:241–247.
Youssef, R A and Chino, M 1988. Development of a new rhizobox. Soil Sci. Plant Nutr. 34:461–465.
Yu, T R and Li, S H 1957. Study on the redox processes in paddy soils. III. Interactions between soil and plant. (in Chinese). Acta Pedol.Sin. 5:166–174.
Zhao, C Z, Zhou. Z D and Dong, B S 1981. Studies on the system of soil tillage of paddy soil in southern Jiangsu Province. (in Chinese). Acta Pedol. Sin. 18:223–233.
Zhu, Z L 1985. Advances in investigations of soil nitrogen supply and fate of fertilizer nitrogen in soils of China. (in Chinese). Soils 17:2–9.
Zhu, Z L, Cai, G X, Simpson, J R, Zhang, S L, Chen, D L, Jackson, A V and Freney, J R 1989. Processes of nitrogen loss from fertilizers applied to flooded rice fields on a calcareous soil in north-central China. Fert. Res. 18:101–115.

9
Ammonia volatilization

CAI GUI-XIN

9.1. Introduction

Ammonia volatilization can now be measured directly without disturbing the environment since the micrometeorological method was developed (Denmead 1983; Denmead *et al.* 1977). Investigations with this method on ammonia volatilization from fertilizers applied to fields have been studied and reviewed by a number of scientists (Cai 1986; Cai 1992; Fenn and Hossner 1985; Fillery and Vlek 1986; Freney *et al.* 1981b; Freney *et al.* 1983; Mikkelsen and De Datta 1979; Nelson 1982; Terman 1979; Ventura and Yoshida 1977). The emission of ammonia following fertilizer application to agricultural fields varies greatly with weather conditions, soil properties and management practices. For instance, ammonia loss from a cotton crop after injection of anhydrous ammonia was negligible (Denmead *et al.* 1977), but it was as high as 47% of the applied N when the fertilizer was applied to the floodwater after transplanting rice, when the weather was sunny and the winds were strong (Fillery *et al.* 1984).

In the present chapter some aspects of ammonia volatilization will be discussed, including the loss process, methods for its determination, the extent of loss after application of fertilizer N, and management practices for reducing loss.

9.2. Loss process

Ammonia volatilization consists of a series of physical and chemical reactions, and the factors controlling these reactions will influence, to different extents, the amount of ammonia loss. The various reactions which govern ammonia volatilization following the application of fertilizer N to a field may be represented as

Nitrogen fertilizer

$$NH_{4\ \text{exchangeable}}^{+} \leftrightarrow NH_{4\ \text{in solution}}^{+} \leftrightarrow NH_{3\ \text{in solution}} \leftrightarrow NH_{3\ \text{gas in soil}} \leftrightarrow NH_{3\ \text{gas in atmosphere}} \quad (1)$$

Factors which promote these reactions towards the right will increase ammonia volatilization. Freney *et al.* (1983) described the above equilibria in detail, and the

Zhu Zhao-liang et al. *(eds.): Nitrogen in Soils of China, 193–213.*

following equation relating ammonia and ammoniacal N in solution at different pH values and temperatures was derived.

$$[NH_3]_{solution} = \frac{[NH_3 + NH_4^+]_{solution}}{1 + 10^{(0.09018 + 2729.92/T - pH)}} \quad (2)$$

T is the absolute temperature (K). The proportion of ammonia-N to ammoniacal N in solution at a given pH and temperature can be calculated from Equation (2) and some values are given in Table 9.1. The proportion of ammonia-N to ammoniacal N increases with pH and temperature. When pH is in the range of 6 to 8, an increase of one unit of pH increases the ratio of ammonia-N to ammoniacal N by 10 times. A further increase in pH from 8 to 9 increases the ratio 5–10 times. Increasing the temperature by 10°C in the range 5–35°C at most doubles the ratio. However, temperature has an additional effect on the partitioning of NH_3 between the liquid and gaseous phases as shown in equation (3):

$$p[NH_3] = \frac{0.00488[NH_3]_{solution}}{10^{(1477.8/T - 1.6937)}} \quad (3)$$

Freney *et al.* (1981) and Denmead *et al.* (1982) used the following equation to relate the rate of ammonia volatilization (F) to the equilibrium ammonia vapour pressure (p_o) and the background ammonia concentration (p_z) at a reference height z.

$$F = k(p_o - p_z) \quad (4)$$

k is an exchange coefficient which is a function of wind speed (u) at the reference height, p_z is the partial pressure of NH_3 at that height, and p_o is the partial pressure of NH_3 at the surface of the water or soil. The greater the difference between the partial pressures, the greater the driving force for ammonia emission. The faster the wind speed, the more rapid is the transport of ammonia away from the surface.

9.3. Factors controlling ammonia volatilization

From the above discussion it is clear that main factors governing ammonia volatilization are ammoniacal N concentration, pH and temperature in soil (upland

Table 9.1. Effect of pH and temperature on the proportion of ammoniacal N present as ammonia (%).[1]

Temperature (°C)	pH				
	6	7	8	9	10
5	0.01	0.12	1.22	11.0	55.2
15	0.03	0.27	2.62	21.2	72.9
25	0.06	0.56	5.32	36.0	84.9
35	0.11	1.11	10.1	52.9	91.8

[1] Calculated from Equation 2.

field) or floodwater (flooded field), and wind speed. Ammoniacal N is the source of ammonia for volatilization, while the other three factors control the direction and/or speed of the reactions. Soil properties, environmental conditions and management practices have large effects on ammonia volatilization by influencing the four main factors.

9.3.1. Cation exchange capacity

Ammonium in solution reacts readily with the cation exchange complex in soil. This reduces the amount of ammonium in solution and therefore the NH_3 concentration in solution is also reduced (Equation 1). Because cation exchange capacity (CEC) is a function of the amount and type of clay minerals and organic matter in soil, these constituents will affect the equilibrium between ammonium in the liquid phase and ammonium on the solid phase, and thereby ammonia volatilization. As demonstrated by incubation experiments ammonia volatilization is negatively correlated with CEC and the amount of clay and/or organic matter in the soil (Qu 1980; Yao and Guan 1983; Yuan and Zhang 1979; Yu and Zhao 1979). Results from a gas-lysimeter experiment with rice showed that ammonia loss amounted to 35% when urea was applied to a sandy loam paddy soil with low CEC, but it was only 10% when urea was applied to a clay soil with high CEC (Vlek and Craswell, 1979).

9.3.2. Soil pH and calcium carbonate

As shown in Table 9.1, when the pH is less than 7, <1% of the ammoniacal N is present in the form of ammonia, thus the potential for ammonia volatilization is low. As the pH increases, the potential for ammonia loss increases sharply. For upland fields, ammonia volatilization increased significantly with an increase in soil pH (Jiang and Liu 1979; Ryan *et al.* 1981). However, this is not always true for flooded paddy fields, because the floodwater pH rather than soil pH directly controls ammonia volatilization. The floodwater pH is in turn influenced by soil pH, irrigation water, applied fertilizers, and especially algal growth.

As ammonia loss proceeds the medium becomes acidified. In soils with high buffering capacity the acidity produced will be neutralized by carbonate or other alkaline materials and the volatilization of ammonia will proceed.

9.3.3. Temperature

Increasing the temperature increases the ratio of NH_3 to ammonium present at a given pH, as well as increasing the ratio of NH_3 in the gaseous phase to NH_3 in the liquid phase. Furthermore, high temperature increases the rate of urea hydrolysis (Mulvaney and Bremner 1981) and the diffusion of ammonium and ammonia in the soil. Therefore, the higher the temperature, the greater the potential for ammonia loss (Liu *et al.* 1980; Qu 1980; Lu *et al.* 1980; Zhao *et al.* 1986; Yuan and Zhang 1979; Vlek and Stumpe 1978). Zhao *et al.* (1986) found that ammonia loss in summer was greater than that in winter when wind speeds were similar; the increased loss was apparently due to the higher temperatures in summer.

9.3.4. *Wind speed*

Increasing wind speed increases the rate of ammonia volatilization by promoting the rapid transport of ammonia away from the surface. This was demonstrated in a maize field by Denmead *et al.* (1982) and in rice paddies by Freney *et al.* (1981a), Fillery *et al.* (1984) and Leuning *et al.* (1984). Leuning *et al.* (1984) observed that the transfer velocity for ammonia increased linearly with wind speeds up to 5 m s^{-1}, but Denmead *et al.* (1982), working in higher wind speeds (up to 8 m s^{-1}), found an exponential relationship between the transfer velocity and wind speed. Denmead *et al.* (1982) suggested that the enhanced volatilization at higher wind speeds was due to better mechanical mixing of the N solution in the furrows, which replenished ammonia in the layer next to the water surface. Bouwmeester and Vlek (1981) studied ammonia volatilization in artificial systems in a wind tunnel. They suggested that at high pH values, the rate of ammonia volatilization reached a maximum at a certain wind speed, and that increasing wind speed beyond that value did not result in increased volatilization because of ammonia depletion in the surface layer. Thus the relationship between wind speeds and ammonia volatilization is complex.

Fillery *et al.* (1984) found that ammonia loss between two sites in the Philippines varied greatly (27% and 47% of the applied N) even though the ammoniacal N concentrations, pH values and temperatures of the floodwaters were similar. They suggested that the difference in ammonia loss was mainly due to the difference in wind speeds between the two sites.

In a sealed enclosure experiment, ammonia loss increased with the increase in frequency of air exchange in the headspace of the enclosure; the greatest loss was observed when the exchange frequency was 15–20 times per minute (Zhu *et al.* 1985; Vlek and Craswell 1979).

9.3.5. *Soil moisture*

The presence of water in soil is a prerequisite for dissolution of the fertilizer N as well as for the hydrolysis of urea. Therefore the soil moisture content is an important factor controlling ammonia loss. Results from an incubation experiment showed that ammonia loss did not occur when soil moisture was too low for urea hydrolysis (Ferguson and Kissel 1986). Results from a field experiment with sunflowers showed that the rate of ammonia loss was limited due to low soil moisture even though the ammoniacal N concentration and pH were high (Smith *et al.* 1989). On the other hand, if the soil moisture content is high, the concentration of ammonia in solution will be low due to dilution and ammonia volatilization should be reduced. Zhao *et al.* (1986), Fenn and Kissel (1976), and Ferguson and Kissel (1986) also found that ammonia loss was low when soil moisture was at the extremes. It seems that moderate amounts of soil moisture are more favourable for ammonia volatilization.

9.4. Methods for measuring ammonia loss

A variety of techniques is available for measuring ammonia volatilization, including micrometeorological methods, enclosure methods, ^{15}N tracer methods (Nommik 1973), Hargrove method (Hargrove *et al.* 1977) and acid-saturated filter papers (Mahendrappa and Ogden, 1973). In the ^{15}N tracer method the deficit in ^{15}N balance is taken as ammonia loss. The results obtained by this method are meaningful only when ammonia volatilization is the sole pathway of N loss. The results obtained by the Hargrove method are difficult to interpret because of ammonium fixation or loss by other processes. Zhao and Zhang (1981) used this method to estimate ammonia loss from applications of ammonium sulfate to calcareous soils in north China and found that the losses were between 4% and 22% of the applied N. Mahendrappa and Ogden (1973) used acid-saturated filter papers to measure ammonia loss in a forest. The method is simple, but only relative differences between the treatments can be obtained.

9.4.1. Enclosure method

In this method an enclosure is placed over the fertilized soil, with or without plants, and the ammonia emitted is absorbed by acid traps and then determined (Datta *et al.* 1971; Kissel *et al.* 1977; Ventura and Yoshida 1977; Liao *et al.* 1982; Zhu *et al.* 1985). This method has commonly been employed because it is simple to use. A great variety of enclosures and trapping techniques has been used with this method. The enclosure may be sealed at all times or at intervals during the period of measurement. In some cases the ammonia emitted from the soil is transferred from the enclosure to acid traps outside by an air stream. In others ammonia is not swept out by an air stream and the acid trap is placed inside the enclosure. Enclosures with an air stream generally result in greater ammonia loss values than those without an air stream and the value obtained increases with the increase in frequency of air exchange. It is difficult to use this method for large plants, and it usually used for small plants only. A special enclosure sealed with water was designed by Datta *et al.* (1971), and used to measure ammonia loss from aquatic plants such as flooded rice. After improvement of the method (Liao *et al.* 1982), it was successfully used to measure ammonia loss after application of N fertilizers to flooded rice (Zhu *et al.* 1985).

Enclosure methods are commonly used for studying the factors controlling ammonia volatilization. It is not suitable for estimating ammonia loss from fertilizers applied in the field because conditions, such as moisture, temperature and air flow rate inside the enclosure are greatly different from those outside. Despite these objections, Black *et al.* (1985) found that the total loss determined by this method was essentially the same as that determined by a micrometeorological method although the patterns of ammonia loss obtained by the two methods were quite different.

9.4.2. *Micrometeorological methods*

Micrometeorological methods are recommended for the estimation of ammonia loss in the field, because they do not disturb the environmental conditions and they integrate the loss over a large area. Two techniques, gradient diffusion and mass balance methods (Denmead 1983) have been used for assessing ammonia volatilization. A large uniform area with a fetch of 150 to 200 m is required for the gradient diffusion method, and this is difficult to get in many crop fields. On the other hand, the mass balance method is most successful when the experimental area is smaller. Circular areas with 25 m radius have been used regularly for this method (e.g. Freney *et al.* 1985a; Fillery *et al.* 1984; Simpson *et al.* 1985; Cai *et al.* 1986; Fillery and De Datta 1986; Zhu *et al.* 1989).

The mass balance method requires that the circular area for investigation be surrounded by a crop at the same stage of growth as that within the circle, and that the background area has not been treated recently with fertilizer N. Following fertilizer application to the circular area, ammonia diffuses upwards and is transported horizontally by the wind. Integration of the horizontal transport of ammonia (product of wind speed and ammonia concentration) through the depth of a vertical plane at the center of the circle is equated with the vertical flux from an upwind area of unit width. To employ this method for evaluating ammonia emission rates in the field, measurements of wind speed and ammonia concentration are required at a number of heights extending through the cloud of ammonia. Since the method requires a supply of power and equipment to measure wind speed, flow rates, ammonia concentration etc., and is laborious, two simplified methods were developed.

The abbreviated method requires measurements of wind speed and ammonia concentration at one height only and thus needs fewer pumps, flow meters, traps, anemometers and labor than the full profile method. As discussed by Wilson *et al.* (1982, 1983) and Denmead (1983), for a given fetch and surface roughness there is a particular height (ZINST) above a circular field emitting ammonia where the horizontal flux density of ammonia is directly proportional to the vertical flux density of ammonia, irrespective of the turbulent state of the atmosphere.

Previous work over flooded rice, with a circular area of radius 25 m, in Australia (Freney *et al.* 1985) and in China (Cai 1988), showed that ZINST was 0.8 m above the water surface, while a value of 0.75 m was obtained by Fillery and De Datta (1986) in the Philippines. Good estimates of ammonia loss from flooded rice fields were obtained using this method. The correlation coefficients for the relationships between flux densities measured by the abbreviated method and full profile methods were high (0.96–0.99), and the total ammonia losses measured by the two methods were essentially the same, the difference between them was only about 5% (Fillery and De Datta 1986; Cai 1988).

There are some limitations to the use of the abbreviated method. The constant relating horizontal and vertical fluxes needs to be determined for a particular location and conditions and the method is not recommended when the plants are tall or the circular area for study is small.

An ammonia sampler was designed to determine the horizontal flux density of ammonia (Leuning *et al.* 1985). It can be used to measure ammonia loss with the full profile method or the abbreviated method. The sampler operates without electric power and there is no need for pumps or separate measurements of wind speed and air flow rate, but a circular fertilized area is still required. The loss measured by this method was not quite as accurate as that measured by the full profile method (Freney *et al.* 1987). Errors may arise if the sampler does not track the wind direction accurately at low wind speeds (Leuning *et al.* 1985), or if the background transport of ammonia is not determined accurately.

9.5. Ammonia volatilization after fertilizer application

Ammonia loss following fertilizer application and the contribution of ammonia loss to total N loss depends on soil properties, environmental conditions and management practices. Fertilizer type, timing and method of application influence the loss greatly. In the following discussion ammonia loss was measured by the micrometeorological mass balance method, and loss by denitrification was calculated as the difference between total loss and ammonia loss.

9.5.1. Flooded rice

Ammonia volatilization from flooded rice fields following fertilizer application has been measured in many countries. The results show that ammonia loss varies greatly with experimental conditions and it is an important pathway of N loss in rice fields in many cases (Fillery *et al.* 1986b; Simpson and Freney 1988; Cai 1992).

In China, four experiments have been conducted in the warm temperate, north-subtropic and mid-subtropic zones (Table 9.2). In all of the experiments, N fertilizer was broadcast into the floodwater and incorporated just before transplanting. The results show that ammonia loss following the application of urea varied greatly, with the loss from the acid soil at Yingtan, in the mid-subtropic zone being the highest (40% of the applied N; Cai *et al.* 1992a). The loss was greater than that from urea applied to a calcareous soil at Fengqiu in the warm temperate zone (30% of the applied N; Zhu *et al.* 1989), and was much larger than the losses obtained when urea was applied to soils of similar pH at Danyang and Fuyang in the north-subtropic zone (9% and 11%, respectively; Table 9.2; Cai *et al.* 1986, 1992b).

Freney *et al.* (1985a) showed that the rate of ammonia loss from flooded rice fields is dependent on the partial pressure of ammonia (pNH_3) in the floodwater and wind speed. Figure 9.1 shows that the maximum pNH_3 value at Yingtan was twice as high as that observed at Fengqiu, while the pNH_3 values at Danyang and Fuyang

Table 9.2. Nitrogen loss (% of applied N) from urea applied to flooded rice[1] at four sites in China.

Location[2]	Weather[3] conditions	Soil pH	Total N loss	NH_3 loss	N_2 loss	Reference
Fengqiu	Sunny	8.8	63	30	33	Zhu *et al.* (1989)
Danyang	Cloudy	5.3	45	9	36	Cai *et al.* (1986)
Fuyang	Cloudy, rainy	6.0	52	11	41	Cai *et al.* (1992b)
Yingtan	Sunny	5.5	56	40	16	Cai *et al.* (1992a)

[1] Nitrogen broadcast into floodwater and incorporated just before transplanting.
[2] Fengqiu located in warm temperate, Danyang and Fuyang in north-subtropic, and Yingtan in mid-subtropic zone.
[3] During the experiments.

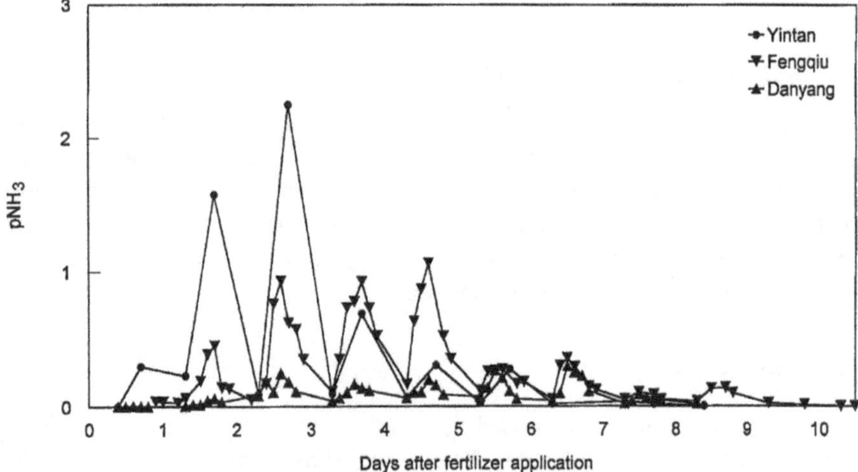

Figure 9.1. Partial pressure of ammonia in the floodwater of flooded rice fields after application of urea (Cai *et al.* 1986; Zhu *et al.* 1989; Cai *et al.* 1992a).

were always low. The wind speeds at 0.8 m height often ranged from 1 to 3 m s^{-1}, and on average were highest at Fengqiu.

As indicated above, the partial pressure of ammonia in the floodwater is a function of ammoniacal N concentration, pH and temperature in the floodwater. The high pNH_3 values at Yingtan resulted primarily from the high ammoniacal N concentrations (maximum 71 g m^{-3}) and the high floodwater temperatures (Figure 9.2). The experiment was conducted in mid-summer with high solar radiation and shallow floodwater (~0.02 m), thus the floodwater temperatures were high (maximum 40°C, minimum 25°C). The high temperature not only increased the rate of urea hydrolysis and subsequent build-up of ammoniacal N in the floodwater, but also promoted the dissociation of ammonium into ammonia, the partitioning of ammonia in the gaseous phase, and the convection in the floodwater. All of these may have contributed to the high rate of ammonia volatilization at Yingtan.

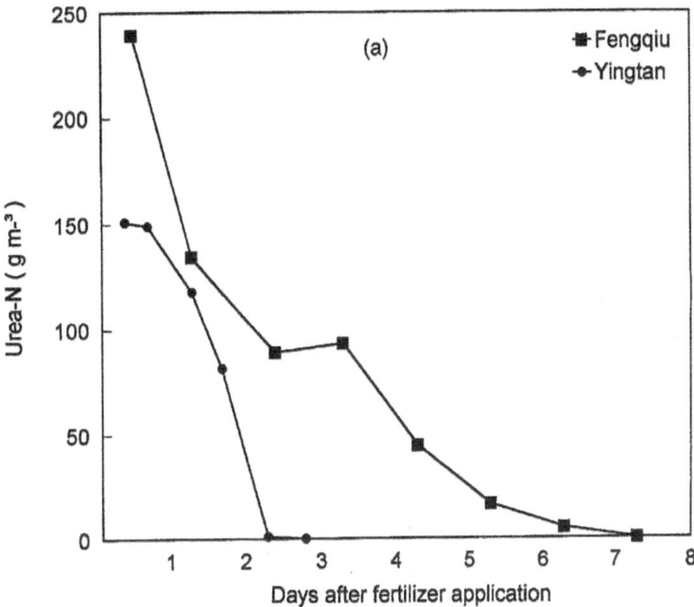

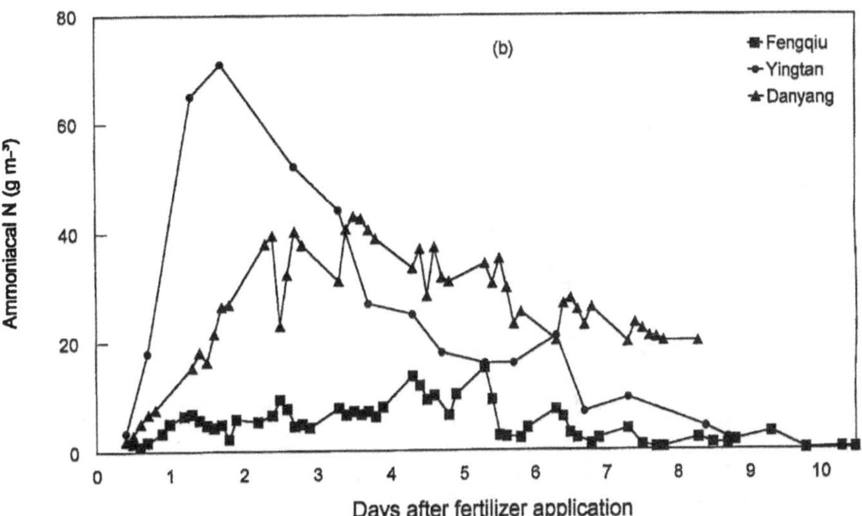

Figure 9.2. Urea (a) and ammoniacal (b) N concentrations, pH (c) and temperature (d) of the floodwater in rice fields after application of urea (Cai *et al.* 1986; Zhu *et al.* 1989; Cai *et al.* 1992a).

In contrast, ammonia loss from applied urea at Danyang was only 9% of the applied N. The main factor limiting ammonia loss in this experiment was the low pH of the floodwater (<7.8; Figure 9.2). This was due primarily to the cloudy days and the consequent poor growth of algae after N application. The low rate of

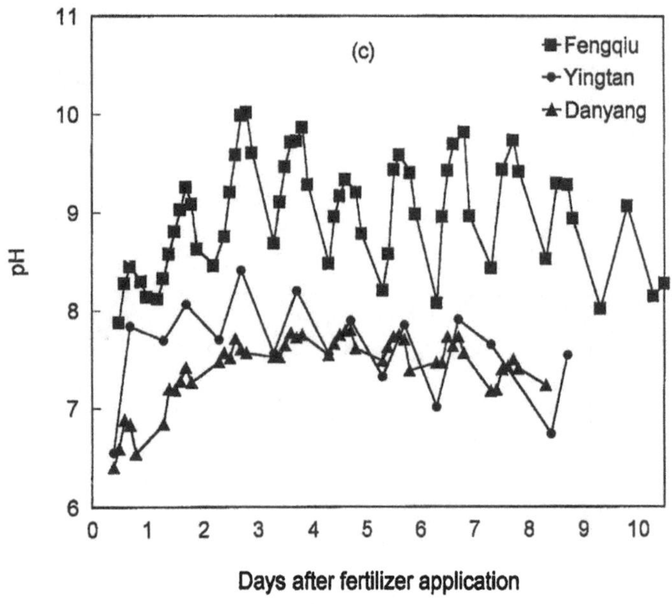

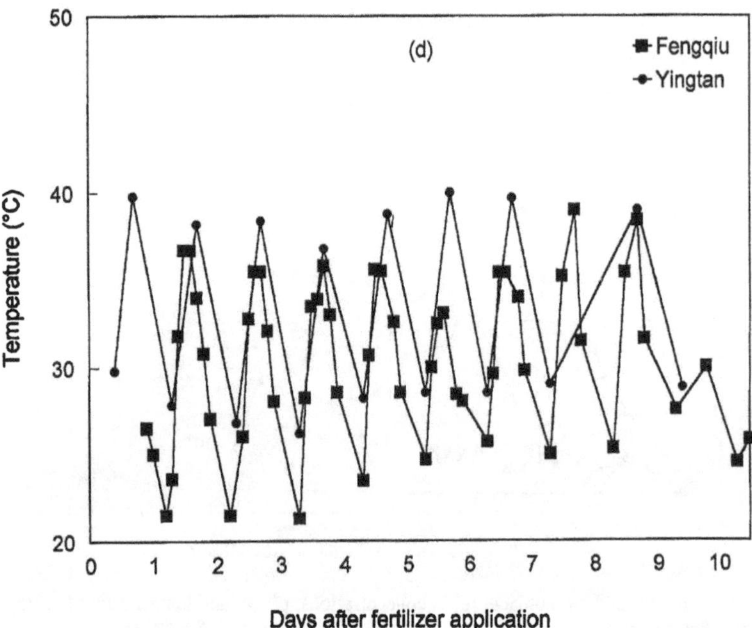

Figure 9.2. (continued)

ammonia emission resulted in comparatively high and steady concentrations of ammoniacal N in the floodwater(20 to 40 g m^{-3}; Figure 9.2). The results obtained at Fuyang (Cai *et al.* 1992a) were similar to those at Danyang.

Differences in ammonia loss between the experimental sites were also found when ammonium bicarbonate was applied (Cai *et al.* 1986; Zhu *et al.* 1989); ammonia loss at Fengqiu was 39% of the applied N, while at Danyang it was 18%. As in the case of urea, the limiting factor for ammonia volatilization at Danyang following the application of ammonium bicarbonate was the low pH of the floodwater.

The preceding discussion suggests that when ammonium based fertilizer or urea is broadcast into floodwater the extent of ammonia loss is largely dependent on weather conditions, in particular solar radiation. Ammonia loss may be high when the days are bright and sunny as was the case for the experiments conducted at Yingtan and Fengqiu, or it may be low when it is raining or cloudy as was the case at Danyang and Fuyang. Thus ammonia loss following application of fertilizer N to late rice might be much greater than that from early rice, at the same site, because it is usually sunny during the late rice crop, and cloudy or rainy days often occur during the early growth stage of the early rice crop.

The extent and pattern of ammonia volatilization following the application of N are influenced by the properties of the fertilizer, such as the form of N and acidity. Results show that ammonia losses from ammonium bicarbonate applied to flooded rice at Fengqiu and Danyang were 39% and 18% of the applied N, respectively, while the corresponding figures for urea were 30% and 9%, respectively (Figure 9.3; Cai *et al.* 1986; Zhu *et al.* 1989). Figure 9.3 also shows that there was little difference between the fertilizers as far as denitrification was concerned, and thus

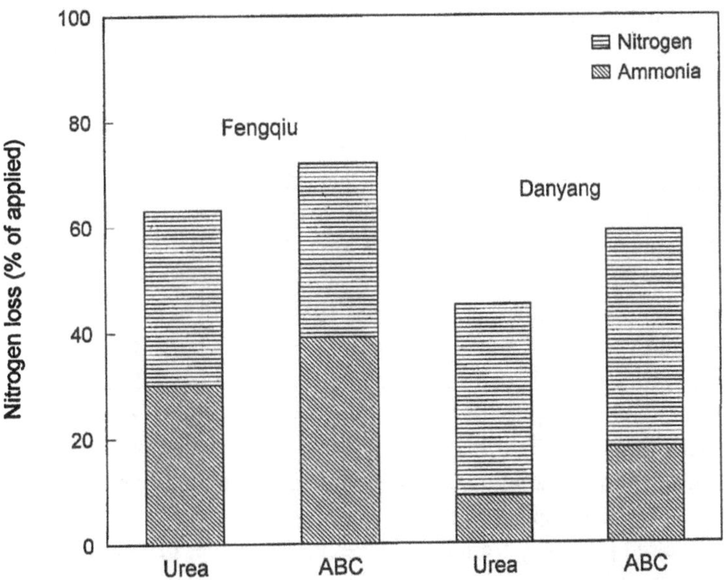

Figure 9.3. Gaseous loss of nitrogen from urea and ammonium bicarbonate (ABC) applied to flooded rice (Cai *et al.* 1986; Zhu *et al.* 1989).

the difference in total N losses between the fertilizers was due mainly to differences in ammonia volatilization.

The pattern of ammonia volatilization from ammonium bicarbonate was markedly different from that obtained with urea (Figure 9.4). Volatilization of ammonia commenced immediately after application of ammonium bicarbonate and proceeded at very high rates just after application. Emission then dropped to very low rates one or two days after fertilizer application. In contrast, ammonia emission from the urea treated plot commenced slowly and never reached the high rates observed in the ammonium bicarbonate treatment, but the emission lasted much longer. The difference in ammonia volatilization resulted from the differences in ammoniacal N concentrations and pH values of the floodwater (Figure 9.5). Fertilization with ammonium bicarbonate provides an immediate source of ammonium ions and alkalinity. However, urea provides alkalinity only after hydrolysis of urea.

Fillery and De Datta (1986) compared the loss of ammonia following application of urea and ammonium sulfate into the floodwater after transplanting rice seedlings. The results showed that under the experimental conditions in the Philippines, i.e. high floodwater pH, losses from urea and ammonium sulfate were comparable, being 36% and 38% of the N applied, respectively. However, the pattern of ammonia loss differed. The pattern of ammonia loss from urea was as described above. With ammonium sulfate volatilization commenced immediately after application and reached its maximum rate soon after; loss rates then declined sharply. The pattern was similar to that observed by Freney *et al.* (1981a), and also to that obtained with ammonium bicarbonate at Fengqiu and Danyang (Zhu *et al.* 1989; Cai *et al.* 1986).

It seems that losses of ammonia following application of urea and ammonium based fertilizers are controlled by floodwater pH and the rate of urea hydrolysis; if both are high there is little difference between the carriers as far as ammonia loss is concerned.

Ammonia volatilization may be modified by the wind speeds which occur after the application of fertilizer N. From the patterns of ammonia volatilization described above it would be expected that loss from ammonium based fertilizers would be enhanced by high wind speeds immediately after application, while loss from urea would be increased by high wind speeds 2 or 3 days after fertilization (Simpson *et al.* 1985; Fillery and De Datta 1986).

The effect of timing of application on total N loss is large. In general, total loss from fertilizer N applied at an early growth stage is greater than that which occurs when the fertilizer is applied at panicle initiation (Cai *et al.* 1986; Nurayama 1979; Vlek and Stumpe 1978). The results of an experiment conducted at Griffith, Australia, showed that ammonia loss from urea applied at tillering was 21% compared with 3% when applied at panicle initiation (Humphreys *et al.* 1988). The difference resulted primarily from differences in floodwater pH and wind speed. The large crop canopy at panicle initiation reduced light intensity at the water surface so that algal photosynthesis (Mikkelsen *et al.* 1978), and consequently pH elevation was restricted. The large crop canopy was also clearly responsible for the attenua-

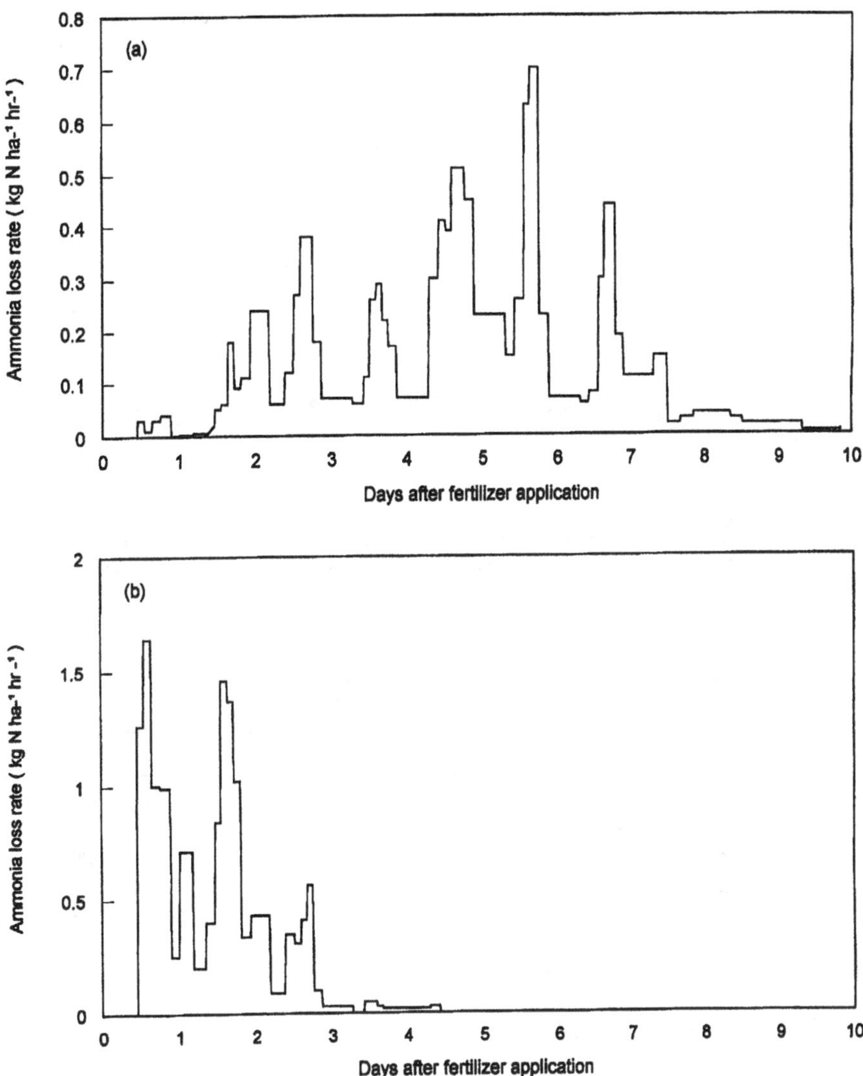

Figure 9.4. Ammonia flux density from urea (a) and ammonia bicarbonate (b) applied to flooded rice at Fengqiu (Zhu *et al.* 1989).

tion in wind speed close to the water surface (Humphreys *et al.* 1988). In addition, much more N is taken up by the abundant rice roots at panicle initiation.

9.5.2. *Upland crops*

Table 9.3 shows the extent of ammonia loss from fertilizer N applied to upland crops in China and Australia. Ammonia loss from urea broadcast onto a maize field at Fengqiu was 30% of the applied N; the loss was considerably reduced (to 12%)

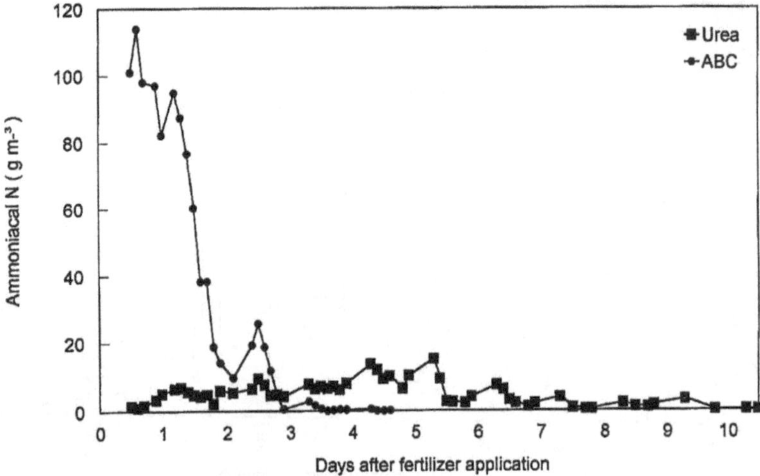

Figure 9.5. Ammoniacal nitrogen concentration of floodwater after application of urea and ammonium bicarbonate (ABC) into the floodwater and incorporation into soil at transplanting at Fengqiu (Zhu *et al.* 1989).

Table 9.3. Ammonia loss from fertilizers[1] applied to upland crops.

Location	Crop	Application method	Ammonia loss (%)	Total N loss (%)	Reference
Fengqiu, China	Maize	Broadcast Deep placed	30 12	45 30	Zhang *et al.* (1992)
Shanghai, China	Cabbage	Broadcast Incorporated	17 11		Xi *et al.* (1987)
Griffith, Australia	Sunflower	Flood irrigation	<2	>20	Freney *et al.* (1985b)
Tatura, Australia	Sunflower	Furrow irrigation	6	35	Smith *et al.* (1989)
Narrabri, Australia	Cotton	Injected	<1		Denmead *et al.* (1977)

[1] Urea was the nitrogen source except at Narrabri where anhydrous ammonia was used.

when the urea was placed 10 cm below the soil surface. The corresponding values for total N loss were 45% and 30%, respectively. Thus, the dominant pathway for N loss from urea broadcast to maize at Fengqiu was ammonia volatilization, and the decrease in total N loss by deep placement was through the reduction in ammonia loss. Studies on ammonia volatilization from a cabbage field in a Shanghai suburb showed that ammonia loss was 17% when urea was broadcast onto the field, whereas it was only 11% when the urea was incorporated (Xi *et al.* 1987). The rates of ammonia emission were low during the first few days after application, due to the low moisture content of the soil, and loss rates increased greatly after irrigation (8 days after fertilization). The maximum rate of ammonia loss after irrigation was twice that before irrigation.

In Australia, when urea was applied to sunflower fields through furrow or flood fertigation after dissolving urea in the water, the losses of ammonia were low (6.1% and <2%, respectively; Freney *et al.* 1985b; Smith *et al.* 1989). However, ammonia loss was rapid and large when water-run ammonia was applied to a maize field through furrow irrigation. With a short maize crop (0.9 m high) the losses approached 30% of the amount applied per hour (Denmead *et al.* 1982). In order to reduce such large ammonia losses the authors recommended substitution of urea for ammonia in the floodwater (Freney *et al.* 1985b).

9.6. Management practices for reducing ammonia loss

Ammonia volatilization is only one of the pathways of N loss from cropland. Therefore, it is important to know whether a management technique adopted to reduce ammonia loss will enhance N loss through other pathways. Results from some experiments indicated that when ammonia volatilization was reduced by an improved management practice, denitrification loss was not promoted significantly so that total N loss was reduced (De Datta *et al.* 1986; Simpson *et al.* 1988). However, in other cases reducing ammonia loss by incorporation of the fertilizer markedly increased loss of N by denitrification and total loss (Freney *et al.* 1990)

9.6.1. Improved methods of application

One of the methods for decreasing ammonia loss is to reduce the amount of fertilizer N remaining in the floodwater (rice field) or on the soil surface (upland crop). Among the proposed methods of application, deep placement is the most effective one for reducing N loss and improving the efficiency of fertilizer N (Chen and Zhu 1982; Cao *et al.* 1984; Craswell and Vlek 1982; Youngdahl *et al.* 1986). Results show that deep placement or incorporation of the fertilizer reduced ammonia loss compared with surface broadcasting irrespective of soil reaction and N source (urea, ammonium bicarbonate or ammonium sulfate; Table 9.4). Ammonia loss decreased with increase in the depth of placement (Chen and Qiao 1950; Yao and Guan, 1983; Fenn and Miyamoto 1981). Irrigation following the surface broadcasting of fertilizer N (particularly urea) on upland crops can reduce N loss to a certain extent.

9.6.2. Modified nitrogen fertilizers

The gradual release of N from coated fertilizers maintains a low concentration of soluble N in the floodwater or soil, and this may reduce ammonia volatilization and total N loss. The most widely tested coated N fertilizer is sulfur-coated urea (SCU). Results show that ammonia loss from SCU was significantly lower than that from prilled urea, though its effectiveness was not as high as deep placement of urea

Table 9.4. Effect of fertilizer form and application method on ammonia loss.

Soil pH	Source	Experimental method	Ammonia loss (%)			Reference
			SB	Inc	DP[1]	
Field experiment						
5.3	ABC[2]	Flooded rice field	12.6	6.3[3]		Cai *et al.* (1986)
	Urea		3.2	2.8[3]		
8.8	ABC	Flooded rice field	19.8	17.3[3]		Zhu *et al.* (1989)
	Urea		19.3	9.2[3]		
7.1	AS	Flooded rice field	3.8	1.6		Ventura and Yoshida (1977)
	Urea		8.2	3.6		
Laboratory experiment						
>7	AS	Aerobic incubation	2.6		0.4	Qu (1980)
	ABC		7.6		0.6	
>7	ABC	Aerobic incubation	2.8		0.1	Lu *et al.* (1980)
>7	AS	Aerobic incubation	7.9		3.1	Chen and Qiao (1950)
>7	urea	Aerobic incubation	7.7		1.9	Jiang (1980)
>7	AS	Column with moist soil	63		52	Feng and Miyamoto (1981)
	ABC		74		52	
	Urea		66		41	
9.0	Urea	Aerobic incubation	79		12	Jiang and Liu (1979)
		Flooded incubation	38		29	
6.6	AS	Flooded incubation	2.5		1.9	MacRae and Ancajas (1970)
	Urea		8.1		3.5	
8.1	AS	Flooded incubation	6.6		5.5	MacRae and Ancajas (1970)
	Urea		19		5.9	

[1] SB = Surface broadcast; Inc = Incorporated; DP = Deep placed.
[2] AS = Ammonium sulfate; ABC = Ammonium bicarbonate.
[3] N was incorporated into drained soil.

supergranules (Chauhan and Mishra 1989). In an international network it was found that SCU gave a significantly better response in yield than split application of urea in 39% of 217 trials, and was equally as good in 56% of the trials (Craswell and Vlek 1982; Youngdahl *et al.* 1986). Ammonia loss from urea could also be reduced by adding nitrophosphate to it; the decreased loss presumably resulted from the reduced pH of the soil (Christianson 1989). Addition of superphosphate also greatly reduced ammonia loss from urea; the effectiveness increased with the amount of superphosphate added (Shi 1986).

9.6.3. *Urease inhibitors*

More than 100 urease inhibitors have been tested. Some of them, such as PPD (phenylphosphorodiamidate), NBPT (N-(n-butyl) thiophosphorictriamide) and hydroquinone are effective. In glasshouse experiments, addition of PPD retarded urea hydrolysis (Zhu *et al.* 1985; Byrnes *et al.* 1983; Rao and Ghai 1986; Vlek *et al.* 1980), and decreased floodwater ammoniacal N concentration; consequently ammonia volatilization and total N loss were reduced (Byrnes *et al.* 1983; Rao and Ghai 1986). In rice field experiments, the ammoniacal N concentration in the floodwater was decreased by addition of PPD or NBPT (Cai *et al.* 1989; Cai *et al.* 1992b), but ammonia loss and total N loss were not always decreased (Simpson *et al.* 1985; Fillery and De Datta 1986; Fillery *et al.* 1986a; Cai 1988; Cai *et al.* 1992b). The effectiveness of six phosphoroamide urease inhibitors were tested in a maize field in the USA. Results showed that ammonia loss was reduced from 19% to between 3 and 12%, and PPD was the most effective inhibitor (Nelson *et al.* 1986). Addition of PPD to a maize field in Canada reduced ammonia loss (Tomar *et al.* 1985), and addition of hydroquinone reduced ammonia loss in a rapeseed experiment (Rodgers *et al.* 1986).

However, the inhibiting effect of some of these urease inhibitors in flooded rice fields was short lived. Retardation of urea hydrolysis by addition of 1% PPD lasted only 1–4 days (Simpson *et al.* 1985; Fillery *et al.* 1986a). Thus a reduction in total N loss by addition of a urease inhibitor was not often found, and yield was not significantly increased in most rice field experiments (Cai 1989). Therefore, conditions for the effective use of urease inhibitors need to be further investigated. Obviously, a reduction in ammonia loss from urea can not be expected when ammonia loss is small. Leaching and runoff may be increased by addition of urease inhibitor due to the retarded hydrolysis of urea.

9.6.4. Surface films

Ammonia loss from a flooded rice field may be reduced by spreading a long chain alcohol on the surface of the floodwater. This technique has been used to control the evaporation of water from surface reservoirs (Frenkiel 1965). It was found that spreading hexadecanol or octadecanol dissolved in ethanol on the water surface greatly reduced rates of ammonia emission in pan experiments (Cai *et al.* 1987).

In a urea fertilized rice crop in Australia, ammonia loss was reduced from 15 to 8.4 kg N ha^{-1} by spreading hexadecanol dissolved in ethanol on the floodwater surface; total N loss was also reduced (Cai *et al.* 1987). In a similar experiment carried out at Yingtan in China, the application of octadecanol on the floodwater reduced ammonia loss from 40% to 23% of the applied N, but total N loss was not reduced. The reasons for the different effects on ammonia loss and total N loss in this experiment are uncertain. It may be that reduction in ammonia volatilization resulted in increased denitrification loss. In these two experiments the effect of the long chain alcohols did not last long; they seemed to be decomposed quickly by microorganisms in the soil-water system. Consequently, addition of the long chain alcohol at regular intervals (say 2 days) needed to be made to maintain the surface film. A cheaper and more stable material is required to drastically reduce ammonia volatilization from ammonium or ammonium-producing fertilizers.

9.6.5. Algicide

Addition of an algicide to a flooded rice field may suppress the elevation of floodwater pH (Bowmer and Muirhead 1987; Simpson *et al.* 1988). A field study with microplots (Simpson *et al.* 1988) showed that addition of algicide depressed the elevation of floodwater pH, and reduced ammonia loss from 20.5% to 10.8%. When floodwater pH was controlled at 7 through addition of algicide and dilute acid, ammonia loss was further reduced to 1.2%, and the total N losses were reduced correspondingly.

9.7. References

Black, A S, Sherlock, R R, Cameron, K C, Smith, N P and Goh, K M 1985. Comparison of three field methods for measuring ammonia volatilization from urea granules broadcast onto pasture. J. Soil Sci. 36:271–280.

Bouwmeester, R J B and Vlek, P L G 1981. Rate control of ammonia volatilization from rice paddies. Atmos. Environ. 15:131–140.

Bowmer, K H and Muirhead, W A 1987. Inhibition of algal photosynthesis to control pH and reduce ammonia volatilization from rice floodwater. Fert. Res. 13:13–29.

Byrnes, B G, Savant, N K and Craswell, E T 1983. Effect of a urease inhibitor phenyl phosphorodiamidate on the efficiency of urea applied to rice. Soil Sci. Soc. Am. J. 47:270–274.

Cai, G X 1986. Ammonia volatilization from fertilizers applied to flooded rice field. In: Agricultural Chemistry and Soil Biology and Biochemistry Committees, Soil Science Society of China (eds.), Advances and Prospects for Soil Nitrogen Research in China. (in Chinese). pp. 55–67. Science Press, Beijing.

Cai, G X 1988. Gaseous loss of nitrogen from fertilizers applied to flooded rice. Ph. D. thesis, University of Queensland, Australia.

Cai, G X 1989. Effect of urease inhibitors on the efficiency of urea. (in Chinese). Progress in Soil Sci 17:1–7.

Cai, G X 1992. Evaluation of gaseous nitrogen losses from fertilizers applied to flooded rice fields in China. Proc. Inter. Symp. on Paddy Soils. pp. 99–106. Science Press, Beijing.

Cai, G X, Freney, J R, Humphreys, E, Denmead, O T, Samson, M and Simpson, J R 1987. Use of surface films to reduce ammonia volatilization from flooded rice fields. Aust. J. Agric. Res. 39:177–186.

Cai, G X, Freney, J R, Muirhead, W A, Simpson, J R, Chen, D L and Trevitt, A C F 1989. The evaluation of urease inhibitors to improve the efficiency of urea as a N-source for flooded rice. Soil. Biol. Biochem. 21:137–145.

Cai, G X, Peng, G H, Wang, X Z, Zhu, J W and Zhu, Z L 1992a. Ammonia volatilization from urea applied to acid paddy soil in southern China and its control. Pedosphere. 2:345–354.

Cai, G X, Yang N C, Lu, W F, Chen, W, Xia, B Q, Wang, X Z and Zhu, Z L 1992b. Gaseous loss of nitrogen from fertilizers applied to a paddy soil in Southeastern China. Pedosphere, 2:209–217.

Cai, G X, Zhu, Z L, Trevitt, A C F, Freney, J R and Simpson, J R 1986. Nitrogen loss from ammonium bicarbonate and urea fertilizers applied to flooded rice. Fert. Res. 10:203–215.

Cao, Z H, De Datta, S K and Fillery, I R P 1984. Effect of placement method on floodwater properties and recovery of applied N (^{15}N-labeled urea) in wetland rice. Soil Sci. Soc. Am. J. 48:196–203.

Chauhan, H S and Mishra, B 1989. Ammonium volatilization from a flooded rice field fertilized with amended urea materials. Fert. Res. 19:57–63.

Chen, S J and Qiao, S H 1950. Theory of ammonia volatilization and the amount of ammonia loss from ammonium fertilizers applied to calcareous soil. (in Chinese). Agric. Res. in China 1:81–87.

Chen, R Y and Zhu, Z L 1982. Studies of fate of nitrogen fertilizer I. The fate of nitrogen fertilizer in paddy soils. (in Chinese). Acta Pedol. Sin 19:122–130.

Christianson, C B 1989. Ammonia volatilization from urea nitrophosphate and urea applied to the soil surface. Fert. Res 19:183–189.

Craswell, E T and Vlek, P L G 1982. Nitrogen management for submerged rice soils. Trans. 12th Inter. Congr. Soil Sci. Vol. II. pp. 158–181. New Delhi.

Datta, N P, Banerjee, N K and Prassada Rao, D M V 1971. A new technique for study of nitrogen balance sheet and an evaluation of nitrophosphate using N-15 under submerged conditions of growing paddy. Proc. Inter. Symp. Soil Fert. Eval. New Delhi. 1:631–638.

De Datta, S K 1986. Improving nitrogen fertilizer efficiency in lowland rice in tropical Asia. Fert. Res. 9:171–186.

Denmead, O T 1983. Micrometeorological methods for measuring gaseous losses of nitrogen in the field. In: Freney, J R and Simpson, J R (eds.), Gaseous Loss of Nitrogen from Plant-Soil Systems. pp. 133–157. Martinus Nijhoff/Dr W Junk, The Hague.

Denmead, O T, Freney, J R and Simpson, J R 1982. Dynamics of ammonia volatilization during furrow irrigation of maize. Soil Sci. Soc. Am. J. 46:149–155.

Denmead, O T, Simpson, J R and Freney, J R 1977. A direct field measurement of ammonia emission after injection of anhydrous ammonia. Soil Sci. Soc. Am. J. 41:1000–1004.

Fenn, L B and Hossner, L R 1985. Ammonia volatilization from ammonium or ammonium-forming nitrogen fertilizers. Adv. Soil Sci. 1:123–169.

Fenn, L B and Kissel, D E 1976. The influence of cation exchange capacity and depth of incorporation on ammonia volatilization from ammonium compounds applied to calcareous soils. Soil Sci. Soc. Am. Proc. 40:394–398.

Fenn, L B and Miyamoto, S 1981. Ammonia loss and associated reactions of urea in calcareous soils. Soil Sci. Soc. Am. J. 45:537–540.

Ferguson, R B and Kissel, D E 1986. Effect of soil drying on ammonia volatilization from surface-applied urea. Soil Sci. Soc. Am. J. 50:485–490.

Fillery, I R P and De Datta, S K 1986. Ammonia volatilization from nitrogen sources applied to rice fields: I. Methodology, ammonia fluxes, and nitrogen-15 loss. Soil Sci. Soc. Am. J. 50:80–86.

Fillery, I R P and Vlek, P L G 1986. Reappraisal of the significance of ammonia volatilization as an N loss mechanism in flooded rice fields. Fert. Res. 9:79–98.

Fillery, I R P, Simpson, J R and De Datta, S K 1984. Influence of field environment and fertilizer management on ammonia loss from flooded soil. Soil Sci. Soc. Am. J. 48:914–920.

Fillery, I R P, De Datta, S K and Craswell, E T 1986a. Effect of phenylphosphorodiamidate on the fate of urea applied to wetland rice fields. Fert. Res. 9:251–263.

Fillery, I R P, Simpson, J R and De Datta, S K 1986b. Contribution of ammonia volatilization to total nitrogen loss after application of urea to wetland rice fields. Fert. Res. 8:193–202.

Freney, J R, Denmead, O T, Watanabe, I and Craswell, E T 1981a. Ammonia and nitrous oxide losses following application of ammonium sulfate to flooded rice. Aust. J. Agric. Res. 32:37–45.

Freney, J R, Simpson, J R and Denmead, O T 1981b. Ammonia volatilization. In: Clark, F E and Rosswall, T (eds.), Terrestrial Nitrogen Cycles. Processes, Ecosystem Strategies and Management Impacts. Ecol. Bull. 33:291–302.

Freney, J R, Simpson, J R and Denmead, O T 1983. Volatilization of ammonia. In: Freney, J R and Simpson, J R (eds.), Gaseous Loss of Nitrogen from Plant-Soil Systems. pp. 1–32. Martinus Nijhoff/Dr W. Junk, The Hague.

Freney, J R, Leuning, R, Simpson, J R, Denmead, O T and Muirhead, W A 1985a. Estimating ammonia volatilization from flooded rice fields by simplified techniques. Soil Sci. Soc. Am. J. 49:1049–1054.

Freney, J R, Simpson, J R, Denmead, O T, Muirhead, W A and Leuning, R 1985b. Transformations and transfers of nitrogen after irrigating a cracking clay soil with a urea solution Aust. J. Agric. Res. 36:685–694.

Freney, J R, Trevitt, A C F, De Datta, S K, Obcemea, W N and Real, J G 1990. The interdependence of ammonia volatilization and denitrification as nitrogen loss processes in flooded rice in the Philippines. Biol. Fertil. Soils. 9:31–36

Freney, J R, Trevitt, A C F, Muirhead, W A, Denmead, O T, Simpson, J R and Obcemea, W N 1988. Effect of water depth on ammonia loss from lowland rice. Fert. Res. 16:97–107.

Freney, J R, Trevitt, A C F, Zhu, Z L, Cai, G X and Simpson, J R 1987. Assessing ammonia volatilization from fertilizer nitrogen applied to flooded rice fields. (in Chinese). Acta Pedol. Sin. 24:142–151.

Frenkiel, J 1965. Evaporation Reduction. UNESCO. Paris. 79 p.

Hargrove, W L and Kissel, D E 1979. Ammonia volatilization from surface applications of urea in the field and laboratory. Soil Sci. Soc. Am. J. 48:359–363.

Hargrove, W L, Kissel, D E and Fenn, L B 1977. Field measurements of ammonia volatilization from surface applications of ammonium salts to a calcareous soil. Agron. J. 69:473–476.

Humphreys, E, Freney, J R, Muirhead, W A, Denmead, O T, Simpson, J R, Leuning, R, Trevitt, A C F, Obcemea, W N, Wetselaar, R and Cai, G X 1988. Loss of ammonia after application of urea at different times to dry-seeded, irrigated rice. Fert. Res. 16:47–67.

Jiang, N H and Liu, G S 1979. Measurement of ammonia emitted from fertilized soil. (in Chinese). Soils (5):193–197.

Jiang, R C 1980. Transformation of urea nitrogen applied to calcareous soil with different methods. (in Chinese). Jiangsu Agric. Sci (3):42–46.

Kissel, D E, Brewer, H L and Arkin, G F 1977. Design and test of a field sampler for ammonia volatilization. Soil Sci. Soc. Am. J. 41:1133–1138.

Leuning, R, Denmead, O T, Simpson, J R and Freney, J R 1984. Processes of ammonia loss from shallow floodwater. Atmos. Environ. 18:1583–1592.

Leuning, R, Freney, J R, Denmead, O T and Simpson, J R 1985. A sampler for measuring atmospheric ammonia flux. Atmos. Environ 19:1117–1124.

Liao, X L, Xu, Y H and Zhu, Z L 1982. Investigation on nitrification-denitrification loss of fertilizer nitrogen in submerged paddy soil. (in Chinese). Acta Pedol. Sin 19:257–263.

Liu, Z H, Luo, Y Y, Xian, Z and Wang, Y S 1980. Investigation on hydrolysis, transformation and movement of urea applied to soils. (in Chinese). Soils and Fertilizers 3:36–39.

Lu, D Q, Liu, X L, Wu, C Z and Hou, L Y 1980. Investigation on factors affecting ammonia volatilization from ammonium bicarbonate applied to calcareous soils and its control. (in Chinese). Shangxi Agric. Sci. (6):7–10.

MacRae, I C and Ancajas, R 1970. Volatilization of ammonia from submerged tropical soils. Plant Soil 33:97–103.

Mahendrappa, M K and Ogden, E D 1973. Patterns of ammonia volatilization from a forest soil. Plant Soil 38:257–265.

Mikkelsen, D S and De Datta, S K 1979. Ammonia volatilization from wetland rice soils. In: Nitrogen and Rice. IRRI, Manila. pp. 135–157.

Mikkelsen, D S, De Datta, S K and Obcemea, W N 1978. Ammonia volatilization losses from flooded rice soils. Soil Sci. Soc. Am. J. 42:725–730.

Mulvaney, R L and Bremner, J M 1981. Control of urea transformation in soils. In: Paul, E A and Ladd, J N (eds.), Soil Biochemistry. 5:153–196. Marcel Dekker, New York.

Nelson, D W 1982. Gaseous losses of nitrogen other than through denitrification. In: Stevenson, F J (ed.), Nitrogen in Agricultural Soils. pp. 327–363. Am. Soc. Agron., Madison, Wisconsin.

Nelson, D W, Beyrouty, C A and Schlegel, A J 1986. Effects of phosphoroamide urease inhibitors on nitrogen transformations in soil and maize yields. Trans. 13th Inter. Congr. Soil Sci. 3:880–881.

Nommik, H 1973. Assessment of volatilization loss of ammonia from surface-applied urea on forest soil by ^{15}N recovery. Plant Soil 38:589–603.

Nurayama, N 1979. The importance of nitrogen for rice production. In: Nitrogen and Rice. pp. 5–23. IRRI, Manila.

Qu, Q X 1980. Investigation of nitrogen loss from ammonium-form fertilizers applied to calcareous soils. (in Chinese). Soils and Fertilizers. 3:31–35.

Rao, D L N and Ghai, S K 1986. Effect of phenylphosphorodiamidate on urea hydrolysis, ammonia volatilization and rice growth in an alkali soil. Plant Soil 94:313–320.

Rodgers, F A, Penny, A and Hewitt, M V 1986 A comparison of the effects of prilled urea, used alone or with a nitrification or urease inhibitor, with those of Nitro-chalk on winter oil-seed rape. J. Agric. Sci. Camb. 106:515–526.

Ryan, J, Curtin, D and Safi, I 1981. Ammonia volatilization as influenced by calcium carbonate particle size and iron oxides. Soil Sci.Soc. Am. J. 45:338–341.

Shi, M H 1986. Ammonia volatilization from urea surface broadcast on calcareous soil – Effect of superphosphate. (in Chinese). M.Sc. Thesis. Nanjing Agric. Univ. Nanjing, China.

Simpson, J R and Freney, J R 1988. Interacting processes in gaseous nitrogen loss from urea applied to flooded rice fields. Urea Technology and Utilization. pp. 281–290. Malaysian Society of Soil Science. Kuala Lumpur, Malaysia.

Simpson, J R, Freney, J R, Muirhead, W A and Leuning, R 1985. Effect of phenylphosphorodiamidate and dicyandiamide on nitrogen loss from flooded rice. Soil Sci. Soc. Am. J. 49:1426–1431.

Simpson, J R, Muirhead, W A, Bowmer, K H, Cai, G X and Freney, J R 1988. Control of gaseous nitrogen losses from urea applied to flooded rice soils. Fert. Res. 18:31–47.

Smith, C J, Freney, J R, Chalk, P M, Galbally, I E, McKenney, D J and Cai, G X 1989. Fate of urea nitrogen applied in solution in furrows to sunflowers growing on a red-brown earth: transformations, losses and plant uptake. Aust. J. Agric. Res. 39:793–806.

Terman, G 1979. Volatilization losses of nitrogen as ammonia from surface-applied fertilizers, organic amendments and crop residues. Adv. Agron. 31:189–223.

Tomar, J S, Kirby, P C and Mackenzie, A F 1985. Field evaluation of the effects of urease inhibitor and crop residues on urea hydrolysis, ammonia volatilization and yield of corn. Can. J. Soil Sci. 65:771–789.

Trevitt, A C F, Freney, J R, Simpson, J R. and Muirhead, W A 1987. Effects of microplots on urea nitrogen reactions in flooded soils. In: IRRI (ed.), Efficiency of Nitrogen Fertilizers for Rice. pp. 177–183. IRRI, Los Banos, The Philippines.

Ventura, W B and Yoshida, T 1977. Ammonia volatilization from a flooded tropical soil. Plant Soil 46:521–531.

Vlek, P J G and Craswell, C T 1979. Effect of nitrogen source and management on ammonia volatilization losses from flooded rice-soil systems. Soil Sci. Soc. Am. J. 43:352–358.

Vlek, P L G and Stumpe, J M 1978. Effects of solution chemistry and environmental conditions on ammonia volatilization losses from aqueous systems. Soil Sci. Soc. Am. J. 42:416–421.

Vlek, P L G, Stumpe, J M and Byrnes, B H 1980. Urease activity and inhibition in flooded soil systems. Fert. Res. 1:191–202.

Wilson, J D, Catchpoole, V R, Denmead, O T and Thurtell, G W 1983. Verification of a simple micrometeorological method for estimating the rate of gaseous mass transfer from ground to the atmosphere. Agric. Meteorol. 29:183–189.

Wilson, J D, Thurtell, G W, Kidd, G E and Beauchamp, E G 1982. Estimation of the rate of gaseous mass transfer from a surface source plot to the atmosphere. Atmos. Environ. 16:1861–1867.

Xi, Z B, Humphreys, E, Shi, X Z, Freney, J R, Huang, W X, Yao, Z and Simpson, J R 1987. Effect of application method on ammonia loss from urea applied to cabbages in the Shanghai district. (in Chinese). Acta Agric. Shanghai 3:47–56.

Yao, R and Guan, S J 1983. Transformation of urea – nitrogen in soil with deep placement and its efficiency. (in Chinese). Soils and Fertilizers (1):15–16.

Youngdahl, L J, Lupin, M S and Craswell, E T 1986. New developments in nitrogen fertilizers for rice. Fert. Res. 9: 149–160.

Yu, S F and Zhao, M Z 1979. Preliminary investigation of nitrogen loss from fertilizers applied to calcareous soils. (in Chinese). Soils 1:31–33.

Yuan, L H and Zhang, Y F 1979. Factors affecting ammonia volatilization from urea applied to soils. (in Chinese). Soils and Fertilizers 2:25–28.

Zhang, S L, Cai, G X, Wang, X Z, Xu, Y X, Zhu, Z L and Freney, J R 1992. Losses of urea-nitrogen applied to maize grown on a calcareous fluvo-aquic soil in North China Plain. Pedosphere. 2:171–178.

Zhao, Z D and Zhang, J S 1981. A study on promotion of utility rate of nitrogen fertilizers (part III). (in Chinese). J. Soil Sci. (1):16–19.

Zhao, Z D, Zhang, J S and Ren, S R 1986. Ammonia volatilization from fertilizers applied to upland crops. In Agricultural Chemistry and Soil Biology and Biochemistry Committees, Soil Science Society of China (eds.), Advances and Prospects for Soil Nitrogen Research in China. (in Chinese). pp. 46–54. Science Press, Beijing.

Zhu, Z L, Cai, G X, Simpson, J R, Zhang, S L, Chen, D L, Jackson, A V and Freney, J R 1989. Processes of nitrogen loss from fertilizers applied to flooded rice fields on a calcareous soil in north-central China. Fert. Res. 18:101–115.

Zhu, Z L, Cai, G X, Xu, Y X and Zhang, S L 1985. Ammonia volatilization and its significance to the losses of fertilizer nitrogen applied to paddy soil. (in Chinese). Acta Pedol. Sin. 22:320–328.

10
Nitrogen in the rhizosphere

LIU ZHI-YU

10.1. Introduction

As a result of its metabolic functions such as exudation of organic and inorganic substances, nutrient and water uptake, and proliferation of microorganisms on the root surface, root growth makes its surrounding microzone differ, both physically, chemically and biologically, from the bulk of the soil. This microzone, which is generally considered to be within 4 mm of the root surface, and accounts for about 1–3% of the volume of the plowed layer, is called the rhizosphere (Liu 1980). However, the zone influenced varies with plant type, the growing period and root development, and may also be influenced by soil properties and nutrient status. Consequently, there is no clearly defined zone for the rhizosphere.

The rhizosphere is a zone where interactions take place between the root, soil, and microorganisms and where nutrients, moisture and various substances, both beneficial and harmful, come into contact with the root. Therefore, the rhizosphere is of vital importance to the growth and yield of plants

Nitrogen is the most active nutrient element in soil. About one half or more of the N taken up during the life of a plant comes from the organic N in soil; so the properties of plant roots and the associated N transformations in the rhizosphere are of special significance. For a long time, soil N has been studied from the standpoint of the whole solum or the plowed layer, but little attention has been given to the rhizosphere. In the last 10 years or so, with the development of experimental techniques, research on the N status of the rhizosphere has progressed somewhat, but even so less data are available compared with other nutrients.

This chapter is devoted to a discussion of topics such as N source at the soil-root interface, dynamics of fertilizer-N within the rhizosphere, effect of absorption of different forms of N on the pH of the rhizosphere, and uptake of N from the rhizosphere in relation to the characteristics of plant roots,

10.2. Nitrogen at the soil–root interface

When the plant root penetrates and extends through the soil, a thin layer of mucilage appears at the interface between the newly growing root and its surrounding soil. This layer consists mainly of secretions from plant roots, microorganisms, sloughed off root tissue and degradation products of microbial residues (Foster

Zhu Zhao-liang et al. *(eds.): Nitrogen in Soils of China, 215–237.*

1981). The mucilage contains organic and inorganic nutrients including N compounds such as amino acids, polypeptides and proteins. These provide the natural media for survival and propagation of microorganisms.

Consequently, in the mucilaginous layer and in the rhizosphere soil there are 10 to 1000-fold as many microorganisms as in the bulk soil, and most of these are present in the cellular interstices of the epidermis and at the base and tips of the root hairs. In general, large quantities of microorganisms are found 15 mm from the root surface; bacteria alone may amount to 36 mg g dry root^{-1} (Haller and Stolp 1985). The increase in weight through propagation of these bacteria at the soil-root interface may amount to, on an average, 0.03 mg dry matter per gram dry root per day and thus a considerable amount of N is immobilized. Furthermore, bacteria of many genera fix atmospheric N_2 non-symbiotically at the root surface of certain crops, thus providing the rhizosphere soil with extra N.

10.2.1. Nitrogen fixation by non-symbiotic organisms

Non-symbiotic fixation of N in the rhizosphere, or associative N fixation, refers mainly to the process in which bacteria at the root surface and in the cellular interstices of the root tissue fix N, using the exudate or material sloughed off root tissue as an energy source. Non-symbiotic N fixation is an important source of N in the soil–plant system and helps to maintain the N supplying capacity of the soil. However, this is often overlooked in agricultural production, due to large quantities of N fertilizers being used.

The non-symbiotic N fixing bacteria in the rhizosphere are mostly *Enterobacter*, *Azospirillum* or *Azotobacter*. They are often associated with specific higher plants, e.g. *Azotobacter beijerinkia* with sugarcane, *Azotobacter paspali* with buckwheat, wheat, or maize (Döbereiner 1983), and the Brazilian *Azospirillum brasilense* with wheat (Mengel and Viro 1978). The C_4 crops (e.g. sugarcane and maize) commonly have a higher rate of nitrogen fixation than the C_3 crops (e.g. barley and wheat). In a growing season the C_3 plants fix ~5–10 kg N ha^{-1}, while the rate of fixation for the C_4 plants may be several times higher than this (Idris *et al.* 1981).

Reports of the rate of non-symbiotic N fixation have been inconsistent, varying from a few to several hundred kg N ha^{-1} y^{-1}. The overestimate seems to have resulted from the extrapolation of data from short-term experiments. The main factors influencing the rate of non-symbiotic N fixation are available N content and partial pressure of O_2. As the mineral N supply increases, the N fixation rate reduces markedly. In field trials, however, the maximum rate of N fixation occurs at a moderate level of N supply. This is because, when N is deficient, photosynthesis is retarded and the exudate decreases in the rhizosphere (Trolldenier 1977). Likewise, the N fixation rate in the rhizosphere is linked to illumination intensity for the plant tops, creating distinct diurnal fluctuations (Sims and Dunigan 1984).

Since the nitrogenase of non-symbiotic N fixing bacteria is highly sensitive to the partial pressure of oxygen, a low oxygen partial pressure and a high soil moisture content will be conducive to N fixation in the rhizosphere. Krotzky *et al.* (1983) showed that, under low oxygen conditions, the phenolic compounds of decomposition products can increase the activity of nitrogenase. Therefore, flooded rice has a higher non-symbiotic N fixation rate in the rhizosphere than upland crops. However, most of the N fixing bacteria in the rice rhizosphere are not anaerobic. There are about 1000 times as many aerobic N fixing bacteria as there are anaerobic bacteria per gram of fresh root. As a result, the N fixing organisms are forced to congregate in the oxidizing zone around the rice root (Balandreau *et al.* 1975). The N fixation rate in the rice rhizosphere measured in situ under field conditions may be as high as 10–15 kg N ha^{-1} $month^{-1}$ (Yoshida and Yoneyama 1980). Unlike upland crops, flooded rice is not concerned with a specific association.

The N fixation rate in the rice rhizosphere varies with growth stage. The rate is very low in the initial stage, then increases considerably at panicle initiation and reaches a maximum at flowering. A similar pattern has been found with other crops such as wheat, maize and sorghum. This is closely associated with an increase in the exudate from plant roots. It has been shown that 15–25% of the N fixed can be taken up by the plant and transported to different parts of the tops within a short time (Yoshida and Yoneyama 1980).

10.2.2. Immobilization

After addition to a soil, mineral N is partly immobilized, i.e. transformed into organic N by microorganisms. Since there are numerous microorganisms at the root surface and around the mucilaginous layer of the root, there is more microbial biomass N in this region than in the bulk of the soil. Döbereiner (1983) reported that the live weight of bacteria alone in the 15 cm thick plough layer of soil amounted to 1500–3750 kg ha^{-1}, or 280–730 kg ha^{-1} on a dry weight basis; the density in the rhizosphere is estimated to be 10–100 times the average of the above values. Assuming the N in bacteria to be 2.0–14.0% (on dry weight basis), it is extrapolated that the N immobilized by rhizosphere microorganisms would account for ~3–10% of the N content of the solum (Zhang *et al.* 1962).

Investigations by ^{15}N techniques show that polypeptides, such as mucopeptides and structural protein, are the main source of amino acid N in the rhizosphere. Mucopeptide is an essential constituent of the cell wall of bacteria. It is an amino sugar (muramic acid) with the main amino acids being alanine, glutamic acid, and either lysine or diaminopimelic acid. The N containing substances in the cell and cell wall residues of these microorganisms are easily degradable. Marumoto (1984) showed that, in incubation studies, mineralization of native soil N was slow and steady, while mineralization of the immobilized N was rapid, and 25–50% of it was re-mineralized within one or two weeks. No doubt, immobilized N is the reserve of potentially available N in the rhizosphere

10.2.3. Root exudates

The organic N compounds released from the growing root, include the low molecular weight substances such as amino acids and amides which diffuse across the electrochemical gradient to the soil, the low and high molecular weight compounds, e.g. colloidal material exuded during the consumption of metabolic energy, and the degradation products of sloughed off cells from the root cap, dead root hairs, and impaired epidermis tissue. It is difficult to differentiate between these sources. The total amount of these exudates account for ~15–20% of the N assimilated by the plant (Foster 1981). The N-containing compounds in root exudates comprise more than 20 kinds of amino acids, peptides, and proteins.

The exudates from different plants vary in type and amount of N-containing compounds. In general, the root exudates from leguminous crops contain a greater variety of N compounds and more N than cereals. For example, the exudate from 21-day-old pea seedlings contained 22 kinds of amino acids compared with 14 in the exudate from oats and had 7 times more N. The exudate of other dicotyledons, such as cucumber, contained twice the N in the exudates of wheat. Differences were also found between the composition of the exudates of cereals. For example, amino acid N formed a greater proportion of the total N in the exudate from wheat than barley (57% and 32%, respectively). However, the exudate from barley contained more protein and peptide N. As the plant aged, the N contained in the exudate varied greatly, and usually decreased (Haller and Stolp 1985). However, because of technical problems associated with the collection of exudates under sterile conditions throughout the growing season, most of the results were obtained during the seedling stage.

The root environment also has noticeable effects on the N content of exudates. For example, the amount of amino acids exuded from the roots of barley and maize was greater when grown in a solid substrate than when grown in solution. This effect seems to be related to the resistance to root growth (Barber and Gunn 1974). In soils of low moisture content, root exudate is usually high and the number of constituents increases. A barley crop grown in soil where the moisture content fell to the wilting point and then increased, exuded more amino acids than a crop grown under constant humidity. The increased release of α-amino acids from the wilted plant appeared to result from the hydrolysis of protein which induced an imbalance of N compounds in the root. Low phosphorus nutrition may reduce the phospholipid content of the plant and alter the permeability of the plasmolemma, thus increasing the release of N compounds such as amino acids from the root (Ratnayake *et al.* 1978). In a saline soil the osmotic pressure of the solution was so high that it appreciably decreased the exudation of free amino acids from barley roots (Polonenko *et al.* 1983).

Nitrogen at the soil root interface comes from many sources apart from the non-symbiotic N fixation and exudation discussed above. Other sources include symbiotic N fixation by legumes, and the activity of fungi, actinomycetes, and algae. In agricultural soils, although the amount of fertilizer N added to each crop usually exceeds the amount of N originally present in the rhizosphere, its availability is still governed by the carbon and nitrogen status at the soil–root interface.

10.3. Gradient of fertilizer nitrogen in rhizosphere

Fertilizer N status in the rhizosphere is affected by the uptake of N and the migration of N towards the root. Investigations using ^{15}N-labelled fertilizer showed that the distribution of the fertilizer N in the soil close to roots was uneven, with a distinct concentration gradient. However, the direction of the gradient varied, depending on the form of fertilizer N (Liu and Qin 1981).

10.3.1. Ammonium and ammonium producing fertilizers

The distribution of fertilizer N in the rhizosphere of rice grown under flooded conditions was investigated with the simulated root–soil interface culture method (PNGISSAS 1988). The results showed that, in contrast to a soil without plants, the rhizosphere soil fertilized with ^{15}N-labelled ammonium sulfate or urea, had a ^{15}N depleted area around the root, with a depletion rate of 30–70%. As the distance perpendicular to root surface increased, the ^{15}N concentration increased gradually. At about 30–40 mm from the root surface, the concentration was the same as that of the initial soil. Such a gradient can be expressed by the equation,

$$y = -74.6 + 12.4x^{1/2},$$

($r = 0.984$; Figure 10.1; Qin and Liu 1984).

The most striking change in gradient occurred within 10 mm of the root surface, indicating that this microzone is the major source of N in the rhizosphere for plant uptake. Calculation shows that the depletion in ^{15}N in the soil within 10 mm of the root surface corresponded to >50% of the ^{15}N taken up by the plant. For instance, on the 6th day after planting the ^{15}N depletion within the 10 mm zone accounted for 70% of the ^{15}N absorbed by plant, while on the 12th day it was only 59%. While the uptake of N by the plant increased with time, the proportion obtained from the soil beyond 10 mm also increased (Table 10.1). Thus, fertilizer N depletion in the zone near the root does not increase proportionately as plant growth proceeds, because the N migrating to the root by diffusion may increase with plant growth. Compared with fertilizer N, total N (soil and fertilizer N) in the rhizosphere soil changed very little, and the maximum depletion rate was no more than 10% of the initial soil N (Figure 10.1).

The depletion of ammonium in the rice rhizosphere is also affected by environmental conditions, and temperature is one of the factors influencing the depletion

Table 10.1. Relationship between fertilizer ^{15}N depletion in rhizosphere and uptake by rice at tillering. [1]

Time after application (days)	Distance from root surface (mm)				^{15}N uptake	^{15}N depletion within 10 mm/^{15}N uptake (%)
	0–2.5	2.5–5.0	5.0–7.5	7.5–10		
	^{15}N depletion					
6	45.2	28.2	15.8	8.4	140	70
12	84.7	71.5	61.4	54.2	460	59
18	89.8	82.2	76.6	69.9	640	49

[1] By simulated root–soil interface culture method (Liu and Qin 1981).

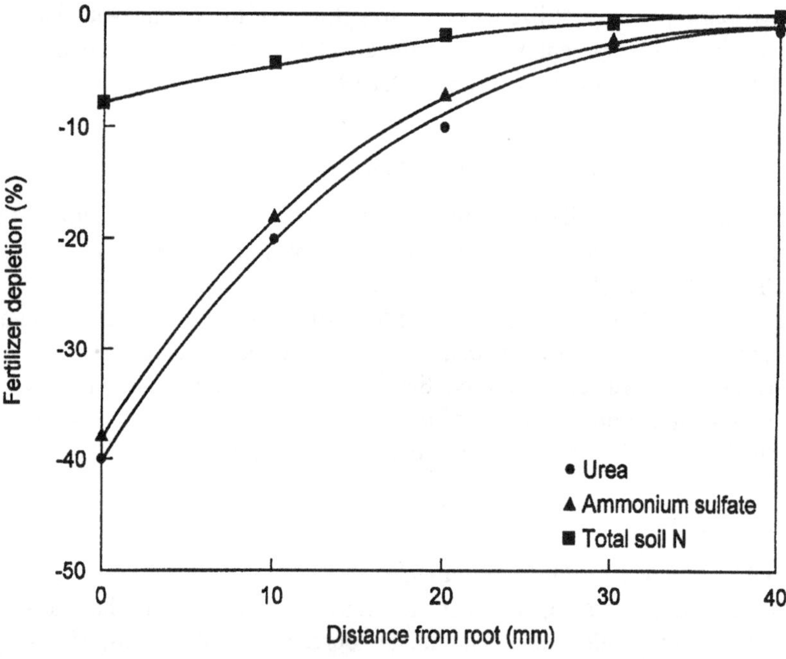

Figure 10.1. Depletion of fertilizer N in rhizosphere soil of flooded rice (Qin and Liu 1984).

(Table 10.2). When the temperature was raised by 10 degrees (from 17°C to 27°C), the depletion rate within 0–2.5 mm of the root surface was increased by 41.6%, whereas within 7.5–10.0 mm, it was increased by about 2–9 times. At 7.5–10 mm from root surface the ammonium depletion was only 2.7% at 17°C, but it increased to 28.8% at 27°C, showing that the area of depletion increased with increasing temperature. It seems that in the rhizosphere the increase in temperature encouraged the uptake of N by rice and enhanced the diffusion of ammonium to the root.

In addition, roots growing at different positions on the plant varied in their ability to absorb N and thus the depletion of fertilizer N in the rhizosphere. As shown by the results obtained at the initial tillering stage of rice (Figure 10.2), the lower positioned root (8–10 cm below the root node) had a lower absorbing capacity than

Table 10.2. Effect of temperature on NH_4^+ depletion in rice rhizosphere at tillering.[1]

Distance from root surface (mm)	Depletion rate of NH_4–^{15}N (%)	
	17°C	27°C
0–2.5	34.1	48.3
2.5–5.0	17.4	38.9
5.0–7.5	8.9	34.4
7.5–10.0	2.7	28.8

[1] Liu and Qin (1981).

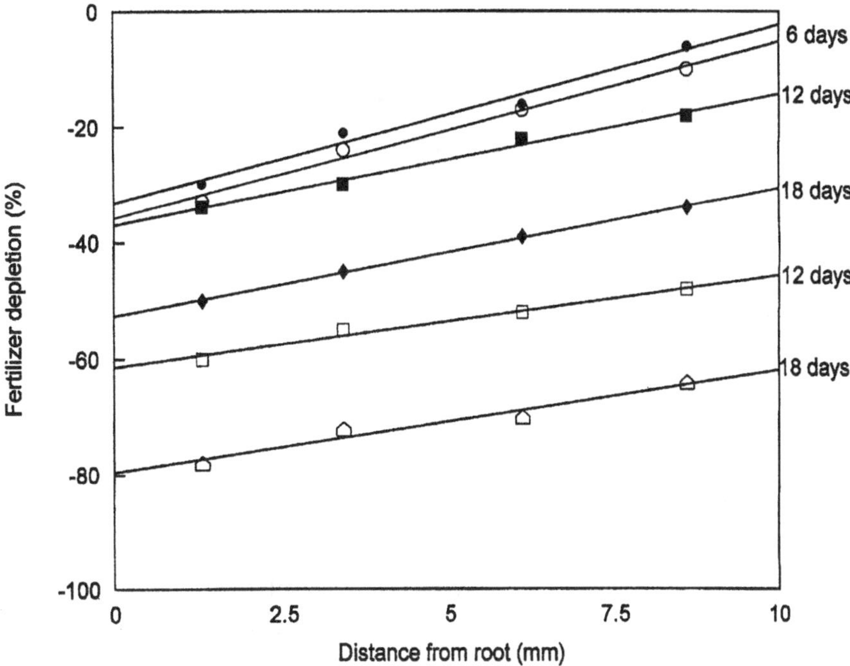

Figure 10.2. Depletion of fertilizer N in rhizosphere soil of flooded rice. Open symbols – upper roots; filled symbols – lower roots (Qin and Liu 1984).

the upper root (4–6 cm below the root node). The difference in the ^{15}N depletion rate by these two roots increased with time (Qin and Liu 1984).

However, the fertilizer N distribution in the rhizosphere of upland crops was not the same as that in the rhizosphere of flooded rice. The results of experiments conducted with upland crops such as maize, barley and rye grass, show that the maximum depletion of ^{15}N occurred at about 2 mm from the root surface. Within the zone 0–2 mm from the root surface ^{15}N accumulated; the nearer to the root surface, the greater the accumulation (Figure 10.3; Qin and Liu 1984). This finding applied whether ammonium sulfate or urea was applied. For all three upland crops tested, the trend was identical. The rate of accumulation was somewhat reduced with time (Figure 10.4). For barley on the 6th day of uptake, the depletion rate in the rhizosphere within 1 mm of the root surface was zero. However, the maximum depletion occurred 2–3 mm further away from the root surface. After that distance, the direction of the concentration gradient was reversed; i.e. the depletion rate decreased with distance from the root surface (Qin and Liu 1984).

It appears that the difference in the distribution of fertilizer ^{15}N in the rhizosphere between the upland crops and rice plants was due primarily to the variation in the quantity of exudates and material sloughed of the root. In another experiment, we let a maize plant absorb ^{15}N ammonium and then placed it in a soil fertilized with non-labelled N. Six days later, it was found that within 1 mm of the root

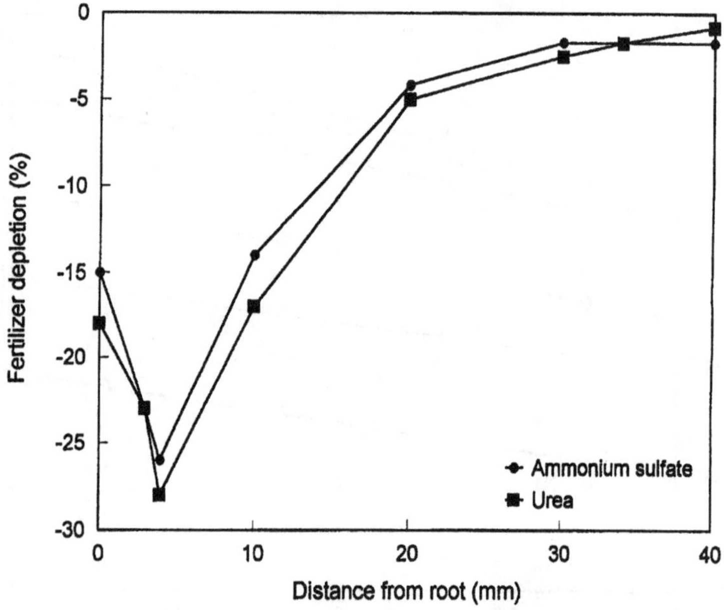

Figure 10.3. Depletion of fertilizer nitrogen in maize rhizosphere soil under upland conditions (Qin and Liu 1984).

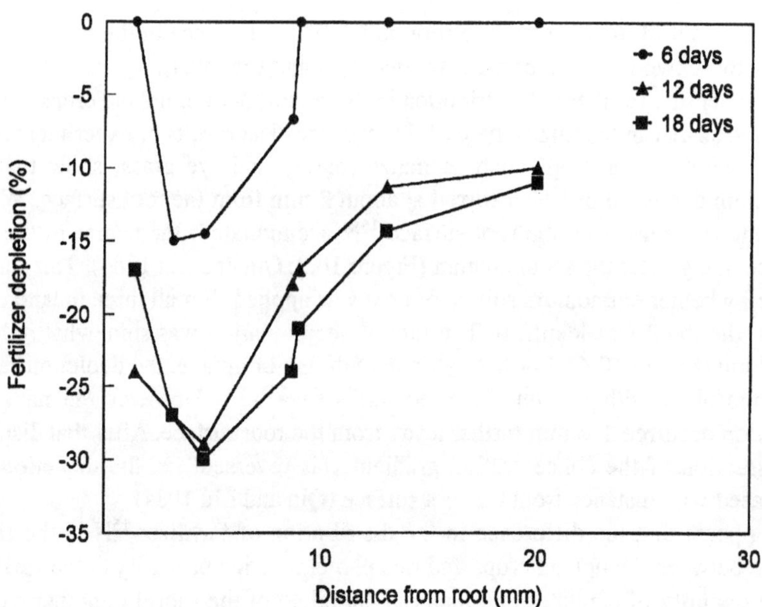

Figure 10.4. Accumulation of fertilizer nitrogen (15N ammonium sulfate) in rhizosphere of barley (Qin and Liu, 1984).

surface the ^{15}N concentration was, on average, 174 μg N 100 g soil^{-1}; it decreased to 45 μg N 100 g soil^{-1} at 2 mm, and was very small at distances beyond 2 mm (Qin and Liu 1989). This proves that the N which accumulates around the roots of upland crops was derived from the root. Other reports indicated that P and K also accumulated within 1 mm of the root surface of upland crops (Jungk *et al.* 1982).

Investigations on the effect of water stress on the distribution of fertilizer ammonium in the rhizosphere of rice and upland crops showed that when rice was grown under upland conditions, there was an accumulation of fertilizer N within 0–2 mm of the root. This was in contrast to rice grown under flooded conditions, but the rate of accumulation was somewhat lower. However, when maize was grown under flooded conditions, such an accumulation was much less noticeable, when compared with maize grown under normal upland conditions (Table 10.3).

These results further indicate that any differences in distribution of fertilizer N in the rhizosphere of plants grown under flooded or upland conditions can be attributed to differences in soil aeration and mechanical resistance. Läuchli and Bieleski (1983) also provide evidence that soil aeration and mechanical resistance greatly influence the rate of root exudation.

10.3.2. Fertilizer nitrate

Nitrate is rarely adsorbed by the soil. It is mainly present in the soil solution as a solute so is subject to leaching and denitrification. Under flooded conditions, the enrichment of denitrifying bacteria in the rhizosphere of rice may enhance the loss of nitrate (Li *et al.* 1984). Mengel and Viro (1978) and Ta and Ohira (1982) found no distinct difference between uptake of ammonium and nitrate by rice. However, nitrate was usually superior to ammonium for the growth of upland crops. It was found that the direction of the concentration gradient for nitrate was opposite to that of ammonium whether in the rhizosphere of rice or upland crops, i.e. with nitrate the concentration decreased as the distance from the root surface increased (Figure 10.5). Moreover, the apparent ^{15}N cumulation usually occurred within 10 mm of the root

Table 10.3. Effect of soil water status on nitrogen depletion (%) in the rhizosphere of rice and wheat after addition of ^{15}N-labelled ammonium.[1]

Distance from root surface (mm)	Rice		Maize	
	Flooded	Upland	Flooded	Upland
1	37.0	13.3	4.2	0.8
2	33.2	22.4	17.7	21.8
3	30.8	27.3	18.3	25.4
4	28.5	23.6	17.5	22.7
5	25.5	22.8	16.3	21.8
10	17.3	12.4	11.4	12.8
20	7.7	3.0	4.1	2.0

[1] Qin and Liu (1989).

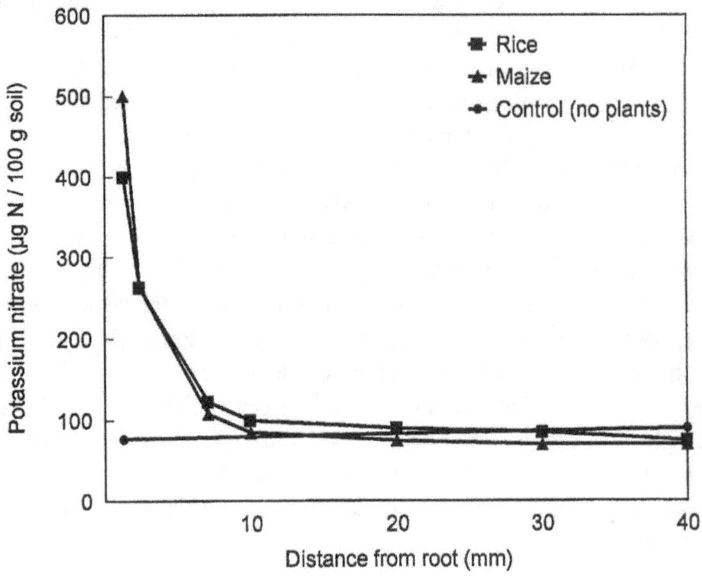

Figure 10.5. Accumulation of potassium nitrate in rhizosphere of maize and rice (Qin and Liu 1989)

surface, a much narrower range compared with ammonium. The maximum concentration close to the root surface was about 5 times that in the initial, uncultivated, soil. Five mm from the root surface it was reduced to twice the initial concentration, and at 10 mm it was almost the same as that of the initial soil.

It is considered that the accumulation of nitrate in the rhizosphere is attributed to the fact that the concentration of nitrate in the soil solution is usually higher than that of ammonium. Analytical data now available show that the average nitrate concentration is 100–20,000 μM, compared with 100–2,000 μM for ammonium (Barber 1984). Assuming the flux of soil water toward the root to be $0.5\text{–}2.0 \times 10^{-6}$ $cm^3\,cm^{-2}\,s^{-1}$, the amount of nitrate reaching the root surface through mass flow is estimated to be several times the amount of N taken up by the plant, thus resulting in an accumulation of nitrate at the root surface.

10.4. Effect of fertilizer on rhizosphere pH

Riley and Barber (1969) showed that when plants were supplied with nitrate, the pH of the soil within 1–2 mm of the roots increased considerably, and the bicarbonate concentration in the soil increased as nitrate was absorbed by the root. In contrast, when ammonium was added, the pH in the rhizosphere decreased. A number of factors influence rhizosphere pH, but the most important appears to be the form of fertilizer N. Other factors, such as pH and buffering capacity of the soil and crop variety, may modify the effect of N on the acidity of the rhizosphere, in both direction and extent (Cunningham 1964).

10.4.1. Release of H^+ or OH^- from plant roots

It can be seen from Table 10.4 (Wu 1986) that OH^- was released from the roots of 7- or 25-day-old maize seedlings when nitrate was added, while H^+ was released when ammonium was added. Of the 3 crops tested, maize, buckwheat and soybeans, the biggest effect of the applied N on acidity or alkalinity, i.e. the ratio m mol H^+/m mol N or m mol OH^-/m mol N occurred with maize. While H^+ or OH^- were released when N was applied to buckwheat, the effect was smaller and the ratios were narrower. With soybeans, OH^- release was found only when nitrate was added to 7-day-old seedlings, and the m mol OH^-/m mol N ratio was only about one tenth of the ratio found for maize seedlings. However, only H^+ was released from the 25-day-old soybean seedlings irrespective of whether they received nitrate or ammonium. In contrast to the large amount of H^+ or OH^- released from maize roots on addition of N, the change in pH of the growth medium per unit N taken up i.e. ΔpH/m mol N, was found to be the lowest with maize (except for the 7-day-old seedlings supplied with nitrate). This suggests that the effect on pH was closely controlled by the buffering effect of the root exudates.

10.4.2. Response of plants to fertilizer form

Both nitrate and ammonium can be taken up rapidly by different plants. If the two forms of N are present in a culture solution, the pH of the medium usually decreases initially and then increases (Figure 10.6). This suggests that plants absorb ammonium preferentially, and then later the proportion of nitrate taken up increases (Wu 1986). This probably occurs because ammonium inhibits the activity of nitrate reductase in plants, thus retarding the uptake of nitrate. However, nitrate does not have a significant effect on ammonium uptake (Mengel and Viro 1978).

Table 10.4. Effect of form of nitrogen on release of H^+ or OH^- from roots and pH of growth medium.[1]

Form		7-day-old seedlings			25-day-old seedlings		
		Soybean	Buckwheat	Maize	Soybean	Buckwheat	Maize
NO_3	N uptake (mmol)	0.531	0.795	1.29	0.780	0.783	1.56
	OH^- release (mmol)	0.022	0.090	0.569	–0.078	0.097	0.860
	mmol OH^-/ mmol N	0.041	0.113	0.441	–0.033	0.041	0.183
	ΔpH/mmol N	+0.584	+0.818	+1.35	–0.310	+0.240	+0.155
	Buffer capacity of medium	0.136	0.278	0.654	0.071	0.115	0.786
NH_4	N uptake (mmol)	0.281	0.351	0.985	1.01	0.660	2.01
	H^+ release (mmol)	0.092	0.076	0.448	0.187	0.052	1.01
	mmol H^+/mmol N	0.327	0.216	0.455	0.062	0.026	0.167
	ΔpH/mmol N	–5.48	–3.83	–2.37	–0.293	–0.295	–0.171
	Buffer capacity of medium	0.120	0.112	0.384	0.141	0.073	0.650

[1] Wu (1986).

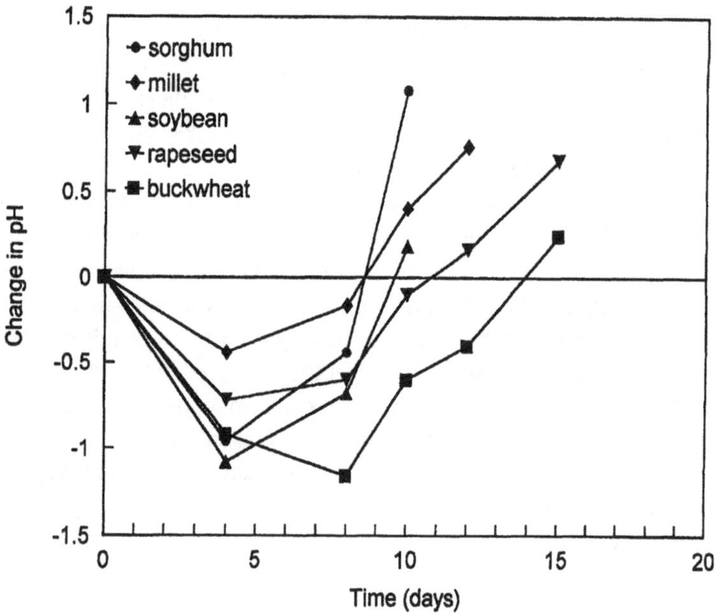

Figure 10.6. Effect of plant type on pH changes in growth medium (water culture containing ammonium and nitrate; Wu 1986).

When nitrate or ammonium is the sole source of N, the pH of the medium may vary with the kind of plant. Plants can be classified into three groups in this regard, viz. cereals, buckwheat, and legumes. When the cereals (rice, wheat, barley, maize and millet) were grown in a medium with ammonium, whether ammonium sulfate or ammonium bicarbonate, the pH of the solution quickly decreased (Figure 10.7). On the 5th day of the experiment the pH of the medium decreased by 1.5–3.0 units and approached its lowest point. However, when nitrate was used as the N source, during the first 15 days the pH of the medium increased with time by 0.5–1.5 units, but at a much lower rate than the pH decrease induced by ammonium. As shown in Figure 10.7, the change in pH of the growth medium was similar for the 4 crops under investigation, and only a slight difference was found between rice and the upland crops. A similar effect of added N on rhizosphere pH was observed when flooded rice and upland crops were grown in soil. However, the pH change with flooded rice, was less pronounced in soil than in water culture, and the affected area was smaller than that when upland crops were grown in soil; this be related partly to the effect of flooding on soil pH (Figure 10.8; unpublished data of Liu Zhi-yu).

The dicotyledon buckwheat belongs to the second group. When ammonium was used as the N source the pH of the medium decreased rapidly, as with cereals. However, when nitrate was used, during the first 12 days the pH increased initially by ~0.3 unit, and then decreased (Figure 10.9). With time of uptake and after every renewal of the culture solution, the pH fluctuated within a certain range, thus an adequate rhizosphere pH was maintained.

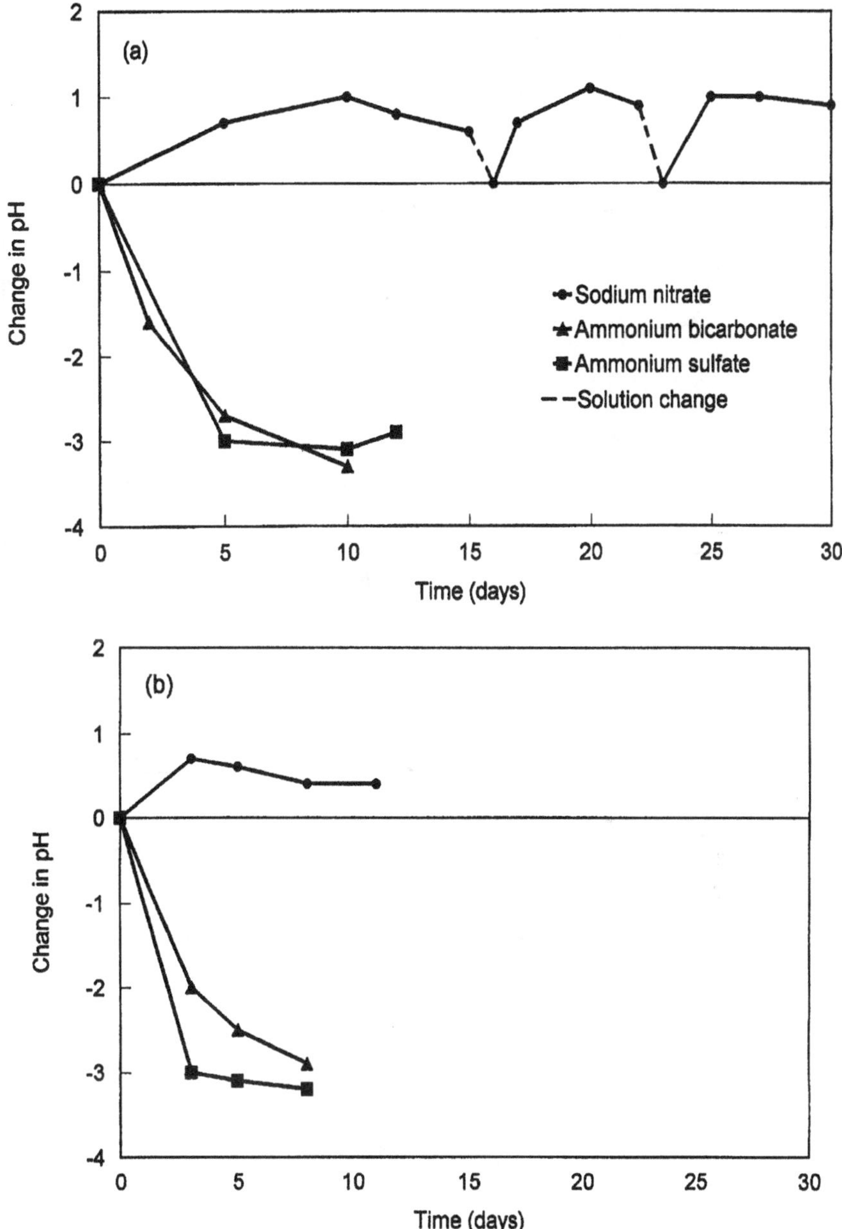

Figure 10.7. Effect of form of nitrogen supplied to (a) rice, (b) maize, (c) barley and (d) millet on pH of growth medium (Liu and Wu 1986).

The legumes respond differently to the form of N than the other two groups. The pH of the medium declined irrespective of whether ammonium or nitrate were supplied and the decrease became more apparent with time (Figure 10.10). One possible explanation for the pH decrease after application of nitrate is that legumes

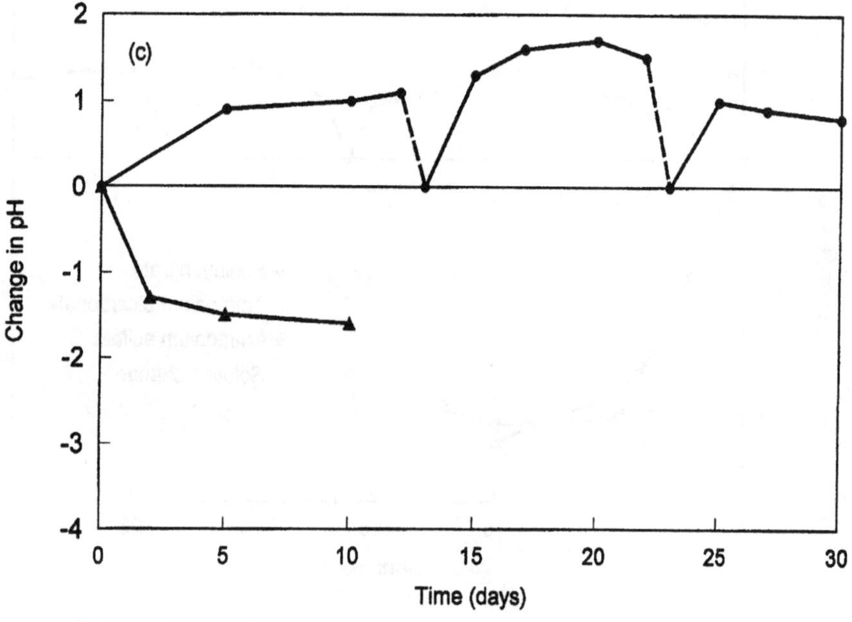

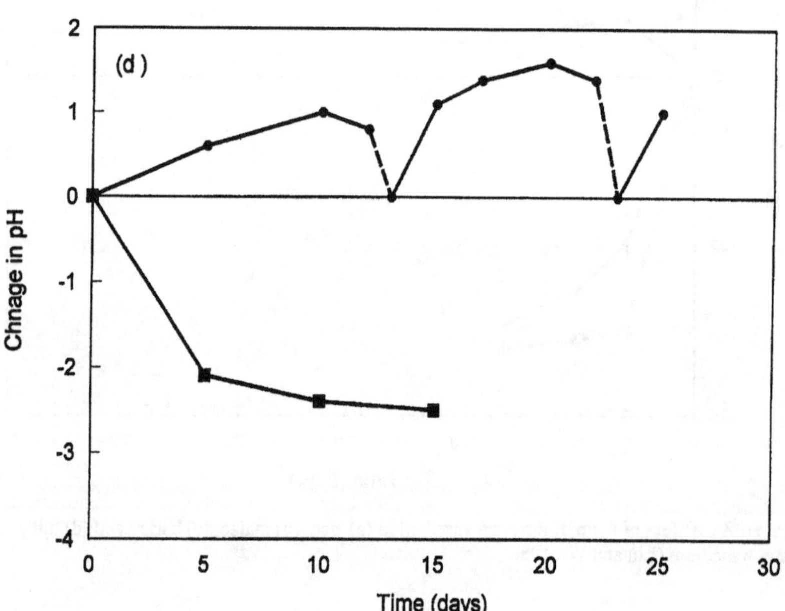

Figure 10.7. (*continued*)

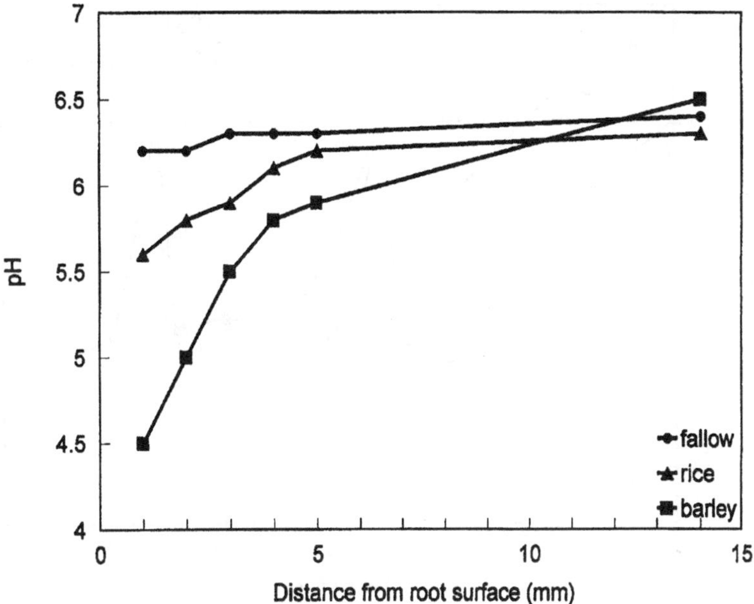

Figure 10.8 Rhizosphere pH of rice and barley grown in submerged paddy soil fertilized with ammonium sulfate (unpublished data of Liu).

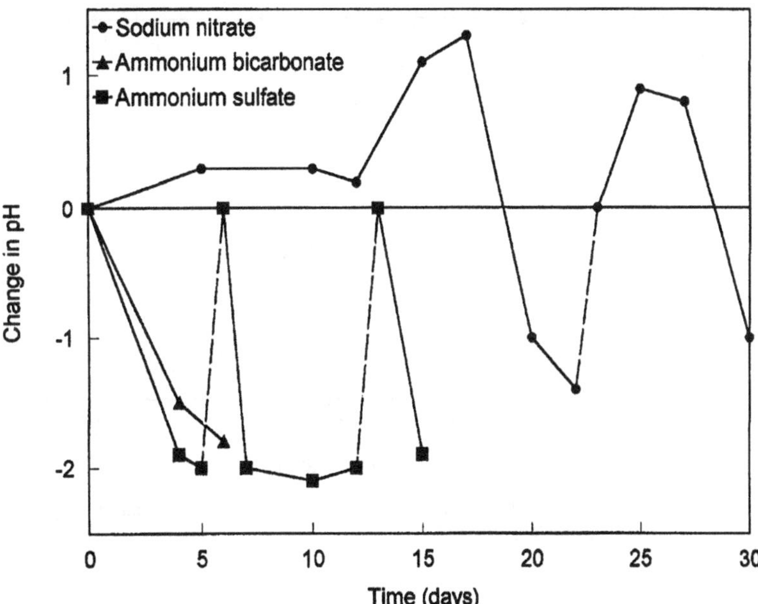

Figure 10.9. Effect of form of nitrogen supplied to buckwheat on pH of growth medium (water culture; Liu and Wu 1986).

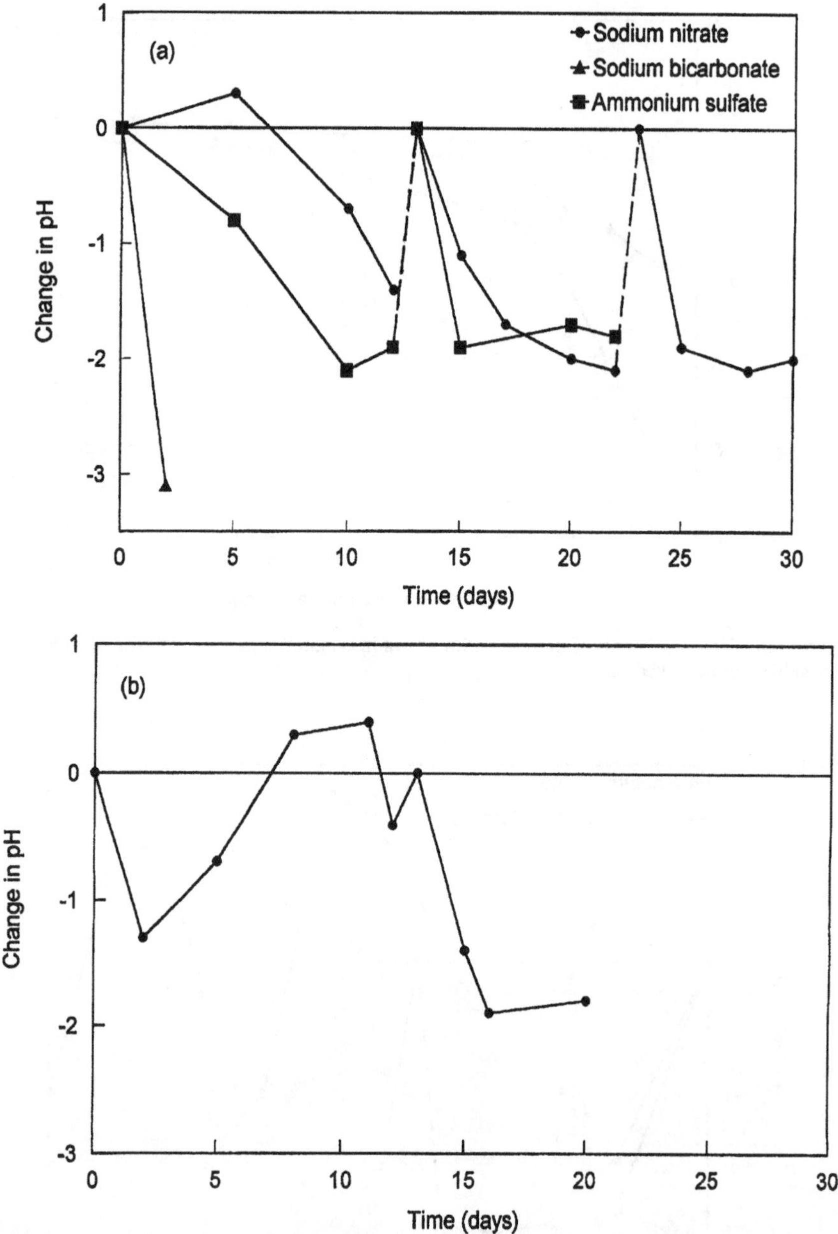

Figure 10.10. Effect of form of nitrogen supplied to soybean (a) and mung bean (b) on pH of growth medium (water culture; Liu and Wu 1986).

absorb more Ca^{2+} which results in an excess of cations over anions in the plant, and excretion of H^+ from the plant to balance the positive and negative charges (Wu 1986). Another possible explanation is that legumes first translocate nitrate from the root to the leaf before assimilation into organic N, whereas cereals transform nitrate into organic N in the root, and then translocate the N to the tops. Thus, legumes have no problems with excess nitrate.

No definite results have been obtained as to whether different crop varieties respond differently to form of N. As shown by Fan (1991) in a comparison of indica and japonica rices (Figure 10.11), when ammonium was supplied, the pH values at different points along the seminal root of both subspecies decreased gradually from the root tip to root base. The pH at the point of the emerging secondary lateral roots (8 cm or more from the tip of the seminal root) was more than one unit lower than the initial soil pH. When nitrate was the N source, the pH values along the roots of both subspecies were higher than the initial soil pH, and the pH from the root tip upward decreased in the same way as with ammonium but to a lesser extent.

The effect of urea on the pH of the rhizosphere soil, was somewhat different from that of ammonium or nitrate, and was related to soil conditions. As shown in Table 10.5, on a fluvo-aquic soil with an initial pH of 8.5, the rhizosphere pH of winter wheat, measured one week after urea application, was considerably lower than the non rhizosphere pH when the rate of application was low, or when a high rate of urea was applied with K_2SO_4. However, there was little difference in pH between rhizosphere and non-rhizosphere soils when urea at a high rate was applied alone. When urea was applied to flooded rice on acid or slightly acid soils, the rhizosphere often had higher pH values than the non-rhizosphere soil, except in the newly reclaimed red soil which had an initial pH as low as 5.0 (Table 10.6). It seems that the rate of urea hydrolysis in soil and the factors controlling it need to be

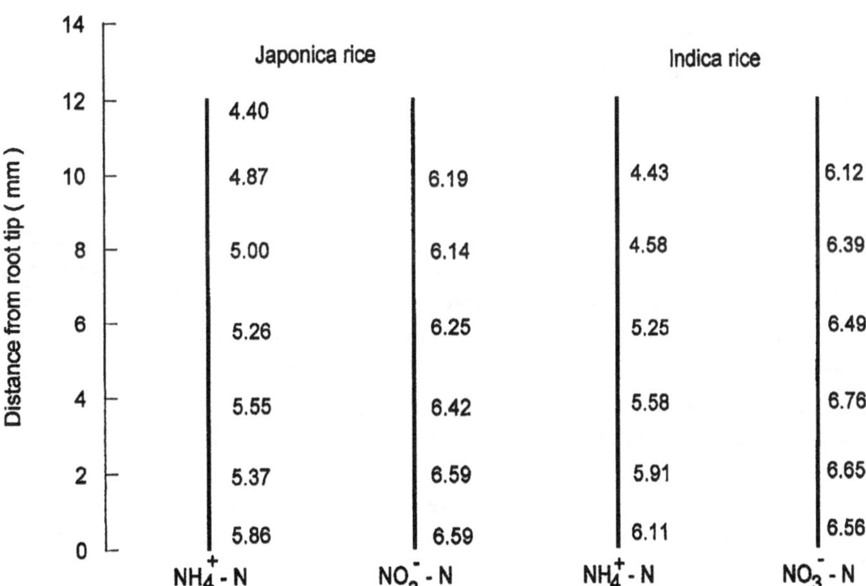

Figure 10.11. pH values along the seminal roots of rice seedlings supplied with different forms of nitrogen (Fan 1991).

Table 10.5. Effect of urea on rhizosphere pH of winter wheat grown on a calcareous soil.[1]

Treatment	Rhizosphere pH	Non-rhizosphere pH
120 kg urea ha^{-1}	7.88	8.07
240 kg urea ha^{-1}	8.37	8.49
240 kg urea + 180 kg K_2SO_4 ha^{-1}	7.69	8.29

[1] Measurements made 7 days after application of fertilizer at tillering (Shi and Liu 1987).

Table 10.6. Effect of urea on the rhizosphere pH of rice grown on acid soils.[1]

Soil	pH		
	Initial soil	Non-rhizosphere soil	Rhizosphere
Paddy soil derived from red soil	6.13	6.48	6.82
Newly reclaimed red soil	5.04	4.71	4.69
Newly reclaimed lateritic red soil	5.26	5.77	5.92

[1] Determinations made one month after urea application (Wang and Liu 1990).

taken into account when considering the effect of urea on pH of rhizosphere and non-rhizosphere soils.

10.5. Uptake of nitrogen by plants

Different crops vary greatly in their demands for N. For instance, rice requires ~1.6–2.5 kg N for every 100 kg of grain produced, whereas wheat, maize and other upland crops need ~2.4–4.3 kg N. In order to meet the N requirement for high yielding crops it is important to study the relationship between plant roots and N uptake.

10.5.1. Uptake by roots

Wada *et al.* (1986) showed that over 50% of the N accumulated in plants at maturity was derived from soil N, and the contribution of soil N to total N in rice is higher than that in upland crops. Moreover, the contribution varied with the growth stage of the plant, and it appears that a vigorous growth of roots enabled better utilization of soil N, and thus increased the relative contribution of soil N to the total N in the plant. For example, at the elongation stage of single cropped late rice (long duration variety), when the roots grew vigorously the contribution of soil N to total N was as high as 73.5% (Table 10.7) However, the contribution of fertilizer N in the reproductive stages increased markedly irrespective of whether the N was applied as a basal or top dressing (Wan *et al.* 1983). However, with early rice (short duration variety) in a triple cropping system, the contribution of soil N at an early growth stage was low, and that of fertilizer N may be as high as 60%. The contribution of soil N increased with growth and was 70% at heading (Xi *et al.* 1978). The

Table 10.7. Uptake of soil and fertilizer nitrogen by rice at different growth stages.[1]

Treatment	Stage of growth	Contribution (%)	
		Fertilizer	Soil
Basal fertilizer	Peak tillering	39.0	61.0
	Jointing	26.5	73.5
	Booting	40.0	60.0
	Heading	45.8	54.2
	Full heading	46.9	53.1
	Maturing	45.5	54.5
Fertilizer applied at jointing	Booting	39.5	60.5
	Heading	45.7	54.3
	Full heading	46.2	53.8
	Maturing	43.6	56.8

[1] Single cropped late rice grown on a submerged paddy soil in Wuxi (Wan *et al.* 1983).

large contribution of fertilizer N at the early growth stage of early rice may be related to the large amount of fertilizer N that was applied at the initial tillering stage. This effect is reinforced by the data in Table 10.8 (STU and SAS 1978). In addition, soil properties and method of application markedly influence the relative contribution of soil and fertilizer N (Zhu *et al.* 1979).

10.5.2. Root growth and uptake

For cereals, root number and weight are related to the number of tillers. However, N uptake is closely related to the age and activity of roots. STU and SAS (1978) showed that N uptake day^{-1} by rice was related to the number of emerging roots and the weight of healthy white roots, but the peak of N uptake occurred later than the peak of root emergence. It was found that young roots had a higher respiratory intensity per unit weight of root, and a higher rate of N uptake. According to early reports, the length and dry weight of adventitious rice roots did not increase after 14 days and after that time the rate of N uptake depended on the number and length of lateral roots. The most intense uptake of N occurred on average, on the 8th–11th

Table 10.8. Effect of rate of application of basal fertilizer on uptake of soil and fertilizer nitrogen by rice.[1]

Rate of application (g N m^{-2})	Total N (g N m^{-2})	Source of N			
		Fertilizer		Soil	
		(g N m^{-1})	(%)	(g N m^{-1})	(%)
0	0.68	0	0	0.68	100
2	1.68	0.62	36.9	1.06	63.1
4	2.85	0.61	56.5	1.24	43.5

[1] STU and SAS (1978).

days of growth, when the rate of ammonium uptake may be 12–13 mg N per 100 mg dry root per hr. However, only 1.8–3.0 mg N per 100 mg dry root per hr was taken up after the 14th day.

As shown in Figure 10.12, the N concentration in roots varied almost in parallel with that in the tops, but the maximum concentration in roots occurred earlier, usually close to the elongation stage. It was reduced to a minimum after panicle initiation (unpublished data of Liu Zhi-yu). The variation in N concentration of rice roots with growth stage was similar for different varieties of rice (Liu and Liu 1962). The N concentration in roots was much lower than that of the leaf blade; the maximum concentration was ~2% and the minimum ~1%, and these were similar to those for the leaf sheath (Figure 10.12).

10.5.3. Toxic effect of ammonia

The concentration of ammonia in solution depends on the pH, ammoniacal N concentration and temperature. Figure 10.13 shows that at a given temperature, the ammonia concentration increases sharply when the pH is higher than 8. At pH 9 about 50% of the ammoniacal N present is ammonia. Liu *et al.* (1987) and Luo *et al.* (1985) found that an ammonia concentration in solution in excess of 5 mg kg^{-1} was toxic to plant roots. The toxicity resulted in the absence of root hairs, fewer and shorter root branches, and a change in colour to yellow-brown.

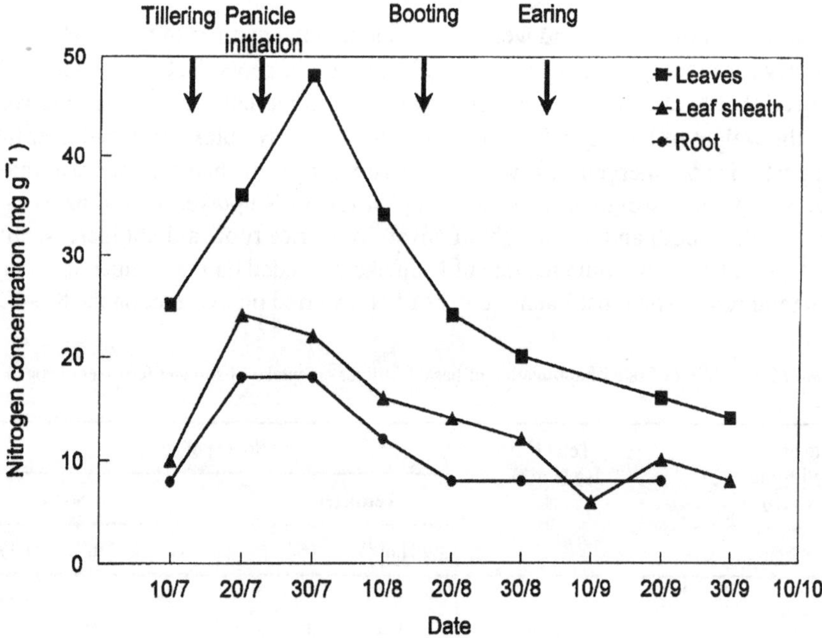

Figure 10.12. Nitrogen in various organs of rice at different growth stages (single-cropped late rice, Nongken 58; unpublished data of Liu).

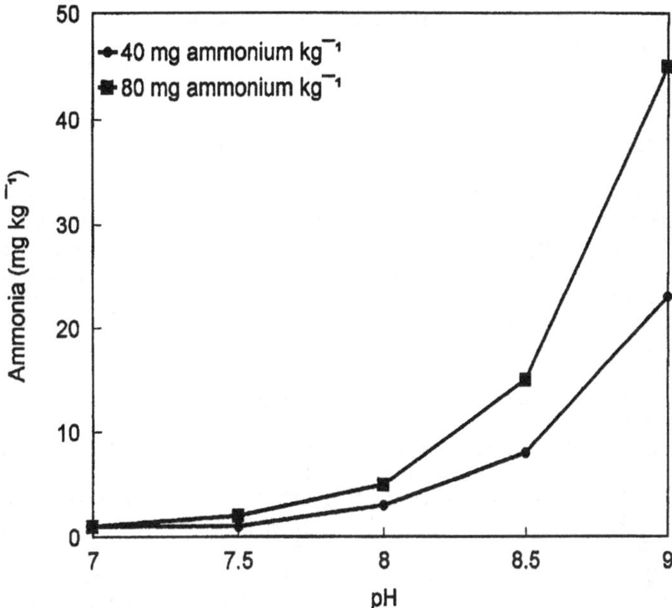

Figure 10.13. Effect of pH on the relationship between ammonium and ammonia concentration in soil solution.

In addition, ammonia affected potassium uptake by plant roots. The injured roots of wheat, maize and rice exuded potassium, so that there was an appreciably higher potassium concentration in the 0–1 mm soil layer around the root than in the control plants. In severe cases, ammonia may cause a physiological potassium deficiency in the plant.

Consequently, it is not advisable to surface broadcast large amounts of ammonium or ammonium producing N fertilizer on calcareous soils. The effective measures to prevent ammonia toxicity are optimisation of the rate of N fertilizer application, deep placement of N fertilizer, and use of mixtures of ammonium fertilizers and acid producing materials.

10.6. References

Balandreau, J, Rinaudo, G, Fares-Hamad, I and Dommergues, Y 1975. Nitrogen fixation in the rhizosphere of rice plants. In: Stewart, W D P (ed.), Nitrogen Fixation by Free-Living Micro-organisms. pp. 57–70. Cambridge Univ. Press.

Barber, D A and Gunn, K B 1974. The effect of mechanical forces on the exudation of organic substances by the roots of cereal plants grown under sterile conditions. New Phytol. 73:39–45.

Barber, S A 1984. Soil nutrient bioavailability: A mechanistic approach. pp. 95–99. John Wiley & Sons, Chichester.

Cunningham, P K 1964. Cation-anion relationships in crop nutrition. III. Relationships between the ratios of sum of the cations: sum of the anions and nitrogen concentrations in several plant species. J. Agric. Sci. 63:109–111.

Döbereiner, J 1983. Dinitrogen fixation in rhizosphere and phyllosphere association. In: Lauchli, A and Bieliski, R L (eds.), Encyclopedia of Plant Physiology, New Series. pp. 330–350. Springer-Verlag, Berlin.

Fan, X H 1991. H^+ exudation from rice and wheat roots in relation to P level in medium. (in Chinese). Acta Phytophysiol. Sin. 17:125–132.

Foster, R C 1981. The ultrastructure and histochemistry of the rhizosphere. New Phytol. 89:263–273.

Haller, T and Stolp, H 1985. Quantitative estimation of root exudation of maize plant. Plant Soil. 86:207–216.

Idris, M, Vinther, F P and Jensen, V 1981. Biological nitrogen fixation associated with roots of field-grown barley. Z. Pflanzenernahr. Bodenkd. 144:385–394.

Jungk, A, Claassen, N and Kuchenbuck, R 1982. Potassium depletion of the soil-root interface in relation to soil parameters and root properties. Proc. 9th Intern. Plant Nutrition Colloq., Vol. 1:253.

Krotzky, A, Berggold, R, Jaeger, D, Dart, P J and Werner, D 1983. Enhancement of aerobic nitrogenase activity by phenol in soil and the rhizosphere of cereals. Z. Pflanze nernahr. Bodenkd. 146:634–642.

Lauchli, A and Bieleski, R L 1983. Encyclopedia of Plant Physiology. New Series. Vol. 12. Inorganic Plant Nutrition. pp. 5–60. Springer-Verlag, Berlin.

Li, L M, Zang, S, Zhou, X R and Pan, Y H 1984. Effect of rice root on N loss. (in Chinese). Soils (16):5–10.

Liu, Z Y 1980. Outline of the study on nutrient environment in the soil-root microzone. (in Chinese). Progress in Soil Science (3):1–11.

Liu, Z Y and Liu, W L 1962. Effect of soil N supply intensity and its duration on coordinate growth of various organs of rice plant. (in Chinese). Acta Pedol. Sin. 10:145–159.

Liu, Z Y, Luo, Z C, Qin, S W Shi, W M and Xu, M L 1987. Study on rational application of urea on calcareous soils. In: Studies on the Regional Management Technology System for Huang-Huai-Hai Plains. (in Chinese). pp. 80–88. Science Press, Beijing.

Liu, Z Y and Qin, S W 1981. The study of nitrogen distribution around rice rhizosphere. In: Institute of Soil Science, Academia Sinica (ed.), Proc. Symp. Paddy Soil. pp. 541–546. Science Press, Beijing.

Liu, Z Y and Wu, W B 1986. The relationship between the status of rhizosphere pH of different crops and the form of nitrogen fertilizer. In: Soil Sci. Soc. of China (ed.), Current Progress in Soil Research in P R China. pp. 244–253. Jiangsu Sci. and Tech. Pub. House, Nanjing.

Luo, Z C, Tang, Y L and Liu, Z Y 1985. Inhibitory action of NH_3 from hydrolyzed urea on plant roots. (in Chinese). Acta Pedol. Sin. 22:56–63.

Marumoto, T 1984. Mineralization of C and N from microbial biomass in paddy soil. Plant Soil. 76:165–173.

Mengel, K and Viro, M 1978. The significance of plant energy status for the uptake and incorporation of NH_4–N by young rice plants. Soil Sci. Plant Nutr. 24:407–416.

Millet, E and Feldman, M 1984. Yield response of a common spring wheat cultivar to inoculation with *Azospirillum brasilense* at various levels of nitrogen fertilization. Plant Soil. 80:255–259.

PNGISSAS 1988. (Plant Nutrition Group of Institute of Soil Science, Academia Sinica) (ed.), Rhizosphere Research Method. (in Chinese). Soil Science Society of China. p. 96.

Polonenko, D R, Dumbroff, E B and Mayfield, C I 1983. Microbial responses to salt-induced osmotic stress. III. Effect of stress on metabolites in the roots, shoots and rhizosphere of barley. Plant Soil. 73:211–225.

Qin, S W and Liu, Z Y 1984. Studies on nutrient status in the soil-root microzone. III. Changes in nitrogen in rice rhizosphere. (in Chinese). Acta Pedol. Sin. 21:238–246.

Qin, S W and Liu, Z Y 1989. Migration regularity of different forms of N in rhizosphere. (in Chinese). Acta Pedol. Sin. 26:117–123.

Ratnayake, M, Leonard, R T and Menge, A 1978. Root exudation in relation to supply of phosphorus and its possible relevance to mycorrhizal infection. New Phytol. 81:543–552.

Riley, D and Barber, S A 1969. Bicarbonate accumulation and pH changes at the soybean root-soil interface. Soil Sci. Soc. Am. Proc. 33:905–908.

Shi, W M and Liu, Z Y 1987. Effect of NH_3 and urea on K distribution in root and rhizosphere of maize. Fert. Res. 14:235–244.

Sims, G K and Dunigan, E P 1984. Diurnal and seasonal variation in nitrogenase activity (C_2H_2 reduction) of rice roots. Soil Biol. Biochem. 16:15–18.

STU and SAS 1978. (Shanghai Teachers University and Shanghai Agricultural School). Physiology of Rice Cultivation. (in Chinese). Shanghai Science and Technology Publishing House, Shanghai. pp. 309.

Ta, T C and Ohira, K 1982. Comparison of the uptake and assimilation of ammonium and nitrate in Indica and Japonica rice plants using the tracer ^{15}N method. Soil Sci. Plant Nutr. 28:79–90.

Trolldenier, G 1977. Influence of some environmental factors on nitrogen fixation in the rhizosphere of rice. Plant Soil. 47:203–217.

Wada, G, Shoji, S and Mae, T 1986. Relationship between nitrogen absorption and growth and yield of rice plants. JARQ 20:135–145.

Wan, C B, Wu, J G, Sun, P L and Pan, Z P 1983. Recovery of basal and top-dressed fertilizer-N by single-cropping late rice and the partial efficiency of added N. (in Chinese). Jiangsu Agricultural Science 12:23–28.

Wang, J L and Liu, Z Y 1990. Chemical behavior of heavy metals in rhizosphere. I. Rhizosphere effect of Cu adsorption in soil. (in Chinese). Acta Scientiae Circumstantiae 11:178–186.

Wu, W B 1986. pH status in rhizosphere soils of different crops and its relation to the form of N fertilizers. (in Chinese). M.Sc. thesis, Institute of Soil Science, Academia Sinica. 51 p.

Yoshida, T and Yoneyama, T 1980. Atmospheric dinitrogen fixation in the flooded rice rhizosphere as determined by the N-15 isotope technique. Soil Sci. Plant Nutr. 26:551–559.

Xi, Z B, Bian, Y J Kuang, A Q, Liu, D B and Liu, M Y 1978. Studies on peak uptake of nutrients by double-cropping rice and on method of applying volatile N fertilizer throughout the plough layer. (in Chinese). Acta Pedol. Sin. 15:113–125.

Zhang, K H, Yu, Y C, Liang, Y K and Wang, S X 1962. Bacterial Physiology. (in Chinese). p. 16. People's Hygiene Publishing House, Beijing..

Zhu, Z L, Chen, R Y, Xu, Y F, Xu, Y H and Zhang, S L 1979. The effect of forms and methods of placement of nitrogen fertilizer on the characteristics of the nitrogen supply in paddy soils. (in Chinese). Acta Pedol. Sin. 16:218–233.

11
Fate and management of fertilizer nitrogen in agro-ecosystems

ZHU ZHAO-LIANG

11.1. Introduction

Since the 1950s the consumption of fertilizer N in China has increased dramatically. It was only 6000 t in 1950, and increased to 13.96 million t in 1987 (ISFCAAS 1986; ECACA 1988), accounting for 19% of the total world consumption in 1986/1987 (FAO 1987). It was forecast that in the year 2000 fertilizer N consumption in China will be further increased to 18 million t (ISFCAAS 1986; Lu 1979). Therefore, the efficient use of fertilizer N is more important than ever before.

The agronomic efficiency of fertilizer N (Ea) depends on (i) the apparent recovery of fertilizer N in plants (Ra), and (ii) the partial physiological efficiency of the increased uptake of N (Ep), expressed as the ratio of the increase in yield to the increase in N uptake as a result of the application of fertilizer N. In this chapter the improvement in Ra and Ea will be discussed in relation to the fate of fertilizer N in the crop–soil system.

11.2. Effects of plants on the fate of fertilizer nitrogen

The fate of fertilizer N applied to a field is the result of N transformations and movement in the soil as affected by plant growth, uptake and environmental conditions. The growth of plants exerts multiple effects on the transformations and movement of fertilizer N. For example, the uptake of mineral N by plants lowers its concentration in the soil, and this reduces the rates of subsequent transformation and transfer processes, such as ammonia volatilization, nitrification, denitrification and leaching. Furthermore, the well-established crop canopy may reduce ammonia emission from the soil surface. On the other hand, the excretion and sloughing off of the roots of plants may increase the activity of microorganisms in the rhizosphere, and enhance denitrification and N loss (Li *et al.* 1984). Table 11.1 shows the net result of the two opposite effects. The results imply that the dominant effect of plant growth is to reduce N loss.

Consequently, the greater the rate of N uptake, the smaller the N loss will be. Tables 11.2 and 11.3 show that topdressing at vigorous growth stages, when uptake is high (such as panicle initiation or elongation), consistently reduced N loss,

Zhu Zhao-liang et al. *(eds.): Nitrogen in Soils of China, 239–279.*

Table 11.1. Effect of rice plants on the fate of fertilizer nitrogen (% of N applied).[1]

Source	Treatment	Recovery in plants	Residual in soil	Loss	Reference
Urea	With rice	64.7	20.1	14.2	Li *et al.* (1984)
	Without rice	0	69.5	30.5	
	With rice	60.3	21.6	18.0	
	Without rice	0	73.1	26.9	
AS[2]	With rice	43.2	47.9	8.9	Liao *et al.* (1982)
	Without rice	0	84.1	15.9	

[1] Pot experiment with ^{15}N, non-calcareous paddy soil.
[2] See list of abbreviations.

Table 11.2. Fate of nitrogen applied to rice at different growth stages (% of N applied).[1]

Soil	N source	Timing of application	Recovery in plants	Residual in soil	Loss	Reference
CFA[2]	Urea	basal	22.3	30.4	47.3	Chen & Zhu (1982a)
		PI[3]	61.8	12.4	25.8	
NCPS	Urea	basal	27.5	18.6	53.9	Chen & Zhu (1982a)
		PI	64.7	5.4	29.9	
NCPS	Urea	basal	39.8	16.4	43.8	Chen & Zhu (1982a)
		PI	54.8	17.3	27.9	
NCPS	AS	basal	54.4	21.6	24.0	Yu *et al.* (1984)
		IT	52.3	11.8	35.9	
		PI	68.9	10.3	20.8	

[1] Microplot experiment with ^{15}N-labelled fertilizers surface broadcast.
[2] See list of abbreviations;
[3] PI = panicle initiation, IT = initial tillering.

Table 11.3. Fate of nitrogen applied to wheat at different growth stages (% of N applied).[1]

Wheat	Soil[2]	N source	Treatment	Recovery in plants	Residual in soil	Loss	Reference
Winter	CFA	Urea	Banded as basal	39.4	22.7	37.9	Zhang *et al.* (1989)
			SB-Ir-R	48.5	28.5	23.0	
			SB-Ir-E	46.6	33.2	20.1	
Spring	AAS	AN	Basal plowed in	42.4	11.3	46.3	Jin *et al.* (1985)
			TD-S	39.1	15.4	45.4	
			TD-E	57.8	21.6	20.6	
			TD-H	15.2	13.3	71.5	

[1] Field microplot experiments with ^{15}N.
[2] See list of abbreviations.

compared with N applied as a basal dressing or at heading, when the rate of N uptake was low. In addition, the elimination of constraints, such as the deficiency of nutrients other than N, will promote the growth and uptake of N by the plants, and reduce N loss (Table 11.4). In soils deficient in P and/or K, the combined application of P and/or K fertilizers with N was found to reduce N loss (Table 11.4). Similarly, the correction of water deficit by irrigation improved plant growth and root extension, thereby reducing ammonia volatilization (Cao 1983).

11.3. Relationship between loss and apparent recovery

Table 11.5 shows that Ra (estimated by difference) was consistently greater than Rt (recovery determined by tracer technique), irrespective of whether the recoveries

Table 11.4. Effect of P and K fertilization on the fate of urea-N applied to upland crops (% of N applied).[1]

Crop	Treatment	Recovery in plants	Residual in soil	Loss
Barley	N	27.5	29.4	43.2
	NP	44.2	25.6	30.3
	NPK	50.9	25.3	23.9
Wheat	N	47.1	14.5	38.4
	NPK	60.7	9.3	30.1

[1] Field microplot experiment with ^{15}N-labelled urea conducted on periodically submerged paddy soils in Taihu Region. (Pan *et al.* 1980).

Table 11.5. Relationship between nitrogen loss and recovery in plant (% of N applied).[1]

Soil	Soil No.	Recovery in plants						Residual in soil	Loss	Loss + Ra
		tops			whole plant					
		Ra[2] (A)	Rt[2] (B)	(A-B)	Ra (C)	Rt (D)	(C-D)	(E)	(F)	(C+F)
NCPS[2]	1	58.2	37.9	20.3	81.8	47.4	34.4	33.1	19.5	101.3
	2	65.9	40.6	25.3	86.6	51.9	34.7	34.0	14.1	100.7
	3	53.5	32.9	20.6	78.0	41.9	36.1	27.7	30.3	108.4
	4	59.9	37.1	22.8	78.4	46.2	32.2	27.5	26.3	107.5
Gleyey	5	50.8	34.7	16.1	69.5	43.3	26.2	27.3	29.4	98.9
	6	43.4	29.1	14.3	65.2	37.3	27.9	27.5	35.2	100.4
	7	50.3	30.6	19.7	75.7	39.6	36.1	29.8	30.6	106.3
Mean		54.6	34.7	19.9	76.5	43.9	32.5	29.6	26.5	103.4
SD		7.4	4.1	3.7	7.2	5.0	4.0	2.9	7.3	3.9

[1] Recalculated from the data of Cai *et al.* (1981). Pot experiment with flooded rice; labelled urea incorporated into 0–5 cm soil from southern Jiangsu Province.
[2] See list of abbreviations.

were determined in the whole plant or in the tops only. The difference between recoveries estimated by the two methods when calculated on the basis of the N in the whole plant (C–D in Table 11.5) was found to be almost equivalent to the residual ^{15}N in the soil (E in Table 11.5), indicating that it resulted primarily from the apparent priming effect (see Chapter 3). In other words, Ra was almost equivalent to the sum of Rt and residual ^{15}N in the soil (C + E close to D). Such a relationship can only be found when organic material is not added. When organic material with wide C/N ratio, such as straw, is added, the increased immobilization disturbs this relationship substantially. It is evident from the agronomic view point that Ra only should be used to evaluate the improvement in N nutrition through N application, whether a real priming effect occurs or not.

Table 11.5 shows that when the recovery of N is based on N in the aerial parts of plant only, as is usually done in field experiments, Ra underestimates the extent of the improved N nutrition through N fertilization. Fortunately, the underestimation of Ra in field experiments is not large, because there is little N in the roots of cereal crops.

There seems to be a complementary relationship between Ra based on N in the whole plant and N loss. As shown in Table 11.5, their sum (i.e C + F) was ~100%. Consequently, the loss of fertilizer N in a particular treatment can be estimated as the difference between 100% and Ra when the latter was derived from N in the whole plant. If such a relationship can be verified in field experiments, then N loss may be assessed from Ra based on N in tops, with a correction for N in roots.

11.4. Fate of nitrogen applied to flooded rice

11.4.1. Immobilization and fixation in relation to loss

Nitrogen applied to flooded rice is subject to a series of interlinked transformations and transfers, such as immobilization, ammonium fixation, nitrification-denitrification and ammonia volatilization. Microplot experiments (Table 11.6) showed that microbial immobilization, ammonium fixation, and total loss approached their maximum values 6–10 days after N application. After that time immobilized and fixed N was released and exchangeable ammonium was taken up by the rice plants, while the extent of N loss was almost unchanged. The immobilization-remineralization of mineral N and the fixation-release of ammonium resulted in a steady and long-lasting uptake of applied N by the plant, and reduced N loss (Wen and Zhang 1986; Zhu *et al.* 1963). It is evident that the rate of these processes may differ with the type of clay mineral and the content and microbial susceptibility of soil organic matter (Cheng *et al.* 1989).

11.4.2. Pathways of loss

Nitrogen can be lost from crop lands by ammonia volatilization, nitrification and denitrification, leaching and runoff. In microplot experiments used to investigate the fate of applied N, runoff is excluded. Leaching was found to be insignificant in

Table 11.6. Dynamics of fertilizer nitrogen applied to flooded rice (% of N applied).[1]

Soil	Treatment[2]	Days after application	Recovery in plants	NH_4^+ex	Immobilized and fixed	Loss	Reference
Non-calcareous							
	Urea, SB	6	9.3[3]	15.3	34.0	41.4	Zhu *et al.* (1979a)
		11	19.3[3]	4.8	28.7	47.2	
		41	28.3[3]	0.5	29.0	42.2	
	Urea, SB	6	8.3[3]	13.5	60.0	18.2	Zhu *et al.* (1979a)
		11	14.0[3]	4.3	45.2	36.5	
		21	20.5[3]	1.0	39.3	39.2	
		41	24.3[3]	0.5	27.3	47.9	
	AS, SB	6	9.5[3]	28.0	26.8	35.7	Zhu *et al.* (1979a)
		11	25.3[3]	6.8	30.5	37.4	
		41	38.0[3]	0.8	26.2	35.0	
	ABC, SB 9		4.0	35.6		60.4	Cai *et al.* (1985)
		65–66	25.4	17.6[4]		57.0	
	Urea, SB	9	5.7	46.2[4]		48.1	Cai *et al.* (1985)
		65–66	37.8	19.5[4]		42.7	
Calcareous							
	Urea, Inc	9–10	6.8	28.4[4]		64.6	Zhu *et al.* (1989b)
		51	27.5	11.6[4]		60.9	
	ABC, Inc	9–10	6.9	17.1[4]		76.7	Zhu *et al.* (1989b)
		51	20.9	11.3[4]		67.8	

[1] Field microplot experiments with ^{15}N, nitrogen applied at transplanting.
[2] See list of abbreviations.
[3] Nitrogen in roots was not included.
[4] Sum of exchangeable and fixed ammonium, and immobilized nitrogen.

most cases, except in soils with high percolation rate. In these soils the downward movement of applied ^{15}N was confined mainly to the plowed layer in the current growing season (Table 11.7).

In preliminary experiments, rice was grown in enclosures to evaluate the effect of soil reaction on ammonia loss from different N sources broadcast and incorporated into soil at transplanting. The enclosure was designed (Liao *et al.* 1982) so that rice could grow normally in the open air, while the air in the headspace of the enclosure was continuously changed 15–20 times min^{-1} (Zhu *et al.* 1985, 1987). Table 11.8 shows that with non-calcareous soils, ammonia loss from ammonium sulfate (AS) and ammonium chloride (AC) accounted for only 19% of the total loss, implying that nitrification-denitrification was the dominant pathway of N loss. However, when urea and ammonium bicarbonate (ABC) were applied, ammonia loss amounted to 42% and 62% of the total loss, respectively, indicating that ammonia loss was comparable with loss by nitrification-denitrification. When N was applied to calcareous soils, total loss and the contribution of ammonia loss to total loss was greater than that obtained with non-calcareous soils. The contribution of ammonia loss to total loss when AS and AC were applied to calcareous soils was similar to that obtained when urea was applied. Ammonia volatilization was the dominant pathway for N loss when ABC was applied to the calcareous soils,

Table 11.7. Distribution of nitrogen applied to flooded rice in the soil profile at maturity.[1]

Source	Texture	Soil depth (cm)	Residual N in soil (%)	Reference
ABC	Clay	0–15	16.4	Cai *et al.* (1985)
		15–30	1.2	
	Sandy loam	0–15	9.5	Zhu *et al.* (1989b)
		15–30	0.2	
		30–45	0.5	
		45–60	1.1	
AS	Clay loam	0–15	17.0; 23.3	Zhu *et al.* (1977)
		15–30	0.9; 2.2	
	Clay loam	0–10	27.6	unpubl. data of Mo, S X.
		10–20	2.9	
	Sandy loam	0–20	16.0-30.9	Li *et al.* (1981)
		20–40	tr.	
	Sandy loam	0–15	14.6	Zhu *et al.* (1988)
		15–30	0.3	
Urea	Clayey	0–15	19.2	Cai *et al.* (1985)
		15–30	0.3	
	Sandy loam	0–15	10.8	Zhu *et al.* (1989b)
		15–30	0.5	
		30–45	0.2	
		45–60	0.1	

[1] Field microplot experiments with ^{15}N.

Table 11.8. Gaseous loss of nitrogen applied to flooded rice at transplanting.[1]

N source[2]	Calcareous soil		Non-calcareous soil	
	Total loss (%)	Ammonia loss/total loss (%)	Total loss (%)	Ammonia loss/total loss (%)
AS, AC	26.7 ± 6.1[3]	42 ± 9 (n = 5)	22.8 ± 8.3	19 ± 15 (n = 5)
ABC	40.6 ± 19.7	72 ± 16 (n = 3)	30.2	62 (n = 2)
Urea	38.1 ± 17.2	45 ± 7 (n = 5)	34.1 ± 6.2	42 ± 22 (n = 4)

[1] Zhu *et al.* (1985, 1987). Rice plants growing in an enclosure with continuous aeration, air flushing frequency 15–20 times/min.
[2] See list of abbreviations.
[3] Values are mean ± standard deviation.

accounting for 72% of the total loss. Needless to say, the results obtained in enclosure experiments are only qualitative and *in situ* field measurements are necessary to quantify ammonia volatilization in a particular location.

Recently, ammonia loss has been determined in the field with micrometeorological techniques after application of N to flooded rice in China and overseas. Loss by nitrification-denitrification, however, can only be determined as the difference between total loss and ammonia loss, when leaching loss is insignificant, because there is no direct method available for its measurement. Obviously, this is only an apparent nitrification-denitrification loss, and the error of the estimation is inevitably high. Table 11.9 shows the relevant results obtained in field experiments with flooded rice at Fengqiu and Danyang, where the soils were calcareous and non-calcareous, respectively. As above with the enclosure experiments, total loss and ammonia loss from ABC and urea was considerably greater with the calcareous soil than with the non-calcareous soil; and the total loss of N from ABC was greater than that from urea. Ammonia loss varied greatly with soil reaction, and the properties of the N sources, ranging from 8.8% to 39.1%, while the variation in nitrification-denitrification loss was quite small (33% to 39%)

There is a linkage between ammonia volatilization and nitrification-denitrification, simply because ammonium is the common source of N for the two processes. This linkage should be considered in the development of new techniques for reducing N loss. For example, ammonia loss may be increased when the nitrification of ammonium is inhibited (Table 11.10; Zhu *et al.* 1985), while the enhancement of nitrification in upland soil may reduce ammonia loss (Fleisker and Hagin 1981).

11.4.3. Extent of loss

Since 1974 a number of microplot experiments have been conducted in the major rice-growing regions of China by different institutions. From the results obtained it is possible to quantify the extent of loss of fertilizer N applied with traditional techniques to flooded rice fields in China. However, it should be kept in mind that the

Table 11.9. Gaseous losses of nitrogen applied to flooded rice at transplanting (% of N applied).[1]

Site	Texture	Soil pH	Source	Total loss	NH_3 loss	Nitrification-denitrification loss[2]	NH_3 loss/ total loss %
Danyang	Clay	5.2–5.4	Urea	45.4	8.8	36.6	19
			ABC	58.7	19.5	39.2	33
Fengqiu	Sandy loam	8.8	Urea	62.9	30.1	32.8	48
			ABC	72.3	39.1	33.2	54

[1] Field measurements (Cai *et al.* 1985; Zhu *et al.* 1989b).
[2] Estimated as the difference between total loss and ammonia loss.

Table 11.10. Effect of nitrapyrin on nitrogen loss from ammonium sulfate broadcast and incorporated into a calcareous paddy soil at transplanting (% of N applied).[1]

Treatment	Recovery in plant	Residual in soil	NH_3 loss	Apparent denitrification	Total loss[2]
AS	48.1	27.3	9.8 a	14.9	24.7 a
AS + Nitrapyrin	46.2	36.1	13.4 a	4.3	17.7 a

[1] Enclosure experiment with rice, air flushing frequency 20 times min^{-1} (Zhu *et al.* 1985).
[2] Values in a column with the same character do not differ significantly at $P = 0.05$.

microplots used in these experiments were small (usually 30 cm in diameter). The cylinder used to isolate the microplots extended above the floodwater surface and this may reduce the speed of the wind across the surface, and produce a shading effect that will suppress ammonia loss (Trevitt *et al.* 1987). According to Cai (1988) the loss obtained in microplots of this size may be ~10% lower than that in microplots 1.2 m in diameter. The results conducted in China with small microplots are summarized in Table 11.11.

Table 11.11 shows that N loss varied greatly in different experiments (range 3–77%). The N losses could be ranked in the following order: calcareous soil > non-calcareous soil; ABC > urea > AS; surface-broadcast as basal or early growth stage ⩾ broadcast and incorporation as basal > deep placement of prilled urea or AS as basal > deep placement of urea supergranules (USG); surface-broadcast as basal or early growth stage > surface-broadcast at panicle initiation stage. It should be noted that N loss from urea surface-broadcast and incorporated or deep placed as a basal dressing varied greatly, presumably due to differences in the extent of urea retained by the soil. It was found that a considerable proportion of fertilizer N remains in the floodwater when the fertilizer was incorporated in the presence of floodwater as traditionally practised. Investigations have shown that the greater the proportion of the N left in floodwater after incorporation the greater the loss of ammonia. A large reduction in N loss can be achieved when the fertilizer is incorporated without water on

Table 11.11. Loss (%) of nitrogen applied to flooded rice.[1]

N source	Treatment[2]	Calcareous soil			Non-calcareous soil		
		No. of data	Mean	Range	No. of data	Mean	Range
ABC	SB-Tr	1	70	–	2	54	51–57
	Inc-Tr	4	63	53–73	10	56	36–77
	SB-PI	–	–	–	1	34	–
	Split	–	–	–	5	53	41–68
AS	SB-Tr	7	47	42–52	5	21	13–29
	Inc-Tr	3	56	50–64	7	35	25–43
	DP-Tr	3	21	3–30	2	19	18–20
	SB-Ti	1	31	–	4	37	34–40
	SB-PI	3	22	20–26	–	–	–
	Split	–	–	–	7	21	3–32
Urea	SB-Tr	2	48	47–48	2	49	44–54
	Inc-Tr	5	56	34–68	10	41	29–59
	DP-Tr	1	51	–	7	10	3–43
	USG-DP-Tr	1	21	–	2	16	13–18
	SB-PI	1	26	–	3	26	21–30
	Split	–	–	–	11	32	10–55

[1] Data obtained in field microplot experiments with ^{15}N in China (Cai and Fu 1985; Cai *et al.* 1985; Chen and Zhu 1982; Chen *et al.* 1986; Guo *et al.* 1980; He *et al.* 1981; Huang *et al.* 1982; Jia *et al.* 1982; Li *et al.* 1981; Liao *et al.* 1983; Mo and Qian 1981, 1983; Xi *et al.* 1978; Yu *et al.* 1984; Zhang *et al.* 1980; Zhu *et al.* 1977, 1988, 1989b.)

[2] See list of abbreviations.

the soil surface (Sun 1987; Zhu *et al.* 1988). Similarly, when prilled urea was deep-placed correctly, little N was left in the floodwater and N loss was small (Cao *et al.* 1984; He *et al.* 1981; Jia *et al.* 1982). However, if urea is not deep-placed correctly, N is left in the floodwater and the loss can be considerable (Chen and Zhu 1982b). It has been shown that among the techniques under investigation deep placement of supergranules is the most effective method for reducing N loss.

Data relevant to nitrate are not shown in Table 11.11. It is well known that loss of N by denitrification following application of nitrate fertilizers to flooded rice fields is large (94–96% of the N applied; Lu *et al.* 1981), and thus nitrate is not recommended as a fertilizer for flooded rice.

As ABC is the major N fertilizer in China, it may be expected that loss of fertilizer N applied by traditional techniques to flooded rice would be as high as 50%.

11.5. Fate of nitrogen applied to upland crops

11.5.1. Immobilization and fixation in relation to loss

Most of the investigations have been conducted with wheat and maize, the two major upland crops grown in China. Tables 11.12 and 11.13 show the results obtained in a winter wheat experiment conducted on a calcareous fluvo-aquic soil in warm temperate, north-central China. Table 11.12 shows that a large proportion of the applied N was present in the 0–20 cm soil layer, therefore it is reasonable to take the N in 0–40 cm soil layer as an estimate of the residual N in the soil profile. The results show that the fertilizer N fixed by clay minerals or immobilized reached a maximum at the first sampling on 8 December (59 days after fertilization). After

Table 11.12. Transformations of urea nitrogen in a calcareous fluvo-aquic soil (% of N applied).[1]

Date sampled	Depth (cm)	Exch NH_4^+	NO_3	Fixed NH_4^+	Organic	Total
8 Dec	0–20	tr[2]	27.1	2.8	32.8	62.7
	20–40	tr	7.6	1.6	8.7	17.9
	0–40	tr	34.7	4.4	41.5	80.6
22 Feb	0–20	tr	tr	1.4	32.0	33.4
	20–40	tr	tr	1.2	11.8	13.0
	0–40	tr	tr	2.6	43.8	46.4
4 April	0–20	tr	tr	0.8	25.7	26.5
	20–40	tr	tr	0.8	6.9	7.7
	0–40	tr	tr	1.6	32.6	34.2
5 May	0–20	tr	tr	1.3	15.9	17.2
	20–40	tr	tr	0.6	3.7	4.3
	0–40	tr	tr	1.9	19.6	21.5
31 May	0–20	tr	tr	1.2	14.1	15.3
	20–40	tr	tr	1.1	1.8	2.9
	0–40	tr	tr	2.3	15.9	18.2

[1] Field microplot experiment with ^{15}N; 75 kg N/ha was banded on 10 Oct. (Zhang *et al.* 1989).
[2] tr = trace amount.

Table 11.13. Fate of urea nitrogen applied to winter wheat on a calcareous fluvo-aquic soil (% of N applied).[1]

Date sampled	Recovery in plants	Residual in soil (0–100 cm)	Loss
Banded 6 cm deep at seeding			
8 Dec.	18.6 b[2]	90.6 a	−9.2 c
4 April	37.3 a	44.4 b	18.3 b
5 May	45.5 a	27.5 c	27.0 b
31 May	39.4 a	22.7 c	37.9 a
Surface-broadcast followed by irrigation at elongation stage			
5 May	40.7 a	34.1 bc	25.2 b
31 May	46.6 a	33.2 bc	20.1 b

[1] Field microplot experiment with ^{15}N (Zhang *et al.* 1989).
[2] Values in a column with the same letter do not differ significantly at $P = 0.05$.

that time the fertilizer N in the fixed ammonium and organic pools decreased with the largest change occurring in the organic pool. The results suggest that immobilization-remineralization of fertilizer N in the soil was more important than fixation-defixation of ammonium in supplying N for plant growth. Similar results were obtained in a pot experiment with wheat growing on a red earth and a lateritic soil. However, owing to the very low capacity of these soils to fix ammonium, fixation of labelled N was not detected (Table 11.14).

Table 11.14. Transformations of urea nitrogen in soil during the growth of winter wheat (% of N applied).[1,2]

Forms	Date sampled				
	10 Dec.	17 Dec.	3 Jan.	3 Feb.	3 March
	Red earth				
Exch $NH_4 + NO_3$	89.9	85.2	61.8	–	–
Fixed NH_4	0	0	0	0	0
Immobilized-N	1.9	3.6	5.0	10.1	20.7
	Red earth, amended with compost[3]				
Exch $NH_4 + NO_3$	91.2	79.8	48.9	0	–
Fixed NH_4	0	0	0	0	0
Immobilized-N	2.7	5.1	14.4	17.1	24.5
	Laterite				
Exch $NH_4 + NO_3$	56.4	88.6	81.0	–	–
Fixed NH_4	0	0	0	0	0
Immobilized-N	0	0.85	2.4	8.7	15.0
	Laterite, amended with compost				
Exch $NH_4 + NO_3$	66.3	86.8	82.5	–	–
Fixed NH_4	0	0	0	0	0
Immobilized-N	0.08	0.61	3.0	8.6	17.5

[1] Pot experiment with ^{15}N; seeded on 3 December, matured on 21 May (Chen *et al.* 1988).
[2] Urea 80 mg N $kg\ soil^{-1}$ and KH_2PO_4 were thoroughly mixed with soil.
[3] 1.3 g compost $kg\ soil^{-1}$.

Table 11.13 shows that when urea was banded 6 cm deep at the time of seeding winter wheat no loss of N was detected 59 days after N application even though the nitrate content of the soil during this period was high. Significant N loss was detected on 4 April (elongation stage), which coincided with significant decreases in nitrate and organic N (Table 11.12). Evidently, part of the nitrate and organic N present in the soil during winter was lost in spring when the temperature increased. When N was topdressed at the elongation stage, loss approached a maximum within about one month. From these results, it may be concluded that loss of fertilizer N applied to winter wheat occurred primarily in spring when the temperature increased. Table 11.12 shows that fixed ammonium was very low and changed only slightly throughout the growth period. Thus it was concluded that fixation-defixation of fertilizer N was not an important process in the fluvo-aquic soil under investigation.

Table 11.15 shows data relevant to summer maize grown on the same soil as the winter wheat discussed above. Although the error in estimating N loss for the summer maize was high (as indicated by the large variation in N loss for the 3 sampling dates), it can still be concluded that loss of fertilizer N proceeded quickly and approached a maximum 20 days after application. This appeared to be mainly due to the high temperatures which occurred during the growth of summer maize. The rapid loss of N which occurred with maize was similar to that obtained when N was applied to winter wheat at the elongation stage in spring. The loss pattern for N

Table 11.15. Fate of urea nitrogen applied to maize grown on a black soil (% of applied N).[1]

Crop	Soil	Treatment[2]	Date sampled	Recovery in plants	Residual in soil	Loss
Summer maize[3]	Fluvo-aquic	SB	21 Jul.	3.8	62.3	33.9
			22 Aug.	16.5	34.9	48.6
			17 Sep.	57.3	15.4	27.3
		UP-DP	21 Jul.	2.8	63.2	34.0
			22 Aug.	16.1	33.2	50.7
			17 Sep.	47.0	16.6	36.4
		USG-1-DP	21 Jul.	3.1	63.3	33.6
			22 Aug.	15.0	31.8	53.2
			17 Sep.	45.8	20.3	33.9
		USG-2-DP	21 Jul.	1.3	59.9	38.8
			22 Aug	12.4	17.4	70.2
			17 Sep.	44.0	18.0	38.0
Spring maize	Black soil	UP-SS (on 15 May)	15 Jun.	1.1	73.9	25.0
			15 Jul.	14.5	51.7	33.8
			15 Aug.	29.3	36.5	34.3
			7 Sep.	30.6	17.7	51.7
		UP-Bs (on 10 May)	17 May	0	71.5	28.5
			24 May	0	70.2	29.8
			31 May	0.02	53.4	46.6
			6 Sep.	38.5	9.0	52.5

[1] Field microplot experiment with ^{15}N (Xu *et al.* 1987, recalculated; Yuan *et al.* 1981).
[2] See list of abbreviations.
[3] Seeded on 10 June, N applied on 1 July.

applied to spring maize grown on a black soil in cool temperate, north-eastern China was somewhat different. In the cool temperate region, loss of fertilizer N proceeded gradually (Table 11.15). The pattern of loss may be related to the comparatively lower temperatures which occurred during the early growth stages at the particular site.

11.5.2. Pathways of loss

Loss of N by leaching can be evaluated from the distribution of fertilizer N in the soil profile at maturity. The depth of sampling for upland crops should be greater than that for flooded rice, primarily because of the high mobility of nitrate, the dominant form of mineral N in upland soils. The relevant results obtained on the fluvo-aquic soil are shown in Table 11.16. It seems that loss of N from the soil by leaching during the growth of winter wheat or summer maize is insignificant.

Ammonia volatilization from urea applied 20 days after sowing summer maize on a fluvo-aquic soil was measured with the full profile micrometeorological technique. The results show that 30% of the surface-broadcast and 12% of the deep placed N was lost as ammonia in 188 hours. These losses accounted for 67% and 40%, respectively, of the total N loss (Zhang *et al.* 1992a). Considerable N was lost by ammonia volatilization when ABC was surface-broadcast on this soil; in this experiment ammonia loss was estimated with the Hargrove method (Zhao *et al.* 1986). The results obtained in a microplot experiment on the same type of soil with winter wheat showed that 33% of the N applied as $^{15}NO_3$ was lost when the fertilizer was banded at sowing at a depth of 6 cm (Table 11.17). These results suggest that denitrification was also an important pathway for N loss.

Table 11.16. Distribution of fertilizer nitrogen (75 kg ha^{-1}) in a fluvo-aquic soil at harvest (% of N applied).[1]

Crop	Treatment[2]	Depth (cm)				
		0–20	20–40	40–60	60–80	80–100
Winter wheat	Urea, Bn-Bs	15.3	2.9	2.8	1.1	0.6
	Urea, TR-Ir	17.4	5.6	2.5	1.5	1.5
	Urea, TE-Ir	23.0	6.9	2.1	1.1	0.2
	Urea, Bn-Bs + TR-Ir	13.8	6.1	1.5	0.7	0.3
	ABC, Bn-Bs + TR-Ir	15.8	2.4	1.4	0.7	0.5
	AS, Bn-Bs + TR-Ir	15.2	2.7	1.3	0.8	0.5
	$^{15}NH_4NO_3$, Bn-Bs + TR-Ir	17.2	5.5	1.3	1.0	0.4
	$NH_4{}^{15}NO_3$, Bn-Bs + TR-Ir	11.2	5.9	2.3	1.0	0.4
Summer maize	Urea, SBn-S	56.8	5.5	0	0	–
	Urea, DP-S	47.4	15.6	0.2	0	–
	USG-1, DP-S	33.9	18.8	1.6	0	–
	USG-2, DP-S	41.2	16.2	1.5	1.0	–

[1] Field microplot experiments with ^{15}N (Xu *et al.* 1987; Zhang *et al.* 1989).
[2] See list of abbreviations.

Table 11.17. Fate of fertilizer N applied to winter wheat on a calcareous fluvo-aquic soil (% of N applied).[1]

Form	Treatment[2]	Recovery in plants	Residual in soil (0–100 cm)	Loss
Urea	Bn-Bs	39.4	22.7	37.9 ab[3]
	TR-Ir	48.5	28.5	23.0 c
	Ir-TR	43.1	27.0	30.0 bc
	TE-Ir	46.6	33.2	20.1 c
	Bn-Bs + TR-Ir	44.4	22.3	33.3 abc
ABC	Bn-Bs + TR-Ir	34.3	20.7	45.1 a
AS	Bn-Bs + TR-Ir	40.3	20.5	39.3 a
$^{15}NH_4NO_3$	Bn-Bs + TR-Ir	37.9	26.0	36.2 ab
$NH_4{}^{15}NO_3$	Bn-Bs + TR-Ir	46.1	20.8	33.1 abc

[1] Field microplot experiment with ^{15}N; sampled at harvest (Zhang *et al.* 1989).
[2] See list of abbreviations.
[3] Values in a column with the same letter do not differ significantly at P = 0.05.

11.5.3. Extent of loss

The relevant results obtained in field microplot experiments in China are summarized in Tables 11.18–11.22. The loss of N applied to wheat and barley ranged from 5% to 72%, but was mostly in the range 14%–55%. The corresponding figures for maize and cotton were 18%–53%. Nitrogen lost from different sources applied to upland crops was generally in the order ABC > AS > urea (Table 11.23; Shi *et al.* 1981; Zuo and Wei 1983; Li *et al.* 1984a; Zhou 1985; Zhang *et al.* 1989), which is different from that obtained with flooded rice. The difference may be attributed to the fact that the upland soils studied were mostly calcareous, and thus ammonia loss was high when ammonium fertilizers were applied. When urea was applied it was easily leached into the soil by rainfall or irrigation water before hydrolysis,

Table 11.18. Loss of ammonium sulfate nitrogen applied to upland crops (% of applied N).[1]

Crop	Soil	Treatment[2]	Loss	Reference
Winter wheat	Drab	SB, –P	42.6	Li *et al.* (1988)
		SB, +P	20.3	
	Drab	SB	30.0	
		DP	28.8	
	Fluvo-aquic	Bn-Bs at 6 cm + TR-Ir	39.2	Zhang *et al.* (1989)
Spring wheat	Albic	–	31.6	Chen *et al.* (1981)
	Black loess	SB-Ir at 3 leaf-stage	27.9	Li *et al.* (1984a)
		DP at 6 cm, Ir at 3 leaf-stage	13.9	
Maize	Fluvo-aquic	SB before tasseling	44.8	Zhao *et al.* (1986)
		Furrow-application at 5 cm	26.4	

[1] Field microplot experiments with ^{15}N.
[2] See list of abbreviations.

Table 11.19. Loss of urea-N applied to wheat and barley (% of applied N).[1]

Crop	Soil	Treatment	Loss	Reference
Winter wheat	Drab	SB	27, 16	Li *et al.* (1988)
		DP at 6 cm deep	23, 24	
	Fluvo-aquic	DP at 6 cm basal + TR-Ir	33.2	Zhang *et al.* (1989)
	Yellow fluvo-aquic	SB in winter	18.8	Chen and Zhu (1982b)
		Side-banded in winter	22.6	
		USG, DP in winter	20.2	
	Non-calcareous	Furrow application in winter	38, 30	Pan *et al.* (1980)
		SB in winter	23.0	Chen and Zhu (1982b)
		Side-banded in winter	35, 25	
		USG, DP in winter	30, 5	
Spring wheat	Black	–	33.1	Chen *et al.* (1981)
	Black loess	DP-Bs-I	21.5	Li *et al.* (1984b)
		DP-I at 3 leaf-stage	23.2	
		SB-I at 3 leaf-stage	20.7	
Barley	Non-calcareous	Split	34, 35, 49,59	Shi *et al.* (1981)
		Furrow-appln in winter, –P	43	
		Furrow appln in winter, +P	25, 30	

[1] Field microplot experiments with ^{15}N.

Table 11.20. Loss of urea nitrogen applied to maize and cotton (% of applied N).[1]

Crop	Soil	Treatment	Loss, %	Reference
Maize	Black	DP at 6 cm at sowing	51.7	Yuan *et al.* (1981)
		Top-dressed	52.5	
	Meadow	DP at 5 cm	33.1	Yao and Guan (1983)
		DP at 15 cm	11.1	
	Fluvo-aquic	SB before tasseling	40.9	Zhao *et al.* (1986)
		Furrow-application at 5 cm before tasseling	17.8	
		Urea, SBn-S	27.3	Xu *et al.* (1987)
		Urea, DP at 10 cm at seedling stage	36	
		USG, DP at 10 cm at seedling stage	34, 38	
Cotton	Non-calc.	Split	15, 23, 30, 41	Tian and Huang (1984)

[1] Field microplot experiments with ^{15}N.

thus ammonia and total N losses were smaller than those obtained with AS (Liu 1986; Li *et al.* 1988; Shi *et al.* 1983; Zhao *et al.* 1986). When the fertilizer was deep-placed there was little difference in the extent of N loss between the sources, presumably because of the depression in ammonia loss.

At present, due to limited data, it is difficult to estimate the extent of fertilizer N loss in the production of upland crops in China. However, from the results given in Tables 11.17–11.23, it appears that losses will be high when N is applied by traditional techniques. There is also the possibility that N loss from upland systems will be underestimated by the microplot technique, as was the case for flooded rice.

Table 11.21. Loss of nitrogen applied as ammonium bicarbonate to upland crops (% of applied N).[1]

Crop	Soil	Treatment[2]	Loss, %	Reference
Winter wheat	Drab	25 kg ha^{-1}, basal	4.6	Li *et al.* (1982)
		50 kg ha^{-1}, basal	16.3	
		75 kg ha^{-1}, basal	22.0	
		SB, –P	10, 45	Li *et al.* (1988)
		SB, +P	5, 21	
	Fluvo-aquic	Bn-Bs + TR-Ir	45.1	Zhang *et al.* (1989)
Spring wheat	Black loess	SB-Ir at 3 leaf-stage	39.7	Li *et al.* (1984a)
		Supergranules, DP-Ir at 3 leaf-stage	13.4	
		Coated supergranules, DP-Ir at 3 leaf-stage	11.3	
Barley	Non-calcareous	Split	42, 54	Shi *et al.* (1981)
Maize	Fluvo-aquic	SB prior to tasseling	62.0	Zhao *et al.* (1986)
		Furrow application at 5 cm prior to tasseling	38.6	

[1] Field microplot experiments with ^{15}N.

Table 11.22. Loss of nitrogen applied as ammonium nitrate, potassium nitrate and ammonium chloride to upland crops (% of applied N).[1]

N source	Crop	Soil	Treatment[2]	Loss, %	Reference
NH_4NO_3	Winter wheat	Drab	SB, –P	29.2	Li *et al.* (1988)
			SB, +P	34.9	
$^{15}NH_4NO_3$	Winter wheat	Fluvo-aquic	Bn-Bs + TR-Ir	36.2	Zhang *et al.* (1989)
$NH_4{}^{15}NO_3$	Winter wheat	Fluvo-aquic	Bn-Bs + TR-Ir	33.1	Zhang *et al.* (1989)
$^{15}NH_4{}^{15}NO_3$	Spring wheat	Anthropogenic-alluvium	60 kg N ha^{-1}	43.7	Hua *et al.* (1982)
			120 kg N ha^{-1}	43.2	
			180 kg N ha^{-1}	39.6	
			Bs, plowed in	46.3	Jin *et al.* (1985)
			TD-3	45.4	
			TD-E	20.6	
			TD-H	71.5	
			Split	54.6	
	Maize	Black	DP, Bs	8.5	Zhou *et al.*, unpubl. data
			DP, Bs & TD	16.3	
		Albic	DP, Bs	12.7	
		Meadow	DP, Bs	21.4	
		Light chernozem	DP, Bs	30.8	
KNO_3	Spring wheat	Black loess	SB-Ir-3	22.1	Li *et al.* (1984a)
			DP-Ir-3	24.4	
NH_4Cl	Maize	Fluvo-aquic	SB prior to tasseling	50.0	Zhao *et al.* (1986)
			Furrow appln at 5 cm prior to tasseling	35.3	

[1] Field microplot experiments with ^{15}N.
[2] See list of abbreviations.

Table 11.23. Nitrogen loss following application of nitrogen to upland crops.

Crop	Soil	Experiment	Treatment	Form	Loss (%)	Reference
Spring wheat	Black loess	Field microplot	SB-3	ABC	39.7	Li *et al.* (1984a)
				Urea	20.7	
				AS	27.9	
				KNO_3	22.1	
			DP-3	ABC-SG	13.4	
				Urea	23.2	
				AS	13.9	
Winter wheat	Drab	Pot	DP	ABC	32.7	Zuo and Wei (1983)
				Urea	25.0	
				AS	28.3	
			SB	ABC	33.1	Zhou (1985)
				AS	30.1	
			DP	ABC	23.2	
				AS	11.3	
		Field microplot	SB	ABC	20.8	Li *et al.* (1988)
				Urea	15.8	
				AS	29.9	
				AN	34.9	
			DP	ABC	4.8	
				Urea	24.0	
				AS	28.8	
Maize	Fluvo-aquic	Field microplot	SB	ABC	62.0	Zhao *et al.* (1986)
				Urea	40.9	
				AS	44.8	
				NH_4Cl	50.0	
			Furrow	ABC	38.6	
			5 cm deep	Urea	17.8	
				AS	26.4	
				NH_4Cl	35.3	

11.6. Plant recovery and agronomic efficiency of fertilizer nitrogen

11.6.1. Apparent recovery

Table 11.24 shows that the apparent recovery of N in crop plants (Ra) varies greatly (range 9%–72% of the applied N) in field experiments conducted in China. The mean Ra values for rice, wheat, barley and naked barley ranged from 28 to 41%. Although the means for different sources of N were calculated, they cannot really be compared, because distinct differences existed, either in soil properties or methods of application in the various experiments. Therefore, the results obtained in field experiments designed specifically for comparing the Ra values of different sources only are given in Table 11.25. No definite conclusion on the effect of N fertilizer source on Ra can be drawn from these results, except that the Ra of nitrate applied to flooded rice was lower than that for ammoniacal N and urea.

The extent of N loss from fertilizers applied to crops in China may also be assessed from the Ra values obtained in field experiments. As indicated earlier in this chapter the extent of N loss may be assessed as the difference between 100%

Table 11.24. Apparent recovery (Ra) of fertilizer nitrogen in crop plants (%).[1]

Crop	Form	Experiments	Site	Mean	Range
Rice	AS	385	Jiangsu	34	30–70
	ABC	18	Shanghai, Jiangsu, Ninxia	33	22–39
	Urea	125	Anhui, Guangxi, Henan, Jiangsu, Jiangxi, Ninxia, Shanghai, Zhejiang,	38	22–62
Wheat, barley	AS	168	Jiangsu	28	23–31
	ABC	28	Jiangsu, Ninxia, Shandong, Shanghai, Shanxi	30	16–38
	Urea	58	Beijing, Gansu, Henan, Jiangsu, Ninxia, Shandong, Shanghai, Sichuan, Zhejiang	41	9–72

[1] Field plot experiments; recovery was estimated by the difference method based on N in plant tops at maturity (Compiled from: Deng *et al.* 1981; Du 1986; Fan and Tao 1984; Guo 1979; ISFJAAS and SAI 1981; Jia *et al.* 1982; Jiang 1980; Jiang and Li 1983; Li *et al.* 1984a; Liu *et al.* 1983; Pan *et al.* 1980, 1982, 1983, 1984; Shi *et al.* 1981, 1983; Wan *et al.* 1983, 1984; Xi *et al.* 1978; Yang *et al.* 1980; Zang 1981; Zhang *et al.* 1988; Zhu 1982; Zhu *et al.* 1988; Zuo 1983a; and unpublished data of, Sun, G. Y. *et al.*; Xia, S. X. *et al.*; Institute of Soils and Fertilizers, Ninxia; Zhang, G. L. and Bao, D. J.; Mao, Z. Y. and Zhou, Z. F.).

and Ra determined from the N in the whole plant. However, the Ra values in Tables 11.24 and 11.25 were estimated from the N in the shoots only, and thus will be lower than those estimated from the N in the whole plant. Fortunately, the N in the roots of cereals generally accounts for less than 10% of the N in the whole plant, thus the Ra in the whole plant may be assessed with only a small error from that determined in the shoots. As shown in Table 11.24, the mean Ra values for the main N fertilizers applied to rice in China ranged from 33 to 38%, and it is estimated that they will be ~40% after allowing for N in roots. This suggests that 60% of the N applied to rice in China is lost. This agrees with the loss estimated from the ^{15}N-microplot experiments.

Nitrogen loss during the production of upland crops in China may be assessed in the same way from the relevant Ra values. As shown in Table 11.24, the mean Ra values for wheat and barley estimated from N in the shoots ranged from 28 to 41%, and they will still be less than 50% when N in roots is taken into account. Consequently, 50% or more of the fertilizer N applied to wheat and barley is apparently lost. As for maize, cotton and other summer crops, definite conclusions cannot be drawn owing to the lack of relevant data, although it is expected that the losses would be close to those of the winter crops. However, such an assessment is not in agreement with the N loss data obtained from the ^{15}N microplot experiments. The mean loss obtained from these was 32% (n = 103) when the values obtained for deep placement of supergranules were excluded (Tables 11.17–11.22). This is much lower than the loss assessed from the amended Ra values (~50%). Therefore, it is impossible, at present, to draw definite conclusions concerning the extent of loss of N applied to upland crops in China. It is most likely that the loss will be ~45%; that is intermediate between the two assessments given. Nevertheless, from

Table 11.25. Apparent recovery of nitrogen in crop plants.[1]

Crop	Soil	Rate (kg N ha^{-1})	Treatment	Ra				
				AS	ABC	Urea	AN	KNO_3
Rice	AAS	95	Split	28	22	22	15	–
		92	Split	36	29	43	20	–
	Fluvo-aquic	90	Inc.	17	13	25	–	–
	NCPS	–	–	39	35	32	–	–
		–		–	27	23	16	–
Winter wheat	Fluvo-aquic	75	Split	54	42	35	33	–
		150	Split	21	19	17	21	–
		150	Basal	–	21	19	–	–
Spring wheat	Black loess	90	SB-Ir	66	59	65	–	73
		90	DP-Ir	63	–	61	–	66
Wheat, barley	NCPS	–	–	33	32	32	–	–
Maize	Fluvo-aquic	75	Split	36	33	31	37	–
		150	Split	21	17	18	20	–

[1] Field plot experiments; recovery was estimated by the difference method based on N in plant tops at maturity (Compiled from Jia *et al.* 1982; Li *et al.* 1984a; Liu *et al.* 1983; Shi *et al.* 1983; Zhu *et al.* 1988; Zuo 1983a, 1983b).

the discussion above it is reasonable to conclude that the recovery of fertilizer N in crops in China is very low.

11.6.2. Agronomic efficiency of fertilizer nitrogen

Table 11.26 shows the Ea values obtained in the national network on fertilizers (ISFCAAS 1986). The results show that Ea decreased as the rate of N applied increased. The agronomic efficiency of fertilizer N in China ranged from 10 to 15 kg grain per kg N applied at the prevailing rates of application (60–150 kg N ha^{-1}), which is comparable to that obtained in the FAO network (12.7 kg grain per kg N applied; Greenwood 1981). Table 11.26 further shows that Ea was also influenced by the yield of the control plot (no N). The agronomic efficiency was low on soils

Table 11.26. The mean agronomic efficiency of fertilizer nitrogen (Ea, kg grain kg N applied^{-1}) in the China national network on fertilizers (1981–1983).

Crop	Experiments	Rate (kg N ha^{-1})						Yield of control plot (t ha^{-1})[1]			
		<60	60–90	90–120	120–150	150–180	>180	<2.25	2.25–3.75	3.75–5.25	>5.25
Rice	829	11.9	11.2	10.2	8.1	6.6	4.8	8.6	10.9	8.8	6.9
Wheat	1260	13.2	11.6	11.3	9.5	5.8	4.5	10.5	10.6	9.3	5.6
Maize	629	18.3	17.8	14.3	11.8	8.9	9.0	16.2	13.8	13.5	10.7
Sorghum	51	4.4	11.5	9.9	6.6	–	–	6.6	10.4	8.9	7.3
Millet	51	18.0	5.9	4.1	–	–	–	5.6	6.2	2.8	–

[1] Without fertilizer application.

of low productivity, presumably due to the presence of limiting factors other than N. However, Ea was also low on soils of high productivity, and this may be related to the high N-supplying capacity of the fertile soils. Ea was highest on soils of medium productivity, where there were no limiting factors other than N, and the N-supplying capacity of the soil was not high. Obviously, these differences in Ea should be taken into account when tailoring N fertilizers to maximize the returns from the application of N.

The agronomic efficiencies of various N fertilizers and AS are compared in Table 11.27 (ISFCAAS 1986). The efficiency of nitrate applied to flooded rice was markedly lower than that of AS and this was apparently due to the severe losses of N by denitrification which occur when nitrate is applied to flooded soils. There was no significant difference between the other ammoniacal and ammonium-producing fertilizers in their agronomic efficiencies. However, the agronomic efficiencies of ABC, aqua ammonia and lime-nitrogen were sometimes lower than that of AS and the agronomic efficiency of urea was usually comparable with that of AS.

In Table 11.27 the efficiencies of the N applications to calcareous and non-calcareous soils were not considered separately, therefore some trends cannot be explained. As indicated above, N losses from the different fertilizers applied to calcareous soils were similar, whereas they were different when applied to non-calcareous soils. Thus it might be expected that the agronomic efficiencies of the different fertilizers would be comparable on calcareous soils but different on non-calcareous soils. Furthermore, differences between the sources may become smaller with high rates of N application, as shown in Table 11.28.

11.7. Strategies for improving plant recovery of nitrogen

As indicated above losses of fertilizer N are high in crop production, and thus the potential for improving N efficiency by reducing N loss is high. The various practices which have been recommended for reducing N loss are discussed in this section.

Table 11.27. Comparisons of the agronomic efficiencies of various nitrogen fertilizers and ammonium sulfate at the same rate of application.[1]

Crop	AA[2]	ABC	AN	CAN	$Ca(NO_3)_2$	NH_4Cl	Urea	$CaCN_2$
	Difference (%)							
Rice	1.9	0.5	−1.1	−4.1	−11.7	2.8	1.4	−4.6
Wheat	–	−29.7	−4.8	−3.4	8.1	−3.8	−8.4	−11.6
Cotton	−2.4	−4.5	−5.7	−1.4	–	−5.1	1.0	−10.0
Maize	−1.0	−19.2	1.5	−0.7	–	−2.7	0	−23.4
Millet	–	–	5.3	−2.3	–	2.4	1.0	−3.7
Sorghum	–	–	−3.8	–	–	−3.7	−4.7	−1.9
Sweet potato	–	3.4	−0.7	–	–	–	–	–
Rapeseed	−16.9	–	4.4	–	–	0	–	–

[1] Means of the results of 92 experiments conducted on 59 sites.
[2] See list of abbreviations.

Table 11.28. Effect of application rate on agronomic efficiency of nitrogen.[1]

Crop	Rate (kg N ha^{-1})	Form	Ea (kg kg N applied^{-1})
Barley	94	AS	12.1
		ABC	9.9
		Urea	9.7
	188	AS	7.2
		ABC	6.9
		Urea	6.0
Rice	75	AS	16.3
		ABC	15.8
		Urea	13.8
	128	AS	10.1
		ABC	9.9
		Urea	9.4
	180	AS	5.9
		ABC	5.7
		Urea	5.8

[1] Field experiments on non-calcareous paddy soils in Shanghai (Shi *et al.* 1983).

11.7.1. Timing and mechanism of loss

Mineral N, including exchangeable ammonium and nitrate, is subject to loss from the soil, and thus it should not be allowed to accumulate in the soil. Some of the techniques suggested to prevent the accumulation of mineral M in soil are use of slow-release fertilizers, split application, and optimization of the rate of N application in intensive agriculture.

Leaching is generally negligible during the growth of rice and runoff is usually controlled. Consequently the important loss processes which need to be controlled are ammonia volatilization and nitrification-denitrification. Losses of N through ammonia volatilization and/or nitrification-denitrification take place immediately after N is broadcast into the floodwater whether incorporated or not, and the bulk of the N is lost within ~10 days of application. Therefore, efforts should be devoted to suppress N loss during this period (Cai *et al.* 1985; Zhu *et al.* 1979a, 1989b).

The relative importance of ammonia volatilization and nitrification-denitrification varies with environmental conditions, and thus the techniques for reducing N loss will also vary. However, the two loss pathways are interlinked, because ammonium is the common source of N for the two processes. Thus, in developing a technique for reducing N loss via one pathway, the possible effect on loss via the other pathway needs to be considered (Zhu *et al.* 1985).

In the Huang-Huai-Hai Plain of north China and the Loess Plateau of northwestern China leaching losses are likely to be negligible due to the dry conditions during the growth of winter wheat. However, because of the high pH of the calcareous soils in these regions, ammonia loss is likely to be an important pathway of N loss when urea and ammonium fertilizers are applied. Therefore, the control of ammonia volatilization is of primary importance. In summer, because heavy rainfall frequently occurs, leaching may be important on occasions when urea is applied (Xu *et al.* 1987; Yuan *et al.* 1981). In southern China, high rainfall may result in considerable leaching of N from upland soils, and thus management practices to reduce leaching loss are required.

11.7.2. Suppression of ammonia volatilization

Recently, it has been established through a series of field investigations that the rate of ammonia volatilization from flooded rice fields is a function of the partial pressure of ammonia in the floodwater and wind speed; the partial pressure of ammonia in turn is a function of the concentration of ammoniacal N, pH and temperature of the floodwater (see Zhu 1990). Consequently, the ammoniacal N concentration or pH of the floodwater need to be reduced to reduce ammonia loss. Several techniques have been proposed for reducing the ammoniacal N concentration in the floodwater, including deep placement, incorporation into drained soil, split application, use of slow-release fertilizers and urease inhibitors. Algicides have been recommended to suppress the elevation of floodwater pH in the daytime. Combinations of acidic N fertilizers with urea or ABC have also been tested for controlling floodwater pH.

Most of the techniques suggested above are also applicable to upland soils. However, since no floodwater is present on upland soils, deep placement of N is the most important way for reducing the partial pressure of ammonia at the soil surface and ammonia loss.

11.7.3. Reducing nitrification-denitrification

Recent investigations show that the nitrification-denitrification process in flooded rice is more complex than previously thought. The process may (i) occur near the soil surface at the junction of the oxidized and reduced layers (Site 1), (ii) take place at the boundary of the oxidized and the reduced layers in the rhizosphere (Site 2; Savant and De Datta 1982), (iii) take place concurrently in the oxidized and reduced layers (Zhou and Chen 1983), (iv) be promoted in the floodwater by green algae (Kimura *et al.* 1986), and (v) be induced by the amorphous iron and manganese oxides acting as electron acceptors to oxidize ammonium to nitrate which is then denitrified (Li *et al.* 1990; Li and Wu 1991). The substantial reduction in N loss which occurs when supergranules are deep-placed suggests that nitrification-denitrification at Site 1 is the most important mechanism. Reddy and Patrick (1986) found that the rate of loss by nitrification-denitrification was governed by the rate of nitrification of ammonium in the surface oxidized layer, and the rate of diffusion of ammonium from the reduced layer to the surface oxidized layer. Consequently, inhibition of either one of these rate-limiting factors is critical for reducing loss by nitrification-denitrification. Deep placement and nitrification inhibitors have been investigated widely for this purpose. In addition, incorporation of the fertilizer into the drained soil, as mentioned above, may also reduce loss by nitrification-denitrification.

Nitrification takes place rapidly in upland soils. The nitrate formed may be denitrified at localized anaerobic sites, and be increased through rainfall or irrigation. Therefore, the techniques under investigation for reducing loss by nitrification-denitrification are the optional management of N application with irrigation and amendment with nitrification inhibitors.

11.7.4. Improve plant uptake of mineral nitrogen

Even though growing plants can stimulate denitrification by providing energy sources (e.g. by exudation), they also assimilate N quickly and thus reduce the pool of mineral N in soil available for denitrifying organisms. The net effect is to reduce N loss, as shown in Table 11.1. Therefore, it is important to ensure plant growth is not limited by deficiencies of elements such as P and K, or water. Application of N at vigorous growth stages may also result in a lower N loss, partly due to the greater uptake rate of N by plants at these growth stages.

11.8. Techniques for improving efficiency of nitrogen

11.8.1. Optimizing rate of application

The decrease in Ea as the rate of N application increased resulted not only from a decrease in Ra, but also from a decrease in the partial physiological efficiency of the absorbed N (Ep), as shown in Tables 11.29 and 11.30. Therefore, optimization of the application rate is essential in intensive agriculture.

Several methods have been proposed for determining the optimum rate of N application in China. The rate of application may be assessed from the balance between requirement and supply. The parameters needed are yield target, amount of N absorbed in crop plants for producing unit weight of grain, content and plant recovery of manure N, Ra of fertilizer N, and soil N supply.

Table 11.29. Effect of rate of application of fertilizer nitrogen on agronomic efficiency (kg kg N^{-1}), apparent plant recovery (%) and partial physiological efficiency of the nitrogen assimilated (Ep, kg kg N^{-1}) for wheat and maize grown on a fluvo-aquic soil.[1]

N source	Rate (kg N ha^{-1})	Winter wheat			Spring maize		
		Ea	Ra	Ep	Ea	Ra	Ep
NH_4Cl	75	13.6	41.8	32.5	6.4	33.2	19.3
	150	6.3	18.8	33.5	4.0	16.9	23.7
AS	75	18.5	53.7	34.5	6.7	36.2	18.5
	150	8.6	21.3	40.4	3.9	20.8	18.8
Urea	75	13.5	35.4	38.1	6.2	30.8	20.1
	150	5.3	16.8	31.5	4.1	17.6	23.3
AN	75	11.7	33.1	35.3	4.8	36.5	13.2
	150	7.0	20.8	33.6	4.0	20.3	19.7

[1] Field experiments (Zuo 1983a, 1983b).

Table 11.30. Effect of rate of application of urea on agronomic efficiency (kg kg N^{-1}), apparent recovery (%) and partial physiological efficiency (kg kg N^{-1}) for rice.[1]

Rate (kg N ha^{-1})	Grain yield (t ha^{-1})	Ea	Ra	Ep
0	5.18	–	–	–
46	5.90	15.6	34.7	45.0
92	6.26	11.7	33.9	34.5
138	6.56	10.1	31.4	32.2
184	6.63	7.9	28.6	27.6
230	6.44	5.5	24.2	22.7

[1] Means of 26 experiments conducted in Taihu Region (Zhang *et al.* 1988).

The yield target is usually determined by experience. Recent investigations show that it can be estimated from the non-linear relationship between the maximum yield (Y_{max}) and the yield of the zero-N treatment (Y_0) (Zhou 1987). Y_{max} can also be calculated from the amount of N mineralized during incubation or the N released by NaOH, because these fractions are non-linearly correlated with Y_0 (Gao *et al.* 1984). Thus, the assessment of yield target becomes reliable. However, the other parameters, e.g. soil N supply (Chapter 3), either vary widely or are difficult to predict with satisfaction. Thus this method has not been practised widely, although the principles involved are the basis of other methods to be discussed below.

The optimum rate of N application may also be assessed from the availability indices. This method was developed for rice in the Taihu region. As mentioned above, Y_0 (i.e., the coefficient 'a' in the quadratic yield-N rate response curve) can be estimated from incubation or NaOH treatment, and Y_{max} can be calculated from the Y_0 value. Then the rate of application (X_{max}) for maximum yield can be derived from the following equations (Gao *et al.* 1984):

$$Y_{max} = a + b\,X_{max} + c\,X_{max}^2 \quad (1)$$

$$b = 2(Y_{max} - a)/X_{max} \quad (2)$$

$$c = -b/2X_{max} \quad (3)$$

This method is simple and only the soil availability index is needed for assessing the optimum rate of N application. However, the coefficients in the empirical equations (1)–(3) can only be used under the conditions where the coefficients were derived. They should be redetermined when the conditions change. Zhang, Y. C. and Guo, Z. F. (unpublished data) showed that this method can also be used to determine the optimum rate of application for cotton. Needless to say, the regression equations obtained were different from those given above.

The optimum rate of application can also be assessed from the yield of the control plot. This method was developed especially for rice. X_{max} can be assessed directly from the equation,

$$X_{max} = 0.017\,(Y_{max} - Y_o)/Ra. \tag{4}$$

Where Y_{max} is derived from Y_0 by an empirical equation, and the N brought in with seedlings and the amount taken up by plants for unit grain yield are combined together and found to be 0.017 under the specific conditions of the experiment (Zhou 1987). Thus, X_{max} can be assessed without soil testing and the inherent error in predicting soil N supply can be eliminated. However, it is not easy to get a value for Yo for every farmers' field.

The optimum split application rates are also based on the balance between the N requirements of the crop and the soil supply. The following steps need to be followed to assess the optimum rate of application: (i) quantify the amount of N accumulated at each growth stage (Np), (ii) determine the N supplied from soil and fertilizer at a particular growth stage (Nsf, the sum of the N in the plants and exchangeable ammonium in soil at that stage), and (iii) predict the N that can be mineralized from soil in the period to the next growth stage (Nsm). Then, the rate of application at the particular stage can be assessed by dividing (Np – Nsf – Nsm) by Ra (Huang 1986). Simplified methods for determining the relevant parameters have been developed so that this method can be used by extension workers. However, it is time-consuming.

In a regional network for a particular crop, the mean optimum rate of application (X_{mean}) can be derived from the specific optimum rates for each field (X_{opt}). Then, the X_{mean} can be used for recommending the rate of application for the whole region. Results obtained in the Taihu region show that when all the rice fields in the network were fertilized with the uniform rate the sum of the yields obtained was only 1–2% lower than the total yield obtained when each field received its optimum rate of application, X_{opt}. With few exceptions, the yields obtained for each field by these recommendations were not significantly different (Zhu 1988). This method is particularly suitable for the Chinese countryside, where soil testing and plant analysis facilities are not available.

11.8.2. Timing of application

Split application is one technique for improving efficiency. Zhu *et al.* (1988) and Gao *et al.* (1984) showed that topdressing at a vigorous growth stage reduces N loss and improves recovery of N by plants. This is primarily due to the capacity of a well-developed root system to compete favourably with soil organisms for soil N. In addition, shading and attenuation of wind speed by the developed plant canopy reduce ammonia volatilization. Furthermore, N nutrition at this stage is critical for yield production. On the other hand, topdressing at a later stage of growth, such as heading, may result in greater N loss and lower N recovery in plants.

Investigations further showed that the Ea of topdressed N at a certain stage of growth depended not only on Ra, but also on the N-supplying pattern of the soil. For example, the Ea of N applied as a basal dressing for rice was strongly correlated with the 'n' value ($r = 0.760$; $P = 0.005$; Cai *et al.* 1981) in the mineralization equation, of cumulative effective temperature, used for characterizing the soil N mineralization pattern. In addition, a positive response to N applied at panicle initiation to the first crop in the rice-rice cropping system was obtained only on those soils with 'n' values less than 1 (Gao *et al.* 1984);

It should be noted that even when N is added at the optimum rate and stage of growth, Ra is still low, when the traditional method of application is used (Table 11.31; Pan *et al.* 1980, 1982, 1983; Wan *et al.* 1984; Zhang *et al.* 1984). Therefore, more research needs to be done to improve N application techniques.

11.8.3. Deep placement

Several methods have been developed for the deep placement of fertilizer N including the use of supergranules, which is most effective (Cao *et al.* 1983). The effect of deep placement of ABC supergranules on yields of rice, wheat and maize is shown in Table 11.32 (ISSAS 1977). Depth of placement and rate of application have a considerable effect on the results obtained with deep placement, and these results are discussed below.

Table 11.31. Agronomic efficiency (kg kg N^{-1}), apparent recovery (%) and partial physiological efficiency (kg kg N^{-1}) of nitrogen applied at optimum rates and growth stages.[1]

Crop	Experiments	Ea	Ra	Ep
Early rice	12	13.4	37.2	36.2
Late rice	10	11.5	27.7	41.5
Single-cropping rice	27	11.7	35.4	33.1
Winter wheat	13	9.5	36.8	25.8

[1] Means of field experiments conducted in Taihu region (Zhang *et al.* 1988).

Table 11.32. Comparison of the efficiency of deep-placed ABC supergranules and surface-broadcast ABC powder.[1]

Crop	Experiments	Treatment	Rate (kg N ha^{-1})	Mean yield (t ha^{-1})	Mean yield (%)
Rice	149	SB-Pd	62	4.68	100
		DP-SG	62	5.30	113
Wheat	20	SB-Pd	63	3.60	100
		DP-SG	63	3.89	109
Maize	4	SB-Pd	92	4.07	100
		DP-SG	92	4.80	118

[1] ISSAS (1977).

Field investigations show that ammonia volatilization was considerably reduced when ABC supergranules were placed in the soil at a depth of 6–10 cm, but the loss was completely eliminated only when the fertilizer was placed at 15 cm (Zhang and Zhao 1983; Jiang and Liu 1980). However, the deeper the placement, the slower the uptake of the N. The results in Table 11.33 show that the best yield was obtained when the fertilizer was placed at 10 cm rather than 15 cm. Furthermore, the deeper the placement, the greater the labor consumption. Judging from these results, placement at 6–10 cm is considered to be adequate.

The optimum rate of N application depends on Ra, which varies greatly with the method of application. Deep placement has the potential to greatly improve Ra, thus leading to lower rates of application (Table 11.34). The optimum rate of application of deep placed USG was 76–93% of that of prilled urea applied by traditional methods. Therefore, the rate of application should be reduced to exploit the maximum efficiency of deep-placed USG.

11.8.4. Management in irrigation agriculture

When urea and ABC are broadcast into the floodwater of lowland rice fields considerable N is lost by ammonia volatilization because of the high ammoniacal N concentrations and high pH values generated (Cai *et al.* 1986; Zhu *et al.* 1989). Attempts to reduce the high ammoniacal N concentrations, and ammonia loss, by incorporating the applied N into the soil by various techniques in the presence of floodwater have not always been successful. Much of the fertilized N was found in the floodwater after incorporation and ammonia losses with and without incorporation were comparable (Cao *et al.* 1984; Zhu 1987). Several methods have been proposed for reducing the N concentration of the floodwater after N application. The recommendation for basal applications is that the soil be drained, then urea or ABC

Table 11.33. Effect of depth of placement of fertilizer nitrogen on yield of summer maize grown on a calcareous soil.[1]

Form	Treatment	Yield ($t\ ha^{-1}$)	Yield (%)
ABC	SB	2.07	100
	Bn 5 cm deep	2.45	118
	Bn 10 cm deep	2.54	122
	Bn 15 cm deep	2.40	116
Urea	SB	2.33	100
	Bn 5 cm deep	2.54	109
	Bn 10 cm deep	2.61	112
	Bn 15 cm deep	2.42	104
AN	SB	1.91	100
	Bn 5 cm deep	2.30	120
	Bn 10 cm deep	2.46	129
	Bn 15 cm deep	2.36	124

[1] Means of 3 field experiments; rate of application, 113 kg N ha^{-1} (Jiao 1983).

Table 11.34. Efficiencies and optimum rates of application (N_{opt}, kg N ha^{-1}) of deep placed or traditionally applied urea.[1]

Crop	Experiments	Treatment[2]	N_{opt}	Yield (t ha^{-1})	Ea (kg kg N^{-1})	Ep (%)
Early rice	5	PU-Tr	87	5.58	12.7	38.3
		USG-DP	69	5.48	15.0	47.6
		DP/Tr	0.79	0.98	1.18	1.24
Late rice	4	PU-Tr	102	5.60	12.8	30.0
		USG-DP	76.5	5.69	16.3	43.6
		DP/Tr	0.76	1.02	1.27	1.45
Single-cropping rice	3	PU-Tr	87	7.2	15.3	42.0
		USG-DP	81	7.46	20.8	59.8
		DP/Tr	0.93	1.04	1.36	1.42
Wheat	4	PU-Tr	132	3.41	8.5	35.4
		USG-DP	118.5	3.48	10.0	42.4
		DP/Tr	0.89	1.02	1.18	1.20

[1] Field experiments conducted in Taihu region.
[2] See list of abbreviations.

is broadcast onto the drained soil and incorporated before reflooding (drained incorporation); alternatively the N can be applied in furrows during plowing (Zhu *et al.* 1987, 1988). For later applications it was suggested that urea be broadcast onto drained soil followed by reflooding (Chen *et al.* 1987; You and Lo 1986).

The results of a pot experiment (Table 11.35) showed that less N was left in the floodwater of the drained-incorporation treatments than in the floodwater of the treatments where urea had been broadcast into the floodwater without incorporation. Zhu (1990) summarized the results of field microplot experiments and indicated that the drained-incorporation treatment reduced total loss by 4–26%. Zhu (1992) concluded that the reduction in ammonia loss by the drained-incorporation treatment was greater when ammonia loss from the unincorporated treatment was large. Tables 11.36 and 11.37 show the relevant results obtained for urea and ABC in experiments conducted in flooded rice growing in a fluvo-aquic soil. Nitrogen

Table 11.35. Effect of draining floodwater and incorporating fertilizer into wet soil on gaseous loss from flooded rice (% of N applied).[1]

Treatment[2]	Nitrogen in floodwater	Total loss (T)	Ammonia loss (A)	T – A[3]	A/T (%)
SB-FW, Urea	74	65.9 a[4]	30.1 b	35.8 a	46
DI, Urea	16	39.8 b	14.6 c	25.1 b	37
SB-FW, ABC	34	57.5 a	42.2 a	15.3 c	73
DI, ABC	7	45.4 b	24.9 b	20.5 bc	55

[1] Pot experiment with ^{15}N; rice was grown on a calcareous paddy soil (Zhu *et al.* 1985).
[2] See list of abbreviations.
[3] (T – A) = apparent nitrification-denitrification loss.
[4] Values in a column with the same letter do not differ significantly at P = 0.05.

Table 11.36. Effect of different methods of incorporation on the fate of nitrogen (90 kg N ha^{-1}) applied as ammonium bicarbonate to flooded rice (% of applied N).[1]

Form	Method of incorporation[2]	Recovery in plants	Residual in soil	Loss[3]
Urea	Inc-F	18.2	14.3	67.5 a
	DI-Ref	26.5	16.9	56.6 b
	DI-Del-Ref	28.1	19.0	52.9 b
ABC	Inc-F	13.8	13.3	72.9 a
	DI-Ref	28.2	18.5	53.3 b
	DI-Del-Ref	27.0	16.4	56.6 b

[1] Field experiment on a calcareous fluvo-aquic soil with ^{15}N applied at transplanting (Zhu *et al.* 1988).
[2] See list of abbreviations.
[3] Values in a column with the same letter do not differ significantly at P = 0.05.

Table 11.37. Effect of method of application of ammonium bicarbonate (90 kgN/ha^{-1}) applied at transplanting on yield of rice.[1]

Treatment	Yield (t ha^{-1})
Control (zero-N)	3.15 d
Inc-F	3.68 bc
DI-Ref	4.25 a
Furrow application	3.84 b

[1] Field experiment on a calcareous fluvo-aquic soil (Zhang *et al.* 1992b).
[2] Values in a column with the same letter do not differ significantly at P = 0.05.

loss was reduced by 11–20% and the yield was increased by 15% by the drained-incorporation treatment. Furthermore, the results obtained in a network extending from the cool temperate to the mid-subtropic regions of China showed that, except on a soil of high fertility, the drained-incorporation treatment increased Ra, Ea and yield by 9–18%, 3.2–8.6 kg kg^{-1}, and 5–16%, respectively (Zhu 1991).

There is the possibility that some of the urea incorporated into the soil will diffuse back to the floodwater, thus reducing the effectiveness of the drained-incorporation treatment. This may be overcome by delaying the reflooding. The results obtained in a field experiment conducted on a black soil in the cool temperate zone confirmed this conclusion (Han *et al.* 1991), but those obtained in an experiment on a calcareous fluvo-aquic soil in the warm temperate zone did not (Table 11.36). Presumably, in the calcareous soil the delay in reflooding permitted ammonia to be lost from the drained surface soil

Although this method is not as effective as the point deep placement of USG for reducing N loss (Zhu 1990), it is a comparatively easy technique for farmers to adopt. It has the disadvantage, however, that in areas with poor water control it may not be possible to replace the water after drainage.

Topdressing fertilizer onto unsaturated paddy soil at early tillering followed by reflooding improved the infiltration of urea into the plowed layer and increased its efficiency (Chen *et al.* 1987). Model experiments conducted with soil columns showed that irrigation after broadcasting urea onto an 80% water saturated soil moved the urea down into the soil. In this treatment the ammonium concentration in the 2 to 7 cm layer was markedly higher than that in the treatments where urea was broadcast onto a saturated soil, or onto a soil with a layer of water on its surface. The results of a field experiment showed that such a method of application significantly improved urea efficiency (Table 11.38).

Waterlogging or drought are the two important factors affecting plant growth on upland soils. Results obtained in pot experiments with wheat showed that plant uptake of N was almost completely inhibited at the wilting point, and that it increased with an increase in the moisture content of the soil, approaching the maximum at the optimum moisture content (Zhao and Zhang 1986). A field experiment with wheat grown on a calcareous soil in Shanxi demonstrated that Ra and Ep were higher in irrigated soil than in dry soil (Table 11.39; Yang *et al.* 1980).

Investigations showed that urea broadcast onto an upland soil surface could be washed into a deeper soil horizon by subsequent irrigation or adequate rainfall. For example, in a field experiment with winter wheat, N loss was lower when urea was surface broadcast onto the soil surface prior to irrigation, than when it was applied after irrigation (Table 11.17; Zhang *et al.* 1989; Li *et al.* 1985). Similar experiments with wheat on black loess soil in Ninxia showed that the loss of urea-N surface

Table 11.38. Efficiency of urea (112.5 kg N ha^{-1}) broadcast onto a drained soil and reflooded.[1]

Treatment[2]	Yield[3] (t ha^{-1})	Ra (%)	Ep (kg kg N^{-1})
Control (zero-N)	6.63 c	–	71.2
SB-Tr	7.17 b	32.0	55.0
SBDr-Ref	7.37 a	46.3	50.7

[1] Field experiment with flooded rice (Chen *et al.* 1987).
[2] See list of abbreviations.
[3] Values in a column with the same letter do not differ significantly at P = 0.05.

Table 11.39. Effect of irrigation on the efficiency of ammonium bicarbonate for wheat.[1]

N application	Treatment	Ra (%)	Ep (kg kg N^{-1})
Furrow application (late June)	Irrigated	40.8	34.2
	Dryland	12.1	28.1
Deep placed (15 Aug.)	Irrigated	38.8	36.2
	Dryland	13.2	29.8
Applied at sowing	Irrigated	30.3	33.6
	Dryland	19.5	28.6
Split	Irrigated	22.7	37.3
	Dryland	19.8	31.3

[1] Field experiment (Yang *et al.* 1980).

broadcast on soil prior to irrigation (20 mm of water) was low and similar to that of point deep placed N (Table 11.19). The yield efficiency of such a technique in a field experiment was comparable with that obtained with deep placement (Liu 1986). The downward movement of ammonium with the infiltration of irrigation water was much less than that obtained with urea; however, irrigation after surface-broadcasting ammonium fertilizer was still helpful for reducing ammonia loss, though to a lesser extent (Dai and Zhang 1983; Li *et al.* 1984a; Liu 1986; Xi *et al.* 1983).

11.8.5. Balanced fertilization

In China a large area of the land used for growing crops is deficient in P (ISFCAAS 1986; Lu 1979), and the area deficient in K is getting larger. The results in Table 11.40 showed that application of P and K to deficient soils markedly reduced the loss of fertilizer N. Table 11.40 further shows that the combined application of N with P for soils deficient in P increased Ra and Ep considerably. The positive interaction between N and P was greater on soils of low productivity and high P deficiency (ISSAS 1978).

In addition, superphosphate (SP) may act as an urease inhibitor (Table 11.41; Shi 1986), and thus ammonia loss from urea may be reduced by adding SP in admixture with urea. In field experiments, the efficiency of the mixture was greater than that obtained when urea and SP were applied separately (Tables 11.42 and 11.43; Lin and Jiang 1985; Tan 1988). The phosphoric acid applied with SP may be responsible for the retardation in urea hydrolysis (Table 11.41).

Table 11.40. Efficiencies of combined applications of N and P to upland crops.[1]

Crop	Site	Treatment	Yield ($t\ ha^{-1}$)	Ra (%)	Ep ($kg\ kg\ N^{-1}$)
Wheat	Henan	Control	2.00	–	47.0
		N	2.13	4.1	43.0
		P	2.99	–	56.8
		NP	3.95	14.6	58.5
Wheat	Henan	Control	1.77	–	45.8
		N	2.01	12.7	43.6
		P	1.80	–	41.3
		NP	2.54	43.6	40.3
Wheat	Shandong	Control	1.50	–	37.5
		N	2.88	21.4	39.8
		P	4.17	–	45.2
		NP	5.07	62.2	38.0
Wheat	Jiangsu	N	–	36.0	–
		NP	–	39.2	–
Maize	Jilin	Control	3.80	–	–
		N	4.71	17.1	–
		NP	6.92	45.1	–

[1] Field experiments (Zhang and Liang 1981; Liu *et al.* 1983; Jiang and Li 1983; Zhang *et al.* 1985; unpublished data of Zhang, S. L., Zhu, Z. L., and Xu, Y. H.).

Table 11.41. Effect of superphosphate on urea hydrolysis (% of N applied).[1]

Soil	Treatment[2]	Urea hydrolyzed	
		3rd day	7th day
Yellow fluvo-aquic, clay	200 mg U-N/kg soil	100	100
	200 mg U-N/kg soil + SP	42.3	79.5
	800 mg U-N/kg soil	71.5	100
	800 mg U-N/kg soil + SP	32.7	58.8
Yellow fluvo-aquic, clay loam	200 mg U-N/kg soil	100	100
	200 mg U-N/kg soil + SP	41.7	89.0
	800 mg U-N/kg soil	44.3	84.3
	800 mg U-N/kg soil + SP	28.2	40.4
Yellow fluvo-aquic, clay loam	800 mg U-N/kg soil	77.5	99.0
	800 mg UPA-N/kg soil	48.3	88.0
Paddy, neutral	800 mg U-N/kg soil	100	100
	800 mg UPA-N/kg soil	18.5	59.2

[1] Incubation at 25°C and 100% WHC (Shi 1986).
[2] See list of abbreviations.

Table 11.42. Effect of combined applications of urea and superphosphate on yield of barley grown on a non-calcareous paddy soil.[1]

Treatment	Yield of barley	
	(t ha^{-1})	(%)
Control (without U and SP)	2.70	100
Supergranules of (U+SP), banded	4.68	173
Mixture of (U+SP), banded	4.37	162
U and SP banded separately	3.68	136

[1] Field experiment (Lin and Jiang 1985).

Table 11.43. Effect of combined applications of urea and superphosphate on yields of maize and wheat grown on a loessial soil.[1]

Treatment	1964–1965		1985–1986		
	Maize	Wheat	Maize	Wheat	Maize
	Increase relative to control (%)				
U	20	14	24	49	6
SP	10	28	14	64	16
(U+SP) mixture	24	84	39	228	42
Supergranules of (U+SP)	28	84	49	256	38

[1] Field experiment (Tan 1988).

11.8.6. Urease inhibitors

Retarding urea hydrolysis, and the subsequent buildup of a high concentration of ammoniacal N in the floodwater of a rice field or on the surface of an upland soil, with urease inhibitors may help to reduce the partial pressure of ammonia in the floodwater, thereby reducing ammonia loss. The advantage of such an amendment is that there is no need to alter the method of application currently practised by farmers. A number of investigations have been conducted on urease inhibitors (Wang *et al.* 1983a; Guan 1985; Sun 1987; Zhu *et al.* 1985; Zhou *et al.* 1988; Cai 1988) but the results (Tables 11.44 and 11.45) showed that the effectiveness of urease inhibitors in reducing N loss and increasing the efficiency of urea N were small or inconsistent.

Many factors control the effectiveness of urease inhibitors to reduce N loss following the application of urea. These include the stability of urease inhibitors and the patterns of wind speed and high ammoniacal N in the floodwater. It was found that the widely investigated urease inhibitor phenyl phosphorodiamidate, PPD, readily decomposes in the solid state before application (Gautney *et al.* 1986). Also urea hydrolysis in field studies is retarded for a few days only (Simpson *et al.* 1985; Fillery *et al.* 1986) because the PPD is hydrolyzed to phenol under the high pH conditions in the floodwater. If the hydrolysis of urea is delayed for a few days only then the extent of loss is determined by the wind speed when ammoniacal N concentrations in the floodwater are high. Fortunately, in most of the relevant experiments, the peaks of ammoniacal N and low wind speeds coincided and so ammonia loss was reduced (Zhu 1992).

Table 11.44. Effect of urease inhibitors on nitrogen loss from urea.[1]

Soil	Crop	Treatment[2]	Recovery in plants	Residual in soil	Loss
Brown earth	Spring wheat	U	48.9	5.8	45.3
		U + 10 mg HQ/pot	52.6	6.8	40.6
		U + 20 mg HQ/pot	52.0	12.5	35.5
		U + 40 mg HQ/pot	35.2	19.2	45.6
Calcareous paddy	Rice	U	46.6	29.4	24.0
		U + 1.6% PPD	48.4	31.1	20.5

[1] In pot experiments (Zhu *et al.* 1985; Zhou *et al.* 1988).
[2] See list of abbreviations.

Table 11.45. Effect of urease inhibitors on yield of maize and rice grown on a brown earth.[1]

Site	Crop	Urease inhibitor	Increase relative to control.		
			Mean	Range	No. of data
Liaoning	Maize	Hydroquinone	5.8	2.3–7.8	5
Shandong	Maize	Phenanthrene quinone, 1–4%	1.5	-2.6–8.5	4
		Chloronitrobenzene	1.6	–	1
Fujian	Rice	PPD, 2–5%	7.6	5.3–11.8	6

[1] Field experiments (Zhou *et al.* 1988; Wang *et al.* 1983a; Guan 1985).

From the data available, it seems that a considerable reduction in N loss and increased Ea of urea N following application of urease inhibitors can only be expected when the conditions are highly conducive to ammonia volatilization.

11.8.7. Nitrification inhibitors

Although nitrification inhibitors suppress ammonium oxidation and subsequent N loss in laboratory incubation experiments (Li *et al.* 1981), they seldom reduce N loss significantly in pot and field experiments (Tables 11.46 and 11.47). At best the total loss was occasionally reduced by around 10% of the applied N. Reasons for the ineffectiveness of nitrification inhibitors may include: (i) volatilization, decomposition or adsorption of the inhibitor by soil (Bremner *et al.* 1978; Hendrickson and Keeney 1979a, b; Guthrie and Bomke 1980; McCall and Swann 1978), (ii) the inhibitor may not move in the soil in synchrony with ammonium (Touchton *et al.* 1978), and (iii) the main pathway of N loss in the experiment is not nitrification-denitrification and/or leaching (Magalhaes and Chalk 1987; Flesker and Hagin 1981; Zhu *et al.* 1985). Further research is needed to develop new inhibitors and to elucidate the factors controlling the effectiveness of these for reducing N loss.

11.8.8. Slow-release fertilizers

Slow-release N fertilizers can drastically reduce N concentrations in floodwaters and soils, and thus help to reduce N losses. In China, most of the investigations were conducted with coated fertilizers, such as sulfur-coated urea (SCU) and ABC coated with calcium-magnesium phosphate. Table 11.48 shows that SCU only reduced N loss appreciably when the rate of N application was high (Zhu 1992; Sun *et al.* 1986). The yield obtained from SCU application was 4–9% higher than that obtained with prilled urea (Table 11.49). It should be noted that N loss from deep placed USG was low, and much less than that obtained with SCU. The yield efficiency of USG was greater than that of SCU. The high cost of slow-release fertilizers is another factor limiting their use for cereal crops.

Table 11.46. Effect of nitrification inhibitors on the fate of fertilizer N applied to wheat, millet and rape (% applied N).[1]

Soil	Crop	N source	Treatment[2]	Recovery in plants	Residual in soil	Loss
Calcareous	Spring wheat-Millet	AS	Control	66.7	20.2	11.1
			Nitrapyrin	65.1	24.8	9.2
			GTU	68.0	20.4	9.3
Non-calcareous	Rape seed	AS	Control	58.7	11.1	30.3
			GTU	63.8	5.9	30.2
		ABC	Control	55.9	9.3	34.7
			GTU	59.1	10.2	30.7

[1] Pot experiments with ^{15}N (Wang *et al.* 1981; Zhang *et al.* 1984).
[2] See list of abbreviations.

Table 11.47. Effect of nitrification inhibitors on the fate of nitrogen applied to flooded rice (% of applied N.[1]

N source	Soil	Method of application[2]	Treatment	Recovery in plants	Residual in soil	Loss
AS[2]	Calcareous	SB, basal	Control	29	23	48
			Nitrapyrin	28	26	46
			Control	23	26	51
			Nitrapyrin	24	27	49
		DP, basal	Control	40	30	30
			Nitrapyrin	39	31	30
		SB, tillering	Control	58	16	26
			Nitrapyrin	61	18	21
	Non-calcareous	SB, basal	Control	54	22	24
			Nitrapyrin	56	23	21
			Control	59	28	13
			Nitrapyrin	55	26	19
		DP, basal	Control	59	22	19
			Nitrapyrin	60	20	20
Urea	Calcareous	SB, basal	Control	22	30	48
			Nitrapyrin	25	28	47
	Non-calcareous	DP, basal	Control	67	29	4
			Nitrapyrin	64	32	4
			GTU	67	27	6
			ATC	68	29	3
		Split	Control	44	23	33
			Nitrapyrin	49	25	26
			GTU	50	24	26
			ATC	46	31	23
			Control	31	48	21
			Nitrapyrin	36	54	10
ABC	Non-calcareous	Split	Control	14	37	49
			Nitrapyrin	14	45	41

[1] Field microplot experiments with ^{15}N (Li *et al.* 1981; Chen and Zhu 1982a; Chen *et al.* 1981; He *et al.* 1981; Guo *et al.* 1980).
[2] See list of abbreviations.

Table 11.48. Fate of modified urea in rice growing on a non-calcareous paddy soil.[1]

Rate	Treatment	Recovery in plant		Residual in soil		Loss	
(kg N ha^{-1})		Flowering	Maturity	Flowering	Maturity	Flowering	Maturity
58	Prilled urea	20.5	25.9	22.0	18.8	57.2	55.3
	SCU	–	28.3	–	20.6	–	51.1
	USG	65.1	67.7	28.8	19.5	5.8	12.9
115	Prilled urea	25.8	26.5	20.3	19.8	53.9	53.8
	SCU	32.9	40.0	18.1	21.3	49.1	38.8
	USG	57.6	53.5	18.9	17.6	23.5	29.0

[1] Field microplot experiment with ^{15}N (Chen *et al.* 1986).

11.8.9. Other approaches

The use of acidic fertilizers such as AC in admixture with urea or ABC to suppress floodwater pH and reduce ammonia loss has been tested in China. Their use did not result in an appreciable reduction in N loss from flooded rice (Zhu *et al.* 1987).

Table 11.49. Effect of modified urea on yield of rice and wheat.[1]

Site	Crop	Treatment[2]	Rate of application (kg N ha^{-1})	Increase relative to control (%)
Fujian	Rice	SCU	30	4.6 (mean of 4 crops)
		USG	30	8.1 (mean of 4 crops)
		SCU	58	4.5 (mean of 4 crops)
		USG	58	8.8 (mean of 4 crops)
		SCU	89	3.6 (mean of 4 crops)
		USG	89	6.7 (mean of 4 crops)
		SCU	117	3.7 (mean of 4 crops)
		USG	117	6.0 (mean of 4 crops)
Gansu	Wheat	WCU[2]	104	4.3
		WCU	87	14.3
		WCU	87	7.7

[1] Field experiments (Deng *et al.* 1981; Ren *et al.* 1987).
[2] See list of abbreviations.

Application of algicide has also been used for reducing floodwater pH, but it is unlikely to be effective in reducing ammonia loss when the soil and weather conditions are unfavourable for the growth of algae (Cai *et al.* 1985) or when ABC is used (Zhu *et al.* 1989b).

Preliminary investigations on the use of surface films to reduce ammonia loss showed that, in spite of the increased floodwater temperature caused by the film, ammonia loss and total N loss were markedly reduced (Cai 1988). Octadecanol and hexadecanol were more efficient than tetradecanol in this respect. The main difficulty encountered was to maintain a continuous film on the floodwater surface during periods of high wind speeds.

11.9. Conclusions

It may be concluded that the loss of fertilizer N in China's agriculture is high, and that there is a high potential for reducing loss. In fact, this is a worldwide challenge that soil scientists are facing. Although the N loss process has been well investigated and several techniques have been developed for reducing N loss and increasing the efficiency of fertilizer N, further basic and applied research into the mechanisms and the factors controlling N loss from plant–soil systems is still required so that better techniques can be developed.

11.10. References

Bremner, J M, Blackmer, A M and Bundy, L G 1978. Problems in the use of nitrapyrin to inhibit nitrification in soils. Soil Biol. Biochem. 10: 441–442.

Cai, D T, and Fu T G 1985. Investigations on the efficiency of split application of nitrogen fertilizers using 15N tracer technique. (in Chinese). J. Nanjing Agric. Univ. No. 2, 82–88.

Cai, G X 1986. Ammonia loss from nitrogen fertilizers applied to flooded rice. In: Soil Agricultural Chemistry and Soil Biology and Biochemistry Committees, Soil Science Society of China (eds.), Advances and Prospects for Soil Nitrogen Research in China. (in Chinese). pp. 55–67. Science Press, Beijing.

Cai, G X 1988. Gaseous Loss of Nitrogen from Fertilizers Applied to Flooded Rice. Ph.D. Thesis, University of Queensland, Australia. 275 pp.

Cai, G X, Zhang, S L and Zhu, Z L 1981. Characteristics of nitrogen mineralization of paddy soils and their effect on the efficiency of nitrogen fertilizer. In: Institute of Soil Science, Academia Sinica (ed.), Proc. of Symp. on Paddy Soil, pp. 793–799. Science Press, Beijing,

Cai, G X, Zhu, Z L, Zhu, Z W, Trevitt, A C F, Freney, J R and Simpson, J R 1985. Nitrogen loss from ammonium bicarbonate and urea applied in flooded rice field. (in Chinese). Soils 17:225–229.

Cao, Y P 1983. Ammonia volatilization from nitrogen fertilizers applied to calcareous soils. In: Shanghai Academy of Chemical Industry (ed.), Selections of Field Evaluation on the Efficiency of Ammonium Bicarbonate. (in Chinese). pp. 41–46. Shanghai.

Cao, Z H and De Datta, S K 1983. Effects of fertilization methods on properties of the water in rice fields and recovery of applied N (15N labelled urea) by lowland rice. (in Chinese). Acta Pedol. Sin. 20:253–261.

Cao, Z H, De Datta, S K and Fillery, I R P 1984. Nitrogen-15 balance and residual effects of urea-N in wetland rice fields. Soil Sci. Soc. Am. J. 48:203–208.

Chen, B H, Ren, Z G, Weng, B Q, Wang, D H, Zhang, W G and Liu, Z Z 1986. The fate of fertilizer nitrogen applied to flooded rice field. (in Chinese). Fujian Agricultural Science and Technology No. 3, 18–20.

Chen, H Q, Cheng, Y, Lin, G F, Gao, M Z and Zhao, Y C 1981. Investigations using 15N on the improvement of plant recovery of fertilizer nitrogen and the soil AN value. (in Chinese). Application of Atomic Energy in Agriculture No. 3, 39–43.

Chen, R Y and Zhu, Z L 1982a. Studies on fate of nitrogen fertilizer. I. The fate of nitrogen fertilizer in paddy soils. (in Chinese). Acta Pedol. Sin. 19:122–130.

Chen, R Y and Zhu, Z L 1982b. Characteristics of the fate and efficiency of nitrogen in supergranules of urea. Fert. Res. 3:63–71.

Chen, R Y, Zhang, J C, Guo, W M and Chen, W 1987. Deep placement of N-fertilizer (urea) topdressed to unsaturated paddy soil with reflooded water. (in Chinese). Chinese J, Rice Sci. 1:184–191.

Cheng, L L, Wen, Q X and Li, H 1988. Fixed ammonium in soils in tropical and subtropical regions of China. (in Chinese). Soils 20:239–242.

Cheng, L L, Wen, Q X and Li, H 1989. Transformation of 15N labelled fertilizer N in soils under greenhouse and field conditions. (in Chinese). Acta Pedol. Sin. 26:124–130.

Dai, Q L, and Zhang, J R 1983. Investigations on the plant recovery and yield efficiency of supergranular ammonium bicarbonate applied to irrigated fields. In: Shanghai Academy of Chemical Industry (ed.), Selections of Field Evaluation on the Efficiency of Ammonium Bicarbonate. (in Chinese). pp. 53–59. Shanghai.

Deng, S Z, Guo, W F and Liang, J L 1981. Investigations on coated fertilizers. (in Chinese). Gansu Agric. Sci. No. 1, 15–18.

Du, Y J 1986. A preliminary investigation on the nitrogen supplying capacity of paddy soils in the lowland along the Changjiang River. (in Chinese). Anhui Agric. Sci. No. 1, 74–80.

ECACA 1988. (Editorial Committee of Almanac of China's Agriculture). Almanac of China's agriculture for 1987. (in Chinese). Beijing.

Fan, Y C and Tao, Q R 1984. Parameters for fertilizing flooded rice. (in Chinese). Jiangxi Agric. Sci. No. 3, 12–14.

FAO. 1987. Fertilizer Yearbook.

Fillery, I R P and De Datta, S K 1986. Ammonia volatilization from nitrogen sources applied to rice fields: 1. Methodology, ammonia fluxes, and nitrogen-15 loss. Soil Sci. Soc. Am. J. 50:80–86.

Fillery, I R P, De Datta, S K and Craswell, E T 1986. Effect of phenyl phosphorodiamidate on the fate of urea applied to wetland rice fields. Fert. Res. 9:251–263.

Fleisker, Z and Hagin, J 1981. Lowering ammonia volatilization losses from urea application by activation of nitrification process. Fert. Res. 2:101–107.

Gao, J H, Zhang, Y, Huang, D M, Wu, J M and Pan, Z P 1984. The application of results of soil nitrogen mineralization as parameters to forecast the maximum dressing rate on nitrogen for early rice plants. (in Chinese). Scientia Agricultura Sinica No. 5, 67–72.

Gautney, J, Barnard, A R, Penney, D B and Kim, Y K 1986. Solid-state decomposition of phenyl phosphorodiamidate. Soil Sci. Soc. Am. J. 50:792–797.

Guan, S Y 1985. The efficiency of using soil urease inhibitors. (in Chinese). J. Soil Sci. 16:232–234.

Greenwood, D J 1981. Fertilizer use and food production: world sense. Fert. Res. 2:33–51.

Guo, W J 1979. The efficiency of deep placement of ammonium bicarbonate. (in Chinese). Soils 1, 20–23.

Guo, Z F, Chen, S S, Wang, Y S and Tang, N X 1980. Investigations on the uptake, immobilization, and loss of fertilizer nitrogen applied to flooded rice field and the improvement of plant recovery. (in Chinese). Application of Atomic Energy in Agriculture 1, 24–29.

Guthrie, T F and Bomke, A A 1980. Nitrification inhibition by N-serve and ATC in soils of varying texture. Soil Sci. Soc. Am. J. 44:314–320.

Han, X Z, Men, K, Liu, H X, Liu, W X and Wu, Y 1991. Fate and efficiency of fertilizer nitrogen applied with different methods to flooded rice field in cool temperate regions. In: Proceedings of Congress of Soil Sci. Soc. of China, 117–119. Soil Sci. Soc. of China, Changsha.

He, C Y, Zhao, Y L, Wang, Q, Li, Z Y, Chen, X L and Gao, J F 1981. The effectiveness of nitrification inhibitors in increasing the efficiency of nitrogenous fertilizer in paddy soils. (in Chinese). J. Soil Sci. No. 3, 1–3.

Hendrickson, L L and Keeney, D R 1979a. Effect of some physical and chemical factors on the rate of hydrolysis of nitrapyrin (N-serve). Soil Biol. Biochem. 11:47–50.

Hendrickson, L L and Keeney, D R 1979b. A bioassay to determine the effect of organic matter as a nitrification inhibitor. Soil Biol. Biochem. 11:51–55.

Hua, X J, Lan, X Q and Wang, H Z 1982. Investigations on the plant uptake of 15N-labelled ammonium nitrate. (in Chinese). Gansu Agricultural Science and Technology 6, 21–24.

Huang, D M, Sun, G Y, Shao, T, Yao, Y C and Zhu, X Q 1982. Investigation on the nitrogen supplying capability of paddy soils in northern plain of Hui River. (in Chinese). Jiangsu Agric. Sci. 10, 1–9.

Huang, J M 1986. Investigations on the regulation technique of nitrogen fertilization for flooded rice. A review. In: Soil Agricultural Chemistry and Soil Biology and Biochemistry Committees, Soil Science Society of China (eds.), Advances and Prospects for Soil Nitrogen Research in China. (in Chinese). pp. 185–194. Science Press, Beijing.

ISFCAAS 1986. (Institute of Soils and Fertilizers, Chinese Academy of Agricultural Sciences). Regionalism of chemical fertilizers in China. (in Chinese). China Agricultural Science and Technology Publishing House, Beijing.

ISSAS 1977. (Institute of Soil Science, Academia Sinica). Super-Granules of Ammonium Bicarbonate. Jiangsu People's Publishing House, Nanjing.

ISSAS 1978. (Institute of Soil Science, Academia Sinica). Soils of China. (in Chinese). Science Press, Beijing.

ISFJAAS 1981. (Institute of Soils and Fertilizers, Jiangsu Academy of Agricultural Sciences and Suzhou Agricultural Institute). Investigations on the efficient application of nitrogen fertilizers in Suzhou District, Jiangsu Province. I. Rate and method of nitrogen fertilizer application for the fist crop of rice in double-cropping system . (in Chinese). Jiangsu Agric. Sci. 2, 1–9.

Islam, M S and Parsons, J M 1979. Mineralization of urea and urea derivatives in anaerobic soils. Plant Soil 51:319–330.

Jia, Y, Li, Z L and Cho, J C 1982. A study on the promotion of utilization rate of nitrogen fertilizer in paddy soils in the Yellow River irrigated area of Ninxia. (in Chinese). J. Soil Sci. 2, 27–29.

Jiang, N H and Liu, G S 1980. Diffusion and loss from ammonium bicarbonate and aqueous ammonia applied to soils. (in Chinese). Acta Pedol. Sin. 17:89–93.

Jiang, R C 1980. Transformation of urea in calcareous soil and the method of application. (in Chinese). Jiangsu Agric. Sci. 3, 42–46.

Jiang, R C and Li, D M 1983. Assessing the rates and ratio of application of nitrogen and phosphorus fertilizers. (in Chinese). Soils and Fertilizers 1, 8–11.

Jiao, G H 1983. Investigation on the depth of nitrogen fertilizer application. Shaanxi Agric. Sci. (in Chinese). 3, 35–36.

Jin, S L, Lan, X Q, Chang, X D, Wang, S L and Zhu, Y B 1985. Investigations on the timing of nitrogen application to spring wheat. (in Chinese). Soils and Fertilizers 6, 25–29.

Kimura, M, Yoshioka, L, Wada, H and Takai, Y 1986. The role of algae for denitrification in paddy soils (part 1). Denitrification by microorganisms grown on filamentous algae. Soil Sci. Plant Nutr. (in Japanese). 57:365–370.

Li, G R, Guo, Y D and Chen, P S 1985. A preliminary investigation on the movement, disintegration and transformation of urea in calcareous soils. (in Chinese). Scientia Agricultura Sinica 1, 73–76.

Li, G R, Chen, P D, Guo, Y D and Zhang, X 1988. Effect of top-dressing ammonium bicarbonate with implement on the fate of fertilizer nitrogen applied for upland crops. A model experiment. (in Chinese). Soils and Fertilizers 2, 15–18.

Li, L M and Wu, Q T 1991. Role of amorphous manganese oxide in nitrogen loss. Pedosphere. 1:83–92.

Li, L M, Zhang, S, Zhou, X R and Pan, Y H 1981. The effect of CP (nitrapyrin) on the nitrification and microbial activity. (in Chinese). Acta Pedol. Sin. 18:58–70.

Li, L M, Zhang, S, Zhou, X R and Pan, Y H 1984b. Influence of rice root system on nitrogen loss. (in Chinese). Soils 16:5–10.

Li, L M, Pan, Y H, Wu, Q T, Zhou, X R and Li, Z G 1988. Investigation on amorphous ferric oxide acting as an electron acceptor in the oxidation of NH_4^+ nitrogen under anaerobic conditions. (in Chinese). Acta Pedol. Sin. 25:184–190.

Li, R G, Wang, S M, Wang, K W, Wu, R Z and Wu, W K 1982. The nitrogen absorption in soil and fertilizer by the winter wheat. (in Chinese). J. Soil Sci. 4, 21–22.

Li, Z L, Li, A R and Cao, Z H 1984a. Effect of application methods of nitrogen fertilizer on the recovery of fertilizer-N by spring wheat in calcareous soil. (in Chinese). Soils 16:134–137.

Liao, X L, Xu, Y H and Zhu, Z L 1982. Investigation on nitrification-denitrification loss of fertilizer nitrogen in submerged paddy soil. (in Chinese). Acta Pedol. Sin. 19:257–263.

Liao, X L, He, D Y, Deng, S L, Wang, C L and Li, X F 1983. Investigations on the fertility of gleyed paddy soil and efficiency of fertilization. II. Nitrogen supplying characteristics of gleyed paddy soil and the nitrogen balance. (in Chinese). Agricultural Modernization 3, 27–33.

Lin, Y X and Jiang, P X 1985. Effect of nitrogen-phosphorus compound fertilizer on growth of crop and conditions of application. (in Chinese). Soils 17:10–14.

Liu, Y Z, Zhang, S M and Chen, Z Q 1983. Efficiency of ammonium bicarbonate and its improvement. In: Shanghai Academy of Chemical Industry (ed.), Selectives of Field Evaluation on the Efficiency of Ammonium Bicarbonate. (in Chinese). pp. 127–132. Shanghai.

Liu, Z H 1986. Investigations on soil nitrogen and nitrogen fertilization in Hebei Provinces. In: Soil Agricultural Chemistry and Soil Biology and Biochemistry Committees, Soil Science Society of China (eds.), Advances and Prospects for Soil Nitrogen Research in China. (in Chinese), pp. 136–143. Science Press, Beijing.

Lu, M L, Wang, Z R, Jiang, B F and Lu, R K 1981. Investigations on the utilization of nitrogen and phosphorus in nitro-phosphate. Chemical Industry 2, 14–17.

Lu, R K 1979. Agricultural modernization and the development of chemical fertilizers. (in Chinese). Chemical Industry 5, 63–66.

Magalhaes, A M T and Chalk, P M 1987. Nitrogen transformation during hydrolysis and nitrification of urea. 2. Effect of fertilizer concentration and nitrification inhibitors. Fert. Res. 11:173–184.

McCall, P J and Swan, R L 1978. Nitrapyrin volatility from soil. Down Earth 34:21–27

Mo, S X and Qian, J F 1981. Uptake of nitrogen by rice plant from straw manure, urea and soil. In: Institute of Soil Science, Academia Sinica (ed.), Proc. of Symp. on Paddy Soil, 800–804. Science Press, Beijing.

Mo, S X and Qian, J F 1983. Studies on the transformation of nitrogen of milk vetch in red earth and its availability to rice plant. (in Chinese). Acta Pedol. Sin. 20:12–22.

Pan, Z P, You, D M and Wan, C B 1982. Investigations on the efficient application of nitrogen fertilizers in Suzhou District, Jiangsu Province. II. Second investigation on the rate and timing of nitrogen fertilizer application for early rice. (in Chinese). Jiangsu Agric. Sci. 5, 15–23.

Pan, Z P, Xu, W Y and Wan, C B 1984. Investigations on the efficient application of nitrogen fertilizers in Suzhou District, Jiangsu Province. VII. The efficiency of nitrogen fertilizer applied to single-cropping late rice and the optimum rate of application. (in Chinese). Jiangsu Agric. Sci. . 12, 1–4.

Pan, Z P, Qian, H C, Wu, J M and Xu, X Q 1980. Investigations on the efficiency of NPK for barley and naked barley. (in Chinese). Jiangsu Agric. Sci. 4, 14–22.

Pan, Z P, Xu, W Y, Wan, C B and You, D M 1983. Investigations on the efficient application of nitrogen fertilizers in Suzhou District, Jiangsu Province. IV. The rate and timing of nitrogen fertilizer application for late rice. (in Chinese). Jiangsu Agric. Sci. 6, 1–8.

Reddy, K R and Patrick, W H Jr 1986. Denitrification losses in flooded rice fields. Fert. Res. 9:99–116.

Ren, Z G, Wang, D H and Zhang, Y Q 1987. Efficiency of modified urea fertilizers. (in Chinese). Soils and Fertilizers No. 4, 33–36.

Savant, N K and De Datta, S K 1982. Nitrogen transformations in wetland rice soils. Adv. Agron. 35:241–302.

Shi, M H 1986. Ammonia volatilization from urea surface broadcast on calcareous soil – Effect of superphosphate. M.Sc. Thesis, Nanjing Agric. Univ. Nanjing, China.

Shi, X Z, Liu, M Y, Xi, Z B and Zhao, D G 1983. Evaluation on the efficiency of ammonium bicarbonate in field experiments. In: Shanghai Academy of Chemical Industry (ed.), Selections of Field Evaluation on the Efficiency of Ammonium Bicarbonate. (in Chinese), pp. 18–25. Shanghai.

Shi, X Z, Ton, Y W, Liu, M Y and Xi, Z B 1981. Characteristics of nitrogen uptake pattern of barley and the technique of application for wheat and barley. In: Sun, X (ed.), Soil nutrients, plant nutrition and improved fertilization. (in Chinese), pp. 328–340. Agriculture Publishing House. Beijing.

Simpson, J R, Freney, J R, Muirhead, W A and Leuning, R 1985. Effect of phenyl phosphorodiamidate and dicyandiamide on nitrogen loss from flooded rice. Soil Sci. Soc. Am. J. 49:1426–1431.

Sun, X T 1987. The fate and efficiency of fertilizer nitrogen (15N-labelled urea) applied in wetland rice soil. (in Chinese). Soils 19:177–182.

Sun, X T, Chen, R Y, Jiang, P X, Pan, Z P, Chen, Q W and Hui, M X 1986. Supplying process and effect of the nitrogen from slow-release urea applied in soil under rice-wheat rotation system. (in Chinese). Acta Pedol. Sin. 23:17–29.

Tan, W L 1988. Efficiency of supergranules of mixed urea-superphosphate fertilizers and the method of application. (in Chinese). Shaanxi Agric. Sci. 4, 22–25.

Tian, W L and Huang, X G 1984. Nitrogen application for cotton in relation to soil properties. (in Chinese). Zhejiang Agric. Sci. No. 3, 143–145.

Touchton, J T, Hoeft, R G and Welch, L F 1978. Nitrapyrin degradation and movement in soil. Agron. J. 70:811–816.

Trevitt, A C F, Freney, J R, Simpson, J R and Muirhead, W A 1987. Effects of microplots on urea nitrogen reactions in flooded soils. In: IRRI (ed.), Efficiency of Nitrogen Fertilizers for Rice, pp. 177–183. IRRI, Manila.

Wan, C B 1986. Nitrogen resource and its efficient use. In: Soil Agricultural Chemistry and Soil Biology and Biochemistry Committees, Soil Science Society of China (eds.), Advances and Prospects for Soil Nitrogen Research in China. (in Chinese), pp. 144–156. Science Press, Beijing.

Wan, C B, Sun, P L, Pan, Z P and Xu, W Y 1983. Investigations on the efficient application of nitrogen fertilizers in Suzhou District, Jiangsu Province. V. Partitioning nitrogen fertilizer application into basal- and top-dressing and the efficiency of organic manures for wheat. (in Chinese). Jiangsu Agric. Sci. 9, 1–4.

Wan, C B, Sun, P L, Pan, Z P and Xu, W Y 1984. Investigations on the efficient application of nitrogen fertilizers in Suzhou District, Jiangsu Province. VI. The optimum rates of nitrogen and phosphate fertilizer application for wheat. (in Chinese). Jiangsu Agric. Sci. 11, 1–4.

Wang, D H, Ren, Z G and Chen, Z C 1983a. Preliminary investigation on the urease inhibitor PPD. Construction of (in Chinese). Soils and Fertilizers 5, 35–37.

Wang, F J, Liu, X S, Peng, G Y, Wen, X F and Wang, B Z 1981. Investigations on the effect of nitrification inhibitors on the uptake of nitrogen with 15N tracer technique. (in Chinese). Application of Atomic Energy in Agriculture 2, 34–39.

Wang, Y H, Jiang, S Z and Gu, Y M 1983b. A study on predicting nitrogen supplying capacity of gleyed soil in the suburbs of Shanghai. (in Chinese). Acta Pedol. Sin. 20:262–271.

Watkins, S H, Strand, R F, De Bell, D S and Esch, J Jr 1972. Factors influencing ammonia losses from urea applied to northwestern forest soils. Soil Sci. Soc. Am. J. 36:354–357.

Wen, Q X and Zhang, X H 1986. Fixed ammonium in soils. In: Soil Agricultural Chemistry Committee and Soil Biology and Biochemistry Committee. Soil Science Society of China (eds.), Advances and Prospects for Soil Nitrogen Research in China. (in Chinese), pp. 34–45. Science Press, Beijing.

Xi, Z B, Bian, Y J, Kuang, A Q, Liu, D B and Liu, M Y 1978. Studies on peak nutrient uptake of double-cropping rice and the method of basal dressing with volatile nitrogen fertilizer to the whole plowed layer. (in Chinese). Acta Pedol. Sin. 15:113–125.

Xi, Z B, Shi, X Z and Liu, M Y 1983. On the agrochemical properties of ammonium bicarbonate. In: Shanghai Academy of Chemical Industry (ed.), Selections of Field Evaluation on the Efficiency of Ammonium Bicarbonate. (in Chinese), pp. 8–17. Shanghai.

Xu, Z H, Cao, Z H and Li, Q K 1987. Studies on the effect of urea supergranules on corn and the fate of nitrogen in calcareous sandy soil. (in Chinese). Acta Pedol. Sin. 24:51–58.

Yang, G, Hu, L H, Jin, Q Y and Wang X L 1980. Investigation on the improved application of ammonium bicarbonate to winter wheat. (in Chinese). Shanxi Agric. Sci. 11, 2–6.

Yao, R H and Guan, S J 1983. Transformation of deep-placed urea in soil and its efficiency. (in Chinese). Soils and Fertilizers 1, 15–16.

You, D M and Lo, D R 1986. Methods for top-dressing nitrogen fertilizers for flooded rice for reducing volatilization loss. (in Chinese). Soils and Fertilizers 5, 43–45.

Yu, J Z, Cai, G X and Zhu, Z L 1984. Fate of 15N-labelled ammonium sulfate in rice fields under different applying methods. (in Chinese). Soils 16:106–107.

Yuan, Z Y, Huang, C Y, Li, S H, Fang, W C, Wang, S Q and Chen, Y S 1981. Investigations using 15N tracer on the movement, loss in soil and the plant uptake of urea. (in Chinese). Application of Atomic Energy in Agriculture 2, 51–56.

Zhang, G L and Liang, G L 1981. A preliminary investigation on the efficiency of combined application of nitrogen and phosphorus fertilizers to wheat. (in Chinese). Henan Agricultural and Forestry Science and Technology 8, 11–13.

Zhang, X Y and Zhao, X C 1983. Investigations on the technology of applying ammonium bicarbonate. In: Shanghai Academy of Chemical Industry (ed.), Selections of Field Evaluation on the Efficiency of Ammonium Bicarbonate. (in Chinese), pp. 47–52. Shanghai.

Zhang, A Z, Liao, P T and Wang, J Y 1980. Investigations on the nutrition of hybrid rice. (in Chinese). Application of Atomic Energy in Agriculture 3, 15–20.

Zhang, K, Zhao, J Y, Wang, X F, Wu, W, Hu, H Y and Yu, T D 1985. Investigations on the rates and proportions of nitrogen and phosphorus fertilizer application. 5. Effect on plant recovery. (in Chinese). Jilin Agric. Sci. 3, 79–84.

Zhang, Q Z, Xi, H F, Ji, Y Q, Shen, H C and Shi, Y H 1984. Investigations using 15N tracer on the improvement of nitrogen efficiency for rape with nitrification inhibitor ASU. (in Chinese). Acta Agriculturae Universitatis Zhejiangensis 10:249–252.

Zhang, S L, Zhu, Z L, Xu, Y H, Chen, R Y and Li A R 1988. On the optimal rate of application of nitrogen fertilizer for rice and wheat in Tai-lake region. (in Chinese). Soils 20:5–9.

Zhang, S L, Zhu, Z L and Xu, Y H 1989. The transformation of urea and the fate of fertilizer nitrogen in fluvo-aquic soil-winter wheat system in flooded plain of Huanghe River. (in Chinese). Acta Agriculturae Nucleatae Sinica 3:9–15.

Zhang, S L, Cai, G X, Wang, X Z, Xu, Y H and Zhu, Z L 1992a. Losses of urea-nitrogen applied to maize grown on a calcareous fluvo-aquic soil in north China. Pedosphere. 2:171–178.

Zhang, S L, Zhu, Z L and Xu, Y H 1992b. Improved techniques for nitrogen fertilizer application for flooded rice. (in Chinese). Soils (in press).

Zhang, Z D 1981. A preliminary investigation on the rate of nitrogen fertilizer application for early rice. (in Chinese). Soils and Fertilizers 4, 25–28.

Zhao, Z D and Zhang, J S 1979. Investigations on the improvement of plant recovery of nitrogen fertilizers. II. Soil moisture status in relation to plant recovery. (in Chinese). J. Soil Sci. 5, 39–40.

Zhao, Z D, Zhang, J S and Ren, S R 1986. Ammonia volatilization from upland soils. In: Soil Agricultural Chemistry and Soil Biology and Biochemistry Committees, Soil Science Society of China (eds.), Advances and Prospects for Soil Nitrogen Research in China. (in Chinese). pp. 46–54. Science Press, Beijing.

Zhou, D C 1985. The effects of 15N-fertilizer on the yields of wheat. (in Chinese). Application of Atomic Energy in Agriculture 3, 53–56.

Zhou, L K, Wu, G Y, Zhang, Z M, Cao, C J and Li, R H 1988. Effect of urease inhibitor hydroquinone on the efficiency of urea fertilizer. (in Chinese). Acta Pedol. Sin. 25:191–198.

Zhou, M Z 1987. Soil testing and fertilizer recommendation in China. (in Chinese). J. Soil Sci. 18:7–13.

Zhou, Q and Chen, H K 1983. The activity of nitrifying and denitrifying bacteria in paddy soil. Soil Sci. 13:531–534.

Zhu, Z L 1982. Parameters for assessing the application rate of nitrogen fertilizer on rice and wheat. (in Chinese). Soils 14:136–140.

Zhu, Z L 1987. 15N balance studies of fertilizer nitrogen applied to flooded rice fields in China. In: IRRI (ed.), Efficiency of Nitrogen Fertilizers for Rice, pp. 163–167. IRRI, Los Baños, Philippines.

Zhu, Z L 1988. On the prediction of soil nitrogen supplying capacity and the use of mean optimal application rate of nitrogen in fertilizer recommendation for rice. (in Chinese). Soils 20:57–61.

Zhu, Z L 1990. Management of nitrogen fertilizers for flooded rice in relation to nitrogen transformations. Trans. 14th Inter. Congress. Soil Sci. IV:337–342. Kyoto.

Zhu, Z L 1991. Investigations on the integrated management of nitrogen fertilizers and irrigation for flooded rice. (in Chinese). Soils 23:241–245.

Zhu, Z L 1992. Efficient management of nitrogen fertilizers for flooded rice in relation to nitrogen transformation processes. Pedology 2: (in press).

Zhu, Z L, Cai, G X and Yu, J Z 1977. A preliminary investigation on the fate of N15-labelled ammonium sulfate in rice field. (in Chinese). Kexue Tongbao 22:503.

Zhu, Z L, Liao, X L and Cai, G X 1978. Soil nutrition status under 'rice-rice-wheat' rotation and the response of rice to fertilizers in Suchow District. (in Chinese). Acta Pedol. Sin. 15:126–137.

Zhu, Z L, Wang, Z Q and Xu, Y H 1963. Investigation of nitrogen supplying regime of soils. II. The transformation of ammonium sulfate in rice soils and its influence on the soil nitrogen supplying status. (in Chinese). Acta Pedol. Sin. 11:185–195.

Zhu, Z L, Zhang, S L and Xu, Y H 1987. Effect of nitrogen fertilizers and application methods on the gaseous loss of nitrogen under rice growing conditions. (in Chinese). Soils 19:5–12.

Zhu, Z L, Cai, G X, Xu, Y H and Zhang, S L 1985. Ammonia volatilization and its significance to the losses of fertilizer nitrogen applied to paddy soil. (in Chinese). Acta Pedol. Sin. 22:320–328.

Zhu, Z L, Chen, R Y, Xu, Y F, Xu, Y H and Zhang, S L 1979a. The effect and forms of placement of nitrogen fertilizers on the characteristics of the nitrogen supply in paddy soils. (in Chinese). Acta Pedol. Sin. 16:218–233.

Zhu, Z L, Cai, G X, Simpson, J R, Zhang, S L, Chen, D L, Jackson, A V and Freney, J R 1989a. Processes of nitrogen loss from fertilizers applied to flooded rice fields on a calcareous soil in north-central China. Fert. Res. 18:101–115.

Zhu, Z L, Simpson, J R, Zhang, S L, Cai, G X, Chen, D L, Freney, J R and Jackson, A V 1989b. Investigations on nitrogen losses from fertilizers applied to flooded calcareous paddy soil. (in Chinese). Acta Pedol. Sin. 26:337–343.

Zhu, Z L, Zhang, S L, Chen, D L, Cai, G X, Xu, Y H, Simpson, J R, Freney, J R and Jackson, A V 1988. Fate and efficiency of nitrogen fertilizers applied with different methods of incorporation to a calcareous paddy soil in Huang-Huai-Hai Plain. (in Chinese). Soils 20:121–125.

Zhu, Z M, Chen, X B and Gao, H B 1986. Investigations on dynamics of ammonium bicarbonate in rice fields. (in Chinese). Hunan Agric. Sci. 6, 35–36.

Zhu, Z M, Wu, T B, Dai, Z Q and Zhen, S X 1979b. A preliminary investigation on nitrogen balance in paddy soil derived from red earth. (in Chinese). Hunan Agric. Sci. 5, 31–34.

Zhu, Z M, Wu, T B, Zhen, S X and Dai, Z Q 1984. Efficiency of split application of nitrogen fertilizer. (in Chinese). Soils and Fertilizers No. 4, 29–32.

Zuo, D F 1983a. Efficiency of nitrogen fertilizers applied to spring maize grown on calcareous soils. In: Shanghai Academy of Chemical Industry (ed.), Selections of Field Evaluation on the Efficiency of Ammonium Bicarbonate. (in Chinese), pp. 31–40. Shanghai.

Zuo, D F 1983b. Comparisons of the efficiency of nitrogen fertilizers applied to winter wheat in field experiments. In: Shanghai Academy of Chemical Industry (ed.), Selections of Field Evaluation on the Efficiency of Ammonium Bicarbonate. (in Chinese), pp. 73–81. Shanghai.

Zuo, D F and Wei, X M. 1983. Investigations on the fate and plant recovery of nitrogen fertilizers using 15N. In: Soil Science Society of China (ed.), Utilization and Fertility Improvement of Soils in China. (in Chinese), Vol. 3, 185–189. Xian.

12
Transformation and management of manure nitrogen

CHENG LI-LI AND WEN QI-XIAO

12.1. Introduction

Before the advent of the fertilizer industry organic manure was the most important source of nutrients for agriculture. The distinctive feature of agriculture in China was the maximum utilization of agricultural and household organic wastes. According to statistics, in the early years of the People's Republic of China, organic manure provided 99.6% of the total input of nutrients (ISF 1986; Zhang 1984). Since then, with the development of the fertilizer industry, consumption of fertilizers has been raised dramatically, while the relative contribution of organic manure to the total input of nutrients, especially N, has declined continuously. Nevertheless, the absolute amount of nutrients applied as organic manure has increased significantly.

Such a change has resulted in the combined use of organic manure and inorganic fertilizers. In 1986, the amounts of N, P and K applied as organic manure in China were estimated to be 4,946, 1,459 and 6,236 thousand tons respectively (Zhu and Xi 1990). Although the amounts were increased by 206%, 419% and 233%, respectively, compared with the amounts applied in 1949, they accounted for only 26.5%, 42.4% and 87.5%, respectively, of the total input of N, P and K. It can be expected that these proportions will be decreased further by the end of this century.

Agricultural and household organic wastes are resources as far as plant nutrients and soil organic matter are concerned. Recently, with the increase in agricultural production, considerable research has been conducted on the transformations of manure-N, and the integrated management of organic manure and inorganic fertilizers, so as to (i) use the resources more efficiently, (ii) overcome environmental problems, (iii) improve soil fertility, and (iv) increase crop production. The results obtained in these studies are summarized in this chapter.

12.2. Transformations of organic nitrogen in manure and its effect on soil fertility

12.2.1. Transformations

In organic manure, N is present largely in organic forms. After applying organic manure to soil, two opposing microbial processes take place continuously and

Zhu Zhao-liang et al. *(eds.): Nitrogen in Soils of China, 281–302.*

simultaneously. One is the liberation (mineralization) of ammonium from various N-containing substances, mainly proteins, nucleic acids and aminopolysaccaride, and the other is the consumption (immobilization) of inorganic N, including ammonium and nitrate, into organic N. Depending on the relative rates of the two processes, net mineralization or net immobilization will occur.

As with fertilizer, the ammonium released from organic manure is partly adsorbed onto soil colloids, and the adsorbed ammonium is in balance with the ammonium remaining in the soil solution. The relative proportion of the two forms of ammonium is largely governed by the CEC of the soil and the nature of the cation exchange complexes. In addition, some of the ammonium may also be transformed from exchangeable into nonexchangeable in soils containing 2:1 type clay minerals, e.g. vermiculite and illite. Ammonium in soil solution may be absorbed by plants directly or after transformation into nitrate, or it may be lost before or after transformation to nitrate by ammonia volatilization, leaching and nitrification-denitrification.

12.2.2. Mineralization

Net mineralization or net immobilization takes place in soil along with the decay of organic materials. As far as methods for determining net mineralization of green manure N is concerned, Bouldin (1988) pointed out that routine incubation procedures gave an underestimate due to the loss of part of the mineralized N during incubation. When N labelled organic material is used mineralization of the organic N can be readily determined, but the biological interchange of mineralized ^{15}N with soil ^{14}N can also lead to an underestimate of the ^{15}N release from the organic materials. Similarly, net immobilization of inorganic N during the decay of organic materials with high C/N ratios will be underestimated by these methods. In our laboratory the pattern of N mineralization from green manure was investigated by a N balance method (Wen *et al.* 1987). Milk vetch or azolla, after mixing with a subsoil extremely low in organic matter, were put into carborundum cylinders for incubation under upland or flooded conditions. The results showed that irrespective of the moisture conditions for incubation and the biodegradability of the green manure, the mineralization of organic N can be clearly divided into rapid and slow release stages. In the initial rapid release stage, which generally lasted 3 months or less, 42–72% of the total N was mineralized. During the subsequent stage the mineralization rate was very slow, and only 10–28% of total N was mineralized over 57 months (Tables 12.1 and 12.2).

Immobilization-mineralization of N during the decomposition of organic materials with high C/N ratios, such as straw, has not been investigated by the N balance method. However, a number of studies on the decomposition of organic materials indicated that mineralization of organic C can also be divided into rapid and slow decomposition stages (Lin *et al.* 1980; Cheng *et al.* 1981). Thus, it can be expected that some of the inorganic N will be immobilized during the rapid decomposition stage of these materials and ammonium will be released slowly in the subsequent slow decomposition stage.

Table 12.1. Fate of nitrogen in milk vetch and azolla under flooded conditions (% of applied N).[1]

Incubation period (months)	Milk vetch					Azolla				
	Organic	Mineralized	Exch NH_4^+	Fixed NH_4^+	Loss	Organic	Mineralized	Exch NH_4^+	Fixed NH_4^+	Loss
					Xiashu loess					
3	30.3	69.7	3.8	17.8	48.1	66.9	33.1	4.6	9.8	18.7
6	32.9	67.1	1.9	14.3	50.9	49.9	52.0	2.0	7.2	40.9
12	30.1	68.9	1.2	12.3	56.4	50.1	49.9	1.1	6.0	42.8
24	24.3	75.7	1.1	11.9	62.7	41.1	58.9	1.2	6.2	51.5
36	23.9	76.1	1.0	11.3	63.8	45.4	54.5	0.8	5.2	48.6
60	19.8	80.2	0.8	10.8	68.6	37.2	62.8	0.8	7.3	54.7
					Quaternary red clay					
3	28.2	71.8	5.1	9.8	56.9	47.8	52.2	3.5	5.8	42.9
6	26.8	73.2	2.0	5.2	65.9	40.6	59.4	1.3	2.3	55.8
12	24.9	75.1	0.7	1.8	72.6	48.8	51.2	0.8	1.7	48.7
24	18.7	81.3	0.9	2.1	78.3	37.5	62.5	0.9	1.6	59.9
36	18.9	81.2	0.7	1.5	79.0	35.5	64.4	0.7	1.3	62.5
60	16.0	84.0	0.5	1.5	82.0	33.8	66.2	0.6	1.5	64.1

[1] Results obtained from incubation experiments in the field.

Table 12.2. Fate of N in milk vetch and azolla under upland conditions (% of applied N).[1]

Incubation period (months)	Milk vetch					Azolla				
	Organic	Mineralized	$NH_4^+ + NO_3^-$	Fixed NH_4^+	Loss	Organic	Mineralized	$NH_4^+ + NO_3^-$	Fixed NH_4^+	Loss
					Xiashu loess					
3	36.5	63.5	2.8	12.6	48.1	58.0	42.1	2.1	5.5	34.4
6	35.6	64.3	2.9	12.5	48.9	54.4	45.6	2.8	4.1	38.6
12	38.2	61.8	3.1	12.5	46.1	52.3	47.7	2.4	3.7	41.5
24	34.8	75.2	1.1	12.6	61.5	33.0	67.0	0.9	5.9	60.2
36	23.2	76.8	0.7	9.5	66.6	29.3	70.7	0.6	4.1	65.9
60	19.2	80.8	0.7	7.6	72.6	23.6	76.4	0.6	3.7	72.0
					Quaternary red clay					
3	31.6	68.3	10.8	6.9	50.6	47.1	52.9	2.0	1.6	49.3
6	32.0	68.0	4.6	6.4	56.9	45.9	54.1	2.8	1.3	50.0
12	27.8	72.4	6.0	2.2	64.0	41.3	58.7	2.1	1.5	55.1
24	23.2	76.8	1.1	2.3	73.4	35.5	64.4	0.7	1.5	62.2
36	19.9	80.1	0.7	1.8	77.6	30.7	69.3	0.6	1.4	67.3
60	15.9	84.1	1.0	1.5	81.6	19.1	80.9	0.6	1.3	79.0

[1] Results obtained from incubation experiments in the field.

Organic manures differ greatly in the stage of decomposition. Manure, such as aerobic compost, waterlogged compost and barnyard manure (SMRWAS 1959; WPM 1959; Liu *et al.* 1959), irrespective of the raw materials used for processing, have undergone a period of decomposition and thus the rapid stage of mineralization or immobilization of N has passed.

12.2.3. Availability

The recovery of manure N by plants is an important index for evaluating the quality of organic manures. Recovery depends not only on the amount of inorganic N originally present in and mineralized from the manure during plant growth, but also on the coincidence of the patterns of mineralization and plant uptake. In addition, it is related to the crop variety and growth period. As indicated above, the amount and pattern of N release from organic manure depends on its chemical composition, total N content and stage of decomposition, as well as soil conditions.

Green manure and crop straw are organic materials which have not undergone decomposition. After addition to soil, the amount of N released or immobilized depends on their lignin and total N content. When different organic manures have the same lignin content, the greatest amount of N is released from the materials with the most N. When the N content of the organic manure is lower than a certain value, net immobilization of soil mineral N rather than net mineralization will occur. On the other hand, for organic manures with the same total N, the higher the lignin content, the less the mineralization (for the high N manures) or immobilization (for the low N manures) will be. Similarly, the release pattern of ammonium from a manure is also affected by its contents of lignin and total N. For manures with similar total N content, the lower the lignin content, the larger the amount of ammonium released (expressed as a percentage of total ammonium released during a cropping season) in the initial stage of decomposition. For instance, Shi *et al.* (1978) showed that N released from milk vetch in the early stage of incubation (20 d) accounted for 64% of the total N release during the growth of rice, whereas the corresponding figure for Azolla which was higher in lignin was only 38%. On the other hand, for manures with similar lignin contents, those with high N contents will release N continuously throughout the whole cropping season, whereas those with a N content lower than a certain critical value will immobilize soil mineral N early and net mineralization will take place only in the late stage of decomposition (Shi *et al.* 1980).

Barnyard manure, aerobic compost and waterlogged compost, which are already partly decomposed, have lower contents of readily decomposable organic C and N than raw plant materials with the same content of total N. The amount of ammonium released from these materials on application to soil depends on the composition of their raw materials and on the degree of decomposition. For instance, with organic manure derived from the materials with low C/N ratios, less N is mineralized from the more decomposed materials. However, for organic manures derived from materials with high C/N ratios, soil mineral N will be immobilized if the degree of decomposition of the manure is low. For organic manure with the same degree of decomposition a greater amount of N is mineralized from materials with high N contents. The pattern of mineralization of N from composted manure is slow and steady, similar to that of organic N in soil, and is strongly affected by temperature.

The degree of decomposition of sewage sludge is intermediate between the plant material and composted organic manure discussed above. In general, the rate of release of N from sludge is quite high due to its low C/N ratio.

The availability of the N of some organic manures estimated by the difference method is given in Table 12.3. The large standard deviations obtained are evidently due to differences in raw materials, degree of decomposition and test crop used. It is interesting to note that the N of all organic manures was less available than that of fertilizer N.

12.2.4. Residual effects

Agricultural practices and field experiments have indicated that application of organic manure is an effective means for maintaining and increasing the content of soil N. In long-term field experiments, soil N in treatments receiving organic manure was greater than that in treatments which received equal amounts of fertilizer N (Table 12.4). The greater increase in soil N with organic manure than fertilizer results from the greater amount of residual N with organic manure. This can be attributed to the following: (i) only part of the organic N in the manure is susceptible to decomposition, (ii) the high energy content in organic manure suppresses net mineralization of N, and (iii) the mineralization of organic N proceeds slowly. The results of a number of ^{15}N-tracer experiments indicate that the residual fertilizer N approximated the increment of N mineralized through the priming effect, thus the net residual N from fertilizer approached zero, no matter what mechanisms are

Table 12.3. Availability of nitrogen in organic manure as determined by a difference method.[1]

Experiment	Organic manure	Studies	Availability (%)
Field	Milk vetch	17	29.4 ± 9.7
	Azolla	4	22.9 ± 3.5
	Pig manure	5	27.5 ± 1.8
	Cake manure	3	24.0 ± 2.9
	Waterlogged compost	19	15.6 ± 4.2
	River sludge	3	5.3 ± 0.6
	Stable manure	23	12.4 ± 6.9
Pot	Milk vetch	11	46.5 ± 16.6
	Azolla	12	31.4 ± 12.7
	Pig manure	4	28.4 ± 4.0
	Activated sludge	4	37.1 ± 1.9
	Rabbit excreta	1	36.9

[1] Adapted from Cheng *et al.* (1986), DSF (1960), Wen (1983), Wu and Ni (1990) and Zhu (1959).

Table 12.4. Effect of application of organic manure[1] on soil N (g kg^{-1}).

Period	Soil	Treatment			Reference
		Control	Fertilizer	Organic manure	
1982–1987	Acid hydromorphic paddy	1.47	1.80	2.03 (Pig manure)	Peng *et al.* (1988)
				1.89 (*A. caroliniana*)	
				1.88 (*A. filiculoides*)	
				1.95 (Milk vetch)	
1978–1985	Coastal salt-affected paddy	0.90	1.06	1.36 (Pig manure)	Li *et al.* (1986)
				1.10 (Rape straw)	
1979–1988	Yellow-chao	0.78	1.00	1.30 (Horse manure)	Zhang *et al.* (1989b)
1981–1986	Yellow-chao	0.69	0.75	1.00 (Horse manure)	Jiang *et al.* (1991)

[1] Samples were collected at the end of the experiments.

involved in the priming effect. Therefore, in long-term field experiments the slightly higher content of soil N in the treatments with fertilizer N is not the result of residual N, but is an indirect result of greater amounts of stubble and roots in the treated soils. The residual N from organic manure can be considerable. For instance, application of *Azolla*, which is difficult to decompose, resulted in a large amount of residual N, accounting for up to 60% of the applied N. Even addition of rabbit excreta, which contains a large amount of easily decomposable organic N, resulted in net residual N amounting to 31% of the applied N. The corresponding values for materials such as leguminous green manures were between 29% and 41% (Table 12.5). Although these values may be overestimates of the actual amount of net residual N because of loss of part of the mineral N, it is certain that the amount of manure N retained in soil after the first crop was considerable.

The availability of the residual N is greater than that of soil N. Results obtained by Mo and Qian (1983) indicated that the recovery of the residual N from milk vetch by a second crop was 17% in a rice-rice double cropping system. Huang *et al.* (1982) reported that the recovery of the residual N from *Crotalaria* by a second crop (rice) was 15.8–16.5 %, by a third crop (wheat) was 2.5–4.0%, and by a fourth crop (rice) was 3.3–5.7%. In a 2 year rotation of rice-rice-barley-rice-wheat, the recovery of the residual N from milk vetch and *Sesbania* by the 2nd, 3rd, 4th and 5th crops were 6.8–7.4%, 1.1–1.2 %, 1.9–2.0% and 0.39–0.41% of applied N, respectively (Shi *et al.* 1991). The common trend of a sharp decrease in the availability of the residual N of the manures mentioned above further confirms that the organic N in manures can be divided into two fractions. The easily decomposable fraction is almost completely mineralized during the growth of the first crop, while the slowly decomposable fraction is largely retained in the soil after the growth of the first crop. Therefore, irrespective of the type of organic manure, the availability of the residual N to plants is low. However, the contribution of the residual N to plant nutrition increases with the long-term application of organic manures (Guo and Shen 1983; Wang and Zhou 1986). It is estimated that depending on the amount of residual N after each application of manure, the contribution of accumu-

Table 12.5. Effect of organic manure and nitrogen fertilizer on soil nitrogen (mg N 100 g soil^{-1}).

Crop	Manure or Fertilizer	Increment of soil N taken up by plant (priming effect) (A)	Residual fertilizer (B)	Net residual N (A–B)	Reference
			Manure		
Rice[1]	Milk vetch	2.11	5.62 (46.8)[4]	+3.53 (29.3)	Gu and Wen (1981)
Rice[2]	Milk vetch	0.62	3.60 (46.2)	+2.98 (38.3)	Gu and Wen (1981)
Rice[1]	Milk vetch	0.42	2.12 (52.8)	+1.70 (40.7)	Mo and Qian (1983)
Rice[1]	Azolla	0.46	7.56 (63.0)	+7.10 (59.3)	Gu and Wen (1981)
Rice[2]	Azolla	0.45	6.03 (77.5)	+5.58 (73.9)	Gu and Wen (1981)
Rice[1]	Crotalaria	0.72	5.46 (46.4)	+4.73 (40.2)	Huang *et al.* (1982)
Rice[1]	Water hyacinth	−1.22	7.84 (65.4)	+9.06 (75.7)	Gu and Wen (1981)
Rice[2]	Sesbania	0.17	3.65 (46.9)	+3.48 (44.8)	Gu and Wen (1981)
Rice[1]	Rabbit excreta	3.10	7.32 (53.7)	+4.22 (30.9)	Wu and Ni (1990)
			Fertilizer		
Incubn[3]	Urea	2.74 ± 0.34	2.53 (29.5)	−0.21 ± 0.27	Cai *et al.* (1981)
Rice[1]	Urea	1.58	1.27 (16.2)	−0.31	Mo and Qian (1983)
Rice[1]	$(NH_4)_2SO_4$	1.75 ± 0.50	1.50 (18.7)	−0.26 ± 0.53	Huang *et al.* (1982)
Rice[2]	$(NH_4)_2SO_4$	1.05	1.52 (32.2)	0.47 (10.0)	Shi *et al.* (1991)
Rice[1]	$(NH_4)_2SO_4$	3.26	3.05 (23.2)	−0.21	Wu and Ni (1990)

[1] Pot experiment, [2] Microplot experiment, [3] Incubation experiment,
[4] Values in parenthesis are % of applied N.

lated residual N to plant nutrition becomes pronounced after 5–10 years of continuous application. Results in some long-term field experiments also indicate that although the crop yield of the organic manure treatment was lower than that of the fertilizer treatment in the first 5 years, it became comparable with or slightly greater than the fertilizer treatment in the period 5–10 years after application (Matsuo *et al.* 1976).

12.3. Transformations of manure and fertilizer nitrogen in combination

Combined use of organic manure and fertilizer, an important feature of the current system of nutrient management in China, is an effective measure for efficient utilization of nutrient resources, maintaining soil fertility and increasing production. Compared with the readily available N from fertilizer applied alone, the pattern of gradual and long-lasting supply of N from the combined application of organic manure and inorganic fertilizer matches more closely with the N uptake pattern by crops. Furthermore, enhanced biological immobilization due to the addition of energy materials in the combined application exerts a considerable effect on other N transformation processes, such as nitrification, denitrification, ammonia volatilization, leaching, fixation and release of ammonium by clay minerals, and N uptake by crop plants, and thus may alter the fate of the applied N. The combined use of organic manure and fertilizer can help to (i) reduce the loss of fertilizer N and increase its availability, and (ii) match the pattern of N supply with uptake and increase crop yield.

12.3.1. Fate of nitrogen

When organic manure and fertilizer N are applied together the fate of the added N depends on the mineralization and immobilization reactions taking place in the soil. It varies with the amounts of easily decomposable organic C and organic N in the manure, as well as the application ratio of manure to fertilizer. Results obtained from combined application experiments, using a ^{15}N cross-labelling technique, demonstrate that, irrespective of whether rice or an upland crop was planted, or whether leguminous green manure or animal excreta was applied, fertilizer N consistently enhanced the mineralization of manure N, and the organic manure promoted the immobilization of fertilizer N (Table 12.6). Consequently, the residual fertilizer N in soil was greater, and plant uptake and loss of fertilizer N were smaller with the combined application than when fertilizer N was applied alone. However, residual manure N was reduced and plant uptake and loss of manure N were increased by the added fertilizer N (Table 12.7).

It is well known that immobilization of inorganic N occurs during the decomposition of straw, which contains a small amount of labile organic N and a large amount of easily decomposable organic C. The amount of N immobilized depends on the N content of the straw and the conditions for decomposition (Cheng *et al.* 1989). Therefore, incorporation of straw in the field has to be supplemented with certain amount of fertilizer N. Investigations with a ^{15}N cross-labelling technique indicate that the interactions which existed between straw N and fertilizer N in the combined application were similar to those which occurred when leguminous green manure or animal excreta were applied in combination with fertilizer N. Viz. in combined application, fertilizer N increased the uptake of straw N by plants and the loss of straw N, but decreased the residual amount of straw N in soil. The added straw increased the residual fertilizer N in soil, but decreased the plant uptake and loss of fertilizer N. The extent of the interaction depends on the rates of application

Table 12.6. Mineralization, immobilization and fixation of nitrogen during rice growth[1] (% of applied N).

Days after	Mineralization of *Crotalaria*-N		Fixation and immobilization of $(NH_4)_2SO_4$	
	1/2 ^{15}N *Crotalaria* + 1/2 $(NH_4)_2SO_4$	^{15}N *Crotalaria*	1/2 ^{15}N $(NH_4)_2SO_4$ + 1/2 *Crotalaria*	^{15}N $(NH_4)_2SO_4$
15	32.5	23.4	61.3	51.1
30	47.6	40.5	47.6	22.1
50	54.7	44.6	42.1	18.1
100	59.0	45.4	37.1	16.2

[1] Huang (1986).

Table 12.7. Effect of application of organic manure and fertilizer nitrogen on the fate of nitrogen[1] (% of applied N).

Treatment	N uptake	Residual N	Lost N	Crop	Reference
AS[2]	54.6 ± 2.9	18.7 ± 3.0	26.7 ± 4.1	Rice	Huang *et al.* (1981)
1/2AS + 1/2 *Crotalaria*	45.2 ± 3.4	34.3 ± 4.4	20.2 ± 2.1		
Crotalaria	37.4 ± 4.6	46.3 ± 5.6	16.3 ± 2.9		
AS	54.6	20.8	24.5	Cotton	Shen *et al.* (1986)
1/2AS + 1/2 Mung bean	41.9	32.7	25.4		
Mung bean	30.3	46.7	23.0		
Urea	43.0 ± 2.7	21.8 ± 3.4	25.2	Rice	Shen *et al.* (1986)
1/2 urea + 1/2 milk vetch	43.5 ± 1.3	35.8 ± 0.5	20.7		
Milk vetch	37.8 ± 1.8	39.3 ± 2.3	23.0		
AS	43.9	23.2	32.4	Rice	Wu and Ni (1990)
1/2 AS + 1/2 excreta	40.5	42.2	17.0		
Rabbit excreta	36.9	53.7	9.1		

[1] Pot experiment, 15N-tracer method.
[2] Ammonium sulfate.

of fertilizer N and straw. The interaction was not significant if the rate of application of fertilizer N was high and that of straw was low, and *vice versa*. As shown in Tables 12.8 and 12.9, when 84 mg fertilizer N kg^{-1} was applied in combination with 0.13% rice straw, the interaction of straw N and fertilizer N, although significant statistically, was not very strong, which is similar to the results obtained by Myers and Paul (1971) in a field experiment. Mo and Qian (1981) reported that when 80 mg urea N kg^{-1} was applied in combination with 0.6% rice straw, residual urea N was significantly increased and N loss was greatly decreased; the reverse was true for rice straw N. Yoshida and Padre (1975) also found that the plant uptake of fertilizer N was greatly decreased and the residual fertilizer N was greatly increased when 125 mg fertilizer N kg^{-1} was applied in combination with 0.5% rice straw.

Mature barnyard manure usually contains less easily decomposable organic N and C than leguminous green manure and fresh animal excreta. It was found in some studies that the combined application of mature barnyard manure and fertilizer N increased yield considerably (Sun *et al.* 1983; Li *et al.* 1982). However, data on the fate of the added N were not given in these publications and thus the relationship between transformation and supply of N and yield with combined application cannot be investigated. Other reports show that mature barnyard manure had little effect on the transformation of fertilizer N. For instance, Li *et al.* (1983) reported that the residual N from an application of ammonium sulfate was not increased by the addition of pig manure. Pomares-Garcia and Pratt (1978) also found that addition of barnyard manure had no significant effect on residual N, plant uptake and N loss from an application of ammonium sulfate irrespective of the ratios of the added materials.

Table 12.8. Effect of application of rice straw and fertilizer nitrogen on residual nitrogen (% of applied ^{15}N).[1, 2]

Soil	^{15}N-urea	^{15}N-urea + rice straw	^{15}N-rice straw	^{15}N-rice straw + urea
Neutral hydromorphic paddy	24.0 ± 1.0	26.3 ± 1.0	57.0 ± 6.3	50.4 ± 5.3
Calcareous hydromorphic paddy	23.0 ± 2.3	36.0 ± 1.3	69.1 ± 7.2	64.8 ± 2.6
Acidic hydromorphic paddy	24.4 ± 0.3	29.0 ± 1.6	72.4 ± 1.5	64.4 ± 1.4
	^{15}N-$(NH_4)_2SO_4$	^{15}N-$(NH_4)_2SO_4$ + rice straw	^{15}N-rice straw	^{15}N-rice straw + $(NH_4)_2SO_4$
Neutral hydromorphic paddy	26.3 ± 1.2	32.3 ± 1.4	57.0 ± 6.3	58.9 ± 2.2

[1] Application rates of fertilizer N and rice straw were 84 mg kg^{-1} and 0.13% (w w^{-1}).
[2] Pot experiment with rice.

Table 12.9. Effect of applications of rice straw and fertilizer nitrogen on nitrogen uptake by rice.[1]

Paddy soil	Treatment	Uptake (mg N pot^{-1})		
		Fertilizer ^{15}N	Straw ^{15}N	Total
Neutral hydromorphic	Urea	165.2 ± 0.69		165.2
	$(NH_4)_2SO_4$	158.9 ± 2.9		158.9
	Straw		5.46 ± 0.09	5.46
	Urea + straw	158.1 ± 1.4	6.84 ± 0.20	164.9
	$(NH_4)_2SO_4$ + straw	147.6 ± 0.73	6.82 ± 0.17	154.4
Calcareous hydromorphic	Urea	138.6 ± 2.3		138.6
	Straw		4.71 ± 0.05	4.71
	Urea + straw	136.7 ± 4.7	5.96 ± 0.19	142.6
Acid hydromorphic	Urea	183.7 ± 3.0		183.7
	Straw		5.00 ± 0.05	5.0
	Urea + straw	174.2 ± 1.4	5.46 ± 0.26	179.7

[1] The application rates of nitrogen and rice straw were 84 mg kg^{-1} and 0.13% (w w^{-1}).

It may be concluded from above discussion that the fate of manure N or fertilizer N in combined application may differ from that when applied alone, depending on the type of manure and its rate of application.

12.3.2. Pattern of supply

The uptake of fertilizer N by plants is rapid but short lasting. Although it is believed that the combined application of fertilizer N and organic manure can improve the pattern of supply of N from soil, little information is available on this topic. In these papers the pattern of supply was evaluated from plant uptake of the added N (manure + fertilizer) at different growth stages, and no further information is available on the amounts of added N present in the soil solution or on the exchange complex, thus it is difficult to make a detailed analysis of the effect of combined application. Huang (1986) reported that the short-acting feature of ammonium sulfate in soil was modified to a comparatively steady and long-lasting feature by the addition of *Crotalaria* (Figure 12.1). In addition the rate of uptake of *Crotalaria* N by rice plants was increased, to different extents at different growth stages, by combined application with ammonium sulfate (Figure 12.2). As stated above, in the whole growth period significantly more *Crotalaria* N and less ammonium sulfate N were recovered by plants in the combined application than when *Crotalaria* and ammonium sulfate were applied alone. The overall uptake of added N in the combined application was approximately equal to the sum of the uptake of ammonium sulfate N and *Crotalaria* N at any growth stage (Figure 12.3). Similar results were obtained by Shen *et al.* (1986).

12.3.3. Effectiveness of combined application

Besides N, organic manure contains various macro and micro nutrient elements such as P, K, Ca and Mg and some biologically active substances. Organic manure is also the major source of soil organic matter. The incorporation of organic manure may improve the physical, chemical and biological properties of soil, which may result in increased yield of the current crop, as with the application of straw on saline and alkaline soils (Wang *et al.* 1988; Chen *et al.* 1987; Xie *et al.* 1987). In studying the effectiveness of combined applications or the interaction between organic manure and fertilizer N, it is impossible to design an experiment in which

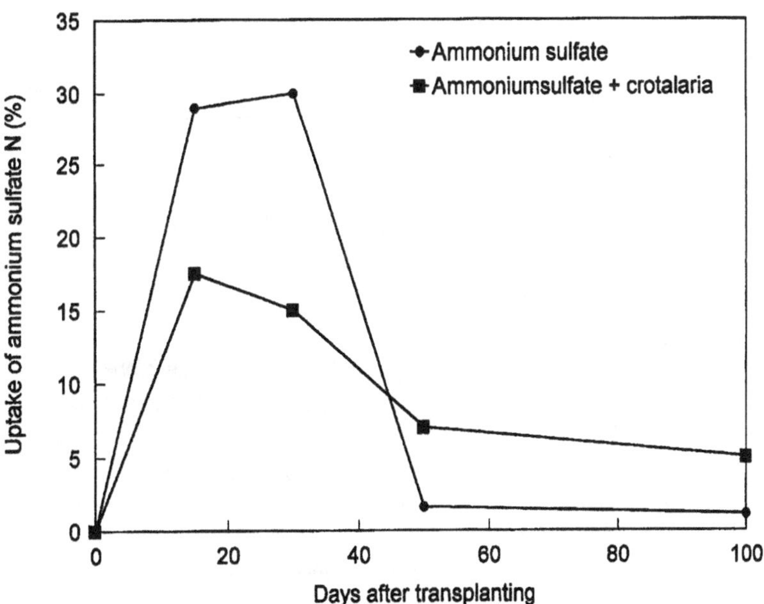

Figure 12.1. Effect of *Crotalaria* residues on the uptake of ammonium sulfate nitrogen by rice.

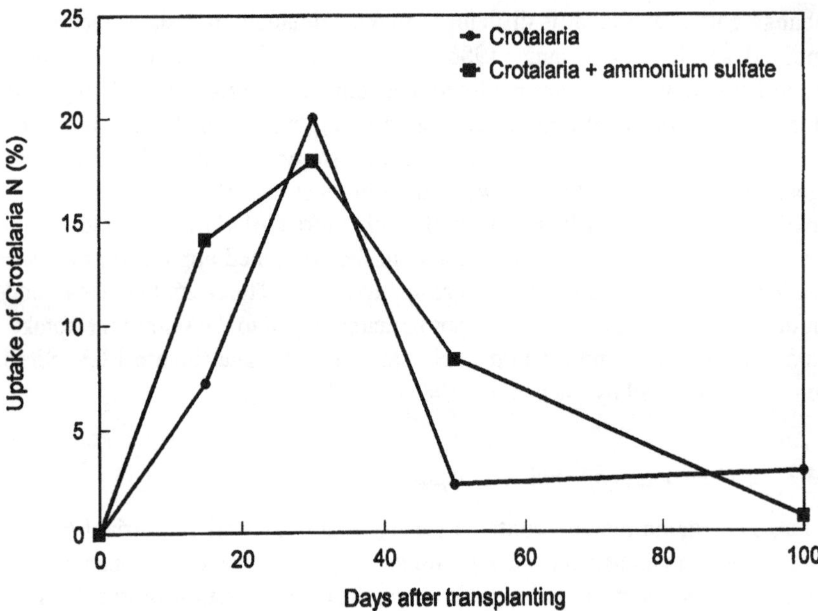

Figure 12.2. Effect of ammonium sulfate on uptake of *Crotalaria* nitrogen by rice.

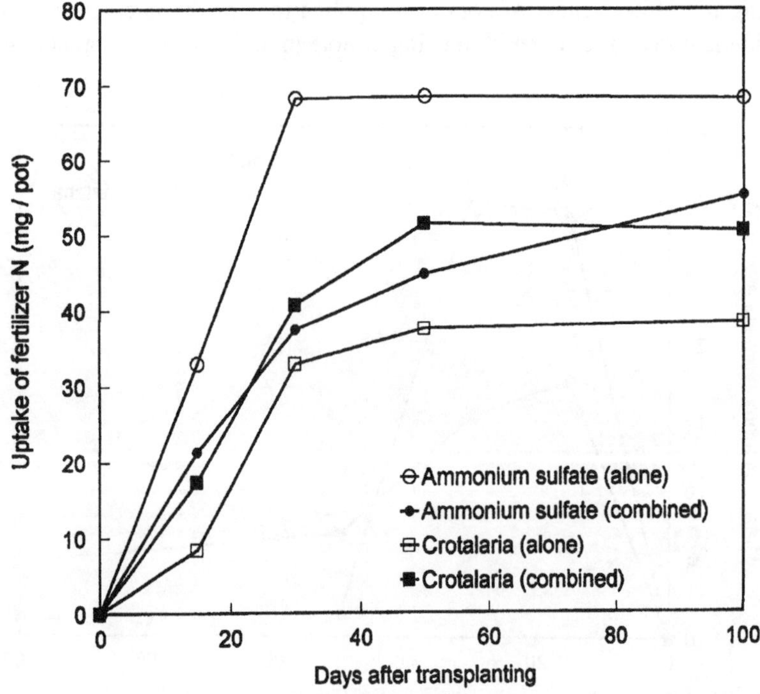

Figure 12.3. Uptake of nitrogen by rice from treatments with ammonium sulfate or *Crotalaria* alone or in combination (silty neutral hydromorphic paddy soil).

all other conditions, except the form of N and the amount of decomposable organic matter, are kept identical. Most scientists designed experiments based on equal rates of N supply with at most further supplements of P and K, and ignored the effect of other nutrients in the organic manure. Thus, on the one hand, the results of almost all experiments cannot be used to determine whether there is an interaction between organic and inorganic forms of N. On the other hand, conclusions on the interaction have been obtained from different experiments, because of differences in soil properties and types of organic manures used in the experiments. In some experiments there was a positive interaction with the combined application of organic manure and fertilizer N on yield; i.e. the increase in yield as a result the combined application was greater than the sum of the increased yields with the fertilizer N or organic manure applied alone. However, in other experiments no interaction was found.

Shen *et al.* (1986) reported that on an acidic hydromorphic paddy soil and a chao soil the yield of rice and cotton with the combined application of urea and green manure (milk vetch or mung bean) was significantly greater than that obtained by an equal amount of N applied as urea or green manure (milk vetch or mung bean) alone. Peng *et al.* (1983) reported that on loessial soils in north-western China, in 14 out of 20 field experiments yield was significantly greater in the treatment with barnyard manure and fertilizer N applied in combination than in the treatments with fertilizer N or barnyard manure applied alone. Liao *et al.* (1986) found that the interaction depended on soil fertility. An effect on yield was found on soils of low fertility, but not with a soil of high fertility. The results obtained by Gong and Li (1986) in pot and microplot experiments with chao soil also indicated that both the yield of ginned cotton and N uptake by cotton plants were higher with the combined application of vetch and ammonium sulfate than in the treatment with vetch or ammonium sulfate alone. In all of these experiments the amount of P and K applied was not the same in all treatments. Thus, the effect of the combined application on yield can be at least partly attributed to the effect of P and K in the organic manure. Also of concern is that most of the soils used in these experiments were deficient in P and/or K.

In only a few cases were the amounts of N, P and K controlled during studies of the effects of combined applications of organic manure and fertilizer N. Sun *et al.* (1986) reported that when the amounts of N, P and K added to all treatments were constant, the yield of barley in the combined treatment was significantly greater than that in the treatments where manure or fertilizer N was applied alone to an acidic paddy soil derived from a red earth, but not with a neutral paddy soil derived from alluvium. Zhang *et al.* (1988) reported that on a yellow-brown earth severely deficient in P, when all treatments received constant amounts of N, P and K, the yield of tomatoes with the combined application of urea and seed cake was comparable with that on the treatment with fertilizer NPK alone, but the quality as determined by the contents of total reducing sugars, and acidity was better (Table 12.10). Shen *et al.* (1988) found that, compared with NPK fertilizer the combined application of organic manure and NPK fertilizer did not significantly increase the

Table 12.10. Effect of application of organic manure and urea on the quality of tomato.[1]

Treatment	Reducing sugar (%)	Acidity (%)	Sugars/ acid
Urea + seed cake	15.1	1.24	3.11
Urea + PK	12.0	1.00	3.33
Seed cake	11.9	0.77	4.23

[1] Zhang *et al.* (1988).

yields of tobacco and watermelon. However, the quality of tobacco as judged by the total sugar content and the ratio of total sugars to protein, and the quality of watermelon, as judged by the amounts of total sugar and soluble solids were increased (Table 12.11). These results further confirm that combined applications of organic manure and fertilizer do not necessarily have a significant effect on yield. Also, the improvement in crop quality with the combined application may not be evidence of the existence of an interaction between manure N and fertilizer N, rather it may be attributed to the effect of other nutrient elements and/or biologically active compounds in the organic manure.

Concerning biologically active compounds, application of seed cake increased the gibberellin content in the growing points of watermelon and tobacco by 2 and 5 times, respectively, as compared with NPK fertilizer. The surface area of the active roots of tobacco was increased 38.6–47.8% more by application of chicken or pig manure than NPK fertilizer, presumably by addition of root growth hormone in the manure.

The above discussion does not mean that the combined application of organic manure and fertilizer to crops will not be beneficial. Treatment with organic manure alone often resulted in lower yields than treatment with fertilizers alone (Li *et al.* 1982; Chen *et al.* 1987). However, the yield with a combined application of organic manure and fertilizer N was always equal to or slightly better than the yield with fertilizer N in the long-term experiments. This suggests that combinations of organic manure and fertilizer N may increase crop yields by improving the pattern of N supply.

Table 12.11. Effect of organic manure on quality of watermelon and tobacco.[1]

Treatment	Watermelon				Tobacco	
	Yield (t ha^{-1})	Total sugars (%)	Vitamin C (mg 100 g^{-1})	Soluble solids (%)	Total sugars (%)	Alkaloid (%)
NPK	14.8 ± 5.24	6.4 ± 0.4	7.6 ± 0.2	7.5 ± 0.3	13.7	2.
Seed cake + NPK	19.5 ± 4.49	8.5 ± 0.2	6.4 ± 0.1	8.4 ± 0.4	16.0	1.7
Pig manure + NPK	20.6 ± 3.77	7.1 ± 0.2	6.3 ± 0.1	8.1 ± 0.6	nd[2]	nd
Chicken manure + NPK	20.2 ± 1.91	7.8 ± 0.2	8.8 ± 0.2	8.2 ± 0.01	17.6	1.9

[1] Shen *et al.* (1988).
[2] nd = not determined.

12.4. Integrated management of manure and fertilizer nitrogen

12.4.1. Cultivation of green manure crops

There are two ways to increase organic manures. One is to maximize the reuse of organic wastes from agriculture, the food processing industry and daily life. The other one is to extend the area sown to green manure and to increase their production. Green manure, especially leguminous green manure, has played an important role in the development of agriculture in China. Since the establishment of the People's Republic of China, many new species of legumes, adapted to different climatic conditions have been screened out. In addition the cultural (Gu and Wen 1981) and fertilization techniques (Zhu 1978) for extending the cropped area and increasing dry matter production have been developed. These include the application of a small amount of N to improve growth and the application of P to improve N fixation.

Recently, the area sown to green manure crops in China has been reduced considerably. This is due to the competition between green manure crops and food crops for land and the low economic benefit of the direct application of green manure to crops. Intercropping, mixed cropping, and the use of green manure crops to link agriculture, animal husbandry and fishery are ways to overcome the reduction in area sown to green manure. Jiao and Sun (1986) reported that more than 15 t ha^{-1} fresh sweet clover could be obtained by intercropping it with maize in the agricultural region of Heilongjiang Province without decreasing, or only slightly decreasing, the yield of maize. The sweet clover was used as fodder for cows and resulted in a significant economic benefit. The barnyard manure produced was applied to the field, and along with the roots and stubble remaining in the field increased the yield of the succeeding crop by 13–30%. Similar results were obtained by Zhang and Jin (1988), who grew a mixed crop of cotton, vetch, rye grass and wheat, and used the green manure to feed domestic animals or fish, and used the animal excreta or fish pond mud as organic manure.

The culture of azolla and other aquatic green manure crops in paddy fields helps to overcome the competition for land use between green manure crops and food crops (Cai *et al.* 1962). The establishment of a rice–azolla–fish system could promote rice production by suppressing the negative effects of weed growth, pathogenic microorganisms and insect pests, increasing soil N and improving soil aeration. Extra benefits were obtained through the production of fish and the saving of pesticides and fertilizers (Liu and Zheng 1986).

12.4.2. Application of straw

The amount of straw available for use in agriculture increases with increasing crop yield. Application of straw may improve soil properties through supplementing soil with nutrients and organic matter. Traditionally in China straw has been used as the raw material for aerobic or anaerobic composting, which is not only labor intensive and time consuming, but also is wasteful of energy. Applying straw directly or after biogas fermentation can in part avoid these disadvantages. Biogas produced during the fermentation of straw can partly meet the needs of fuel for the farmer, and the produced sludge can be used as manure, thereby eliminating the losses of C, N and other nutrients which occur when straw is burnt.

In comparison with composting, biogas fermentation of straw (along with animal excreta and/or green manure) can improve the utilization of the energy of straw and reduce the losses of C and N in the straw. Wen (1984) reported that 0.11–0.38 m^3 of biogas could be produced from one kg of organic material. Biogas produced from 4 t raw material (2.7 t pig manure and 1.3 t rice straw) could supply sufficient biogas to meet the energy requirements for lighting and cooking of a 4–5 person household over a whole year (Wang 1985). The sludge produced could be used as organic manure as its efficiency (in terms of increased yield) was almost equivalent to that of pig manure (Jin 1989; Li 1981).

Two techniques of biogas fermentation were developed, water sealed fermentation and solid state fermentation. The rate of fermentation by the water sealed fermentation was about 10% higher than the solid state fermentation, but the high water content of the sludge creates problems in emptying the fermentation pool, and transportation and distribution of the sludge. In addition, the losses of organic matter and nutrients during solid state fermentation were lower than that during water sealed fermentation. Therefore, solid state fermentation is more promising in spite of its low production rate of biogas.

SIRWAS (1959) found that the grain yield of rice when the raw materials (including straw) used for anaerobic composting were applied was only 22.5–37.5 kg ha^{-1} less than that obtained when the compost was applied, and the yield of straw was even slightly greater. Similar results have been obtained by other scientists. The results in Table 12.12 show that provided the amount of organic material and fertilizer was kept constant, the effectiveness of direct application of the raw materials used for composting was not lower than that obtained by applying aerobic or anaerobic compost. This applied irrespective of whether the organic material used was green manure or the straw of leguminous or cereal crops.

The direct application of straw may not be beneficial for the growth of the current crop because of immobilization of mineral N and production of toxins during the decomposition of straw. This detrimental effect varies with soil type and fertility, and the rate and method of straw application. Cheng *et al.* (1992) found that an application of rice straw 0.13% (w/w) to calcareous and acidic hydromorphic paddy soils had no effect on grain yield of rice, but the same application to a neutral hydromorphic paddy soil markedly decreased yield (Table 12.13). Liu *et al.* (1984) found that the changes in Eh values in a flooded acidic hydromorphic paddy soil receiving less than 3 t rice straw ha^{-1} were similar to those obtained in the absence of straw. The Eh value of the plow layer soil declined and reached its lowest value 8 days after flooding. It then rose gradually and leveled off ~20 days after flooding. However, at higher rates of straw application (4.5–6.0 t ha^{-1}), the soil Eh value declined rapidly and fell to below zero 10 days after flooding. The time when the Eh value tended to be constant was delayed 10 days compared with

Table 12.12. Effect of straw and green manure applied directly or after composting on grain yield (t ha^{-1}).

Soil	Crop	Manure	(t ha^{-1})	Fertilizer (kg ha^{-1})	Method of application		Reference
					Composted	Raw material	
Shajiang black[1]	Wheat	Wheat straw	3.8	N 138, P_2O_5 138	4.29[1]	4.43	Zhou and Wang (1989)
		Legume stalk	3.8	″	4.74	4.92	
		Maize stalk	3.8	″	4.35	4.43	
		Sesbania	3.8	″	4.32	4.38	
		Amorpha fruticosa	3.8	″	4.77	4.92	
Calcareous hydromophic paddy	Rice	Rice straw + pig manure	3.75+ 0.38	N 105, P_2O_5 95	2.97[2]	2.97	Jin (1989)
	Wheat	″	″	″	2.48[2]	2.25	
Neutral hydromorphic paddy	Rice	Rice straw	4.5	N 450	7.01[3]	7.11	Jin *et al.* (1984)
	Rice	Rice straw	″	″	7.35[1]	7.35	

[1] Aerobic composting.
[2] Anaerobic fermentation for producing biogas.
[3] Anaerobic composting.

Table 12.13. Grain yield (g pot^{-1}) of rice in three hydromorphic paddy soils as affected by application of rice straw.[1]

	Neutral	Calcareous	Acidic
Control	15.3 ± 0.3	10.0 ± 0.9	12.4 ± 0.5
Straw	13.5 ± 0.4	9.1 ± 0.4	12.0 ± 0.4
t test[2]	6.56**	1.61	1.03
Urea	27.7 ± 1.0	19.2 ± 0.3	19.3 ± 1.4
Urea + straw	26.5 ± 1.0	19.2 ± 0.9	17.1 ± 0.8
t test	0.73	0.22	2.41
$(NH_4)_2SO_4$	26.5 ± 3.1		
$(NH_4)_2SO_4$ + straw	25.4 ± 0.7		
t test	0.60		

[1] The rate of addition of straw and nitrogen to pots were 0.13% (w w^{-1}) 84 mg N $k^{-1}g$.
[2] $t_{0.05}$ = 2.78, $t_{0.01}$ = 4.60; ** significant at 1% level.

the control treatment (no rice straw), and thus the development of rice roots, and nutrient uptake was suppressed in the initial stages of growth. Huang and Zhu (1988) showed that when wheat straw was applied on the soil surface the content of reducing substances varied within 300–500 c M during the 14 weeks after flooding; this was similar to the values obtained in the treatment to which ammonium sulfate was added. However, when wheat straw was incorporated into soil at the same rate of application, the content of reducing substances was between 900–1200 c M. Improvement of soil drainage and the application of the appropriate amount of fertilizer N are the main measures for eliminating the detrimental effects of direct application of straw. The results obtained in pot and field experiments show that on a well drained soil of low fertility, the application of 0.45% rice straw (w/w) did not affect the grain yield of rice, provided 120 mg N kg^{-1} was added (Table 12.14). Zhang *et al.* (1989a) also found that on a well drained light textured chao soil, even though the rate of application of straw was as high as 9 t ha^{-1}, the total yield of upland crops was not affected by the addition of straw, irrespective of whether N

Table 12.14. Effect of rates of addition of straw and urea on grain yield of rice in the greenhouse. (g pot^{-1}).

Urea (mg N kg $soil^{-1}$)	Rice straw (% of soil)			
	0	0.15	0.30	0.45
0	13.3 A[1]	nd[2]	nd	nd
80	24.5 B	24.1 a	26.4 A	25.1 A
100	27.8 C	28.0 ab	28.6 AB	29.9 B
120	29.5 D	29.5 bc	32.6 C	32.3 BC

[1] DMRT test. Values followed by the same lower case or capital letters are not significant at 5% and 1%, respectively.
[2] not determined.

was applied or not (Table 12.15). However, the application of straw and N had a residual effect; the yields of the treatments which had previously received straw were significantly greater than those which had not received straw.

12.4.3. *Ratio of manure nitrogen to fertilizer nitrogen*

In all of the studies on the ratio of manure N to fertilizer N the experiments were designed to apply equal amounts of N to the treatments with different manure N/fertilizer N ratios, but the amounts of P and K applied were different. Consequently the effects on yield were largely due to differences in the amounts of P and K applied, rather than to differences in manure N/fertilizer N ratios. This was confirmed by the results of some experiments which showed that the effect of combined applications differed with soil fertility, especially P and K status, as well as the rate of N application, kind and quality of organic manure, and manure N/fertilizer N ratio. For example, on soils deficient in N and P or N, P and K, when the rate of N application (195–225 kg N ha^{-1} $crop^{-1}$) was higher than usually applied, and leguminous green manure or pig manure were applied, the optimum ratio was 50:50. Zhang *et al.* (1989b) also found that on a sandy loam chao soil, the average crop yield over 17 seasons was highest in the treatment with a manure N/fertilizer N ratio of 50:50. The average yield with the 50:50 ratio was 4.4%, 13.6% and 45.5% higher than the yields with manure N/fertilizer N ratios of 30:70, 70:30 and 100:0, respectively. Similar results were obtained by You and Lou (1983). On the other hand, it was found that the proportion of manure N should be low if the soil

Table 12.15. Crop yield (t $ha^{-1}yr^{-1}$) as affected by addition of straw and fertilizer nitrogen.[1]

	Rice straw (t ha^{-1})			
Fertilizer	0	2.25	4.5	9
	Average yield from 1982 to 1984			
N_0P_0[2]	4.19	3.80	4.40	4.37
$N_{90}P_{60}$	6.80	7.19	6.56	6.89
$N_{180}P_{120}$	8.19	8.12	7.92	7.79
$N_{360}P_{180}$	8.85	8.19	8.75	8.46
	Average yield from 1985 to 1986[3] (Residual effect)			
N_0P_0	4.32	4.52	4.80	5.01
$N_{90}P_{60}$	5.15	5.31	5.58	5.79
$N_{180}P_{120}$	5.94	5.81	6.35	6.12
$N_{360}P_{180}$	8.39	8.48	8.37	8.97

[1] Zhang *et al.* (1989a). Maize-millet, double cropping system on light textured chao soil with clayey bottom layer.
[2] Subscripts refer to the rate of application of N and P_2O_5 (kg ha^{-1}).
[3] No straw or fertilizer was applied during this period.

had (i), a low N status, (ii) a moderate P or K status, (iii) a low P and K status, but P and K fertilizers had been applied, or (iv) the soil had a moderate N status and a low rate of fertilizer N was applied. Liu *et al.* (1990) reported that on a neutral hydromorphic paddy soil, the average yield of 7 crops in 4 years was highest and the yield variation was lowest in the treatment with a manure N/fertilizer N ratio of 30:70. The average yield in the treatment with manure N/fertilizer N ratio of 0:100 was the lowest and the variation was high. Fan *et al.* (1987) showed that on a neutral hydromorphic paddy soil derived from alluvium with a rate of application of 150–300 kg N ha^{-1}, the grain yield of rice in the treatment with a manure N:fertilizer N ratio of 30:70 was higher, in most cases, than that in other treatments. Obviously, in these experiments the effect of N with P and K, rather than the manure N/fertilizer N ratio, was under investigation.

12.5. References

Bouldin, D R 1988. Effect of green manure on soil organic matter content and nitrogen availability. In: Green Manures in Rice Farming. pp. 151–163. IRRI, Los Banos, Philippines.

Cai, D J, Tai, J C and Cheng, L L 1962. Study on cultivation and utilization of azolla in paddy field. (in Chinese). Chinese J. Soil Sci. (4), 49–53.

Cai, G X, Zhang, S L and Zhu, Z L 1981. Characteristics of nitrogen mineralization of paddy soils and their effect on the efficiency of nitrogen fertilizer. In: ISSAS (ed.), Proc. of Symp. on Paddy Soil. pp. 793–799. Science Press, Beijing.

Chen, X H, Xiao, H J, Zhu, Q and Zhang, X Y 1987. Experiment on improvement of medium- and low-yield paddy soil derived from yellow soil by application of fertilizers. (in Chinese). Soils and Fertilizers (6), 13–15.

Cheng, L L, Wen, Q X and Li, H 1989. Transformation of ^{15}N labelled fertilizer in pot and field experiments. (in Chinese). Acta Pedol. Sin. 26:124–130.

Cheng, L L, Wen, Q X and Li, H 1992. Effects of rice straw return to fields on soil nitrogen and rice plant growth. (in Chinese). Soils 24:234–238

Cheng, L L, Wen, Q X and Shi, S L 1986. Supply and transformation of nitrogen of combined application of organic and inorganic fertilizers. In: Advances and Perspectives in Soil Nitrogen Research in China. (in Chinese). pp. 104–115. Science Press, Beijing.

Cheng, L L, Wen, Q X, Wu, S L and Xu, N 1981. The effect of chemical composition and decomposition conditions of plant materials on the newly-formed humus. (in Chinese). Acta Pedol. Sin. 18:360–367.

DSF 1960. (Department of Soil and Fertilizer, Jiangsu Branch of Chinese Academy of Agricultural Sciences). Summary of experiments on effects of farm manures applied into early rice field. (in Chinese). Huadong Agricultural Science Bulletin (5), 226–238.

Fan, Y C, Tao, Q X and Zhang, M H 1987. Study on the application of chemical fertilizers alone and in combination with organic manures. (in Chinese). Jiangxi Journal of Agricultural Sciences and Technology (10), 22–24.

Gong, G Y and Li, G Z 1986. Absorption of different forms of nitrogen by cotton and the effect of combined applications of organic manures and inorganic fertilizers. (in Chinese). Chinese J. Soil Sci. 17:209–211.

Gu, R S and Wen, Q X 1981. Cultivation and application of green manure in paddy fields of China. In: Proc. of Symp. on Paddy Soil. pp. 207–209. Science Press, Beijing.

Guo, Y T and Shen, Z Q 1983. Efficiency of different forms of nitrogen fertilizer in paddy soils and its effect on soil nitrogen supply. (in Chinese). Soils (6), 206–211.

Huang, D M 1986. Contribution of organic and inorganic fertilizers to soil N and their combination. In: Advances and Perspectives in Soil Nitrogen Research in China. (in Chinese). pp. 92–103. Science Press, Beijing.

Huang, D M and Zhu, P L 1988. Soil fertility of no-tillage rice-based cropping system in southern China. In: 1st Intern. Symp. on Paddy Soil Fertility. Chiangmai, Thailand. Vol. 2, pp. 683–693.

Huang, D M, Gao, J H and Zhu, P L 1981. Distribution of N of organic and inorganic fertilizers in rice-soil system. (in Chinese). Acta Pedol. Sin. 18:107–121.

Huang, D M, Zhu, P L and Gao, J H 1982. Residual effect of N of organic and inorganic fertilizers in paddy field and upland field. (in Chinese). Scientia Sinica Series B. 10:907–912.

ISF 1986. (Institute of Soil and Fertilizer, Chinese Academy of Agricultural Sciences) (ed.), Regionalization of Chemical Fertilizers of China. (in Chinese). Agricultural Science Press, Beijing. 90 p.

Jiao, B and Sun, C F 1986. Experiences on improving soil fertility by planting fodder and raising livestock in Suihua region. (in Chinese). Soils and Fertilizers (1), 46–47.

Jiang, R C, Zhang, H Y, Li, Z H, Zhang, M P, Li, D M and Liu, M Z (1991). Effect of fertilizers and manure application in different combination on fertility of fluvo-aquic soil. Jiangsu Provincial Symposium on Improvement of Soil Fertility and Rational Application of Fertilizers, Peixian, China. (in Chinese).

Jin, Y K 1989. Experiment on application of straw after anaerobic fermentation for producing biogas. (in Chinese). Soils and Fertilizers (1), 23–26.

Jin, Z R, Chen, H Y and Huang, Z J 1984. Ways of straw application into soil and their yield-increasing effects. (in Chinese). Jiangsu Journal of Agricultural Sciences (5), 26–29.

Li, H Z, Han H R, Wu, Z C, Yang, J C and Ge, L M 1986. A study on the results of increasing the fertility of paddy soil with organic manures. (in Chinese). Chinese J. Soil Sci. 17:252–258.

Li, S J 1981. Yield efficiency of biogas sludge. (in Chinese). Jiangsu Journal of Agricultural Sciences (1), 62–63.

Li, S Y, Wang J Y and Kong F G 1982. Study on the characteristics of nitrogen supply in paddy soils. II. Effect of fertilization on the soil nitrogen supply and grain yield of double-cropping-rice. (in Chinese). Acta Pedol. Sin. 9:13–20.

Li, S Y, Wang, J Y, Kong, F G and Zhu, X Q 1983. Study on the characteristics of nitrogen supply in paddy soils: III. Characteristics of uptake of different forms of nitrogen by double-cropping-rice in various gleyed paddy soils. In: Chinese Society of Soil Science (ed.), Reasonable Utilization and Fertilization of Soils of China. (in Chinese). Xian. Vol II. p. 229.

Liao, S Z, He, C F and Guan, Y L 1986. Role of nitrogen fertilizer in agriculture of Sichuan province. In: Advances and Perspectives in Soil Nitrogen Research in China. (in Chinese). pp. 173–182. Science Press, Beijing.

Lin, X X, Cheng, L L, Shi, S L and Wen, Q X 1980. Characteristics of decomposition of plant residues in soils of southern part of Jiangsu province. (in Chinese). Acta Pedol. Sin. 17:320–327.

Liu, D H, Ding, R X, Wu, S M, Sun, Y H and Rang, W 1990. Improvement of soil fertility by combined application of organic manure and chemical fertilizer. (in Chinese). Soils and Fertilizers (2), 1–4.

Liu, J R, Leu, Y H, Zhang, D Y and Guo, C Z 1984. Effect of incorporation of rice straw on paddy soil fertility and rice plant growth. (in Chinese). Chinese J. Soil Sci. 15:49–53.

Liu, Q S, Qiu, F Q, Chen, E F, Li, F Z, Zhang, L S, Leu, D Y, Zhou, H M and Li, J Z 1959. Study on the advantages of organic manures: I. Increasing the effectiveness of compost. (in Chinese). Acta Pedol. Sin. 7:159–179.

Liu, Z Z and Zheng, W H 1986. Nitrogen fixation by azolla and its utilization. In: Advances and Perspectives in Soil Nitrogen Research in China. (in Chinese). pp. 195–211. Science Press, Beijing.

Matsuo, H, Hayase, T, Yokoi, H and Onikuram, Y 1976. Results of long-term fertilizer experiments on paddy rice in Japan. Ann. Agron. 27:957–968.

Mo, S X and Qian, J F 1981. Uptake of nitrogen by rice plant from straw manure, urea and soil. In: ISSAS (ed.), Proc. of Symp. on Paddy Soil. pp. 800–804. Science Press, Beijing.

Mo, S X and Qian J F 1983. Studies on the transformation of nitrogen of milk vetch in red earth and its availability to rice plant. (in Chinese). Acta Pedol. Sin. 20:12–21.

Myers, R J K and Paul, E A 1971. Plant uptake and immobilization of ^{15}N-labelled ammonium nitrate in a field experiment with wheat. In: Nitrogen-15 in Soil-Plant Studies. pp. 55–64. IAEA, Vienna.

Peng, L, Peng, X L and Yu, C Z 1983. Efficiency of combined application of organic manures and chemical fertilizers in loessial soil region. (in Chinese). Soils and Fertilizers (5), 14–15.

Peng, K L, Tang, R B and Yang, K L 1988. Effect of organic manure on fertility of paddy soil derived from red earth. (in Chinese). Second National Symposium on Soil Organic Matter and Organic Manure, Taian, China.

Pomares-Garcia, F and Pratt, P F 1978. Recovery of ^{15}N-labelled fertilizer from manured- and sludge-amended soil. Soil Sci. Soc. Am. J. 42:717–720.

Shen, Z Q, Guo, Y T, Leu, L X and Cai, Q X 1986. Fate of nitrogen in organic manure and inorganic nitrogen fertilizer in combined application. (in Chinese). Chinese J. Soil Sci. 17:107–110.

Shen, Z Q, Guo Y T, Yuan, J F and Fan, X B 1988. Effect of organic manure on the quality of crop products. (in Chinese). Second National Symposium on Soil Organic Matter and Organic Manure, Taian, China.

Shi, S L, Wen, Q X and Liao, H Q 1980. The availability of nitrogen of green manures in relation to their chemical composition. (in Chinese). Acta Pedol. Sin. 17:240–246.

Shi, S L, Cheng, L L, Lin, X X Shu, Z L and Wen Q X 1978. Effect of azolla on the fertility of paddy soil. (in Chinese). Acta Pedol. Sin. 15:54–60.

Shi, S L, Liao, H Q, Wen, Q X, Xu, X Q and Pan, Z P 1991. Fate of N from green manures and ammonium sulfate. Pedosphere 1:219–227.

SMRWAS 1959. (Section of Manure Research, Institute of Soil Science, Academia Sinica and Wuxi Agricultural School, Jiangsu). Investigation on the process of decomposition and yield efficiency of waterlogged compost. (in Chinese). Acta Pedol. Sin. 7:190–202.

Sun, X, He, N Z and Zhang, Y S 1983. Role of organic manures in improving soil fertility. In: Chinese Society of Soil Science (ed.), Reasonable Utilization and Fertilization of Soils of China. (in Chinese). Xian. Vol. II. p. 226.

Sun, X, Zhang, Y S, Yi, Q H and Tang, C X 1986. Effect of organic manure on soil fertility and crop production. In: Current Progress in Soil Research in People's Republic of China. pp. 197–206. Jiangsu Science and Technology Press, Nanjing.

Wang, T G 1985. Dry fermentation and its effect on biogas and fertilizer production. (in Chinese). Jiangsu Agricultural Sciences 1:24–25.

Wang, Y H and Zhou, D X 1986. Yield efficiency and efficient utilization of organic manures in the suburbs of Shanghai. (in Chinese). Acta Agriculture Shanghai 2:75–82.

Wang, Y Q, Gao J L, Ma W P and Xie S X 1988. The effect of wheat straw and maize stalk in ameliorating saline wasteland. (in Chinese). Chinese J. Soil Sci. 19:274–275.

Wen, Q X 1983. Role of organic manure in supplying nutrients and maintaining the content of soil organic matter. In: Chinese Society of Soil Science (ed.), Reasonable Utilization and Fertilization of Soils of China. (in Chinese). Xian. Vol. II. pp. 169–173.

Wen, Q X 1984. Utilization of organic materials in rice production in China. In: IRRI (ed.), Organic Matter and Rice. pp. 45–56. IRRI, Los Banos, Philippines.

Wen, Q X, Cheng, L L and Shi, S L 1987. Decomposition of azolla in the field and availability of azolla nitrogen to plants. In: IRRI (ed.), Azolla Utilization. pp. 241–254. IRRI, Los Banos, Philippines.

Wu, S M and Ni, M J 1990. Transferring and cycling of organic and inorganic N in micro-agroecosystem. (in Chinese). Journal of Applied Ecology 1:67–74

WPM 1959. (Wuhan Laboratory of Microorganisms, Academia Sinica). A method of rapid composting. (in Chinese). Soils 3:10–12.

Xie, C T, Yan, H J and Xu, J X 1987. Experiment on improvement of saline and alkaline soils with organic manures. (in Chinese). Chinese J. Soil Sci. 18:97–99.

Yoshida, T and Padre, Jr B C 1975. Effect of organic matter application and water regimes on the transformation of fertilizer nitrogen in a Philippine soil. Soil Sci. Plant Nutr. 21:281–292.

You, D M and Lou, D R 1983. Long-term experiment on application rate and combination ratio of mineral nitrogen and organic N. (in Chinese). Jiangsu Agricultural Sciences (2), 26–28.

Zhang, B Y, Li, J S, Zhang, S F, Qu, S L and Wu, R D 1989a. A preliminary study on the effect of combined use of organic manure and inorganic fertilizer on crop yield and soil organic matter. (in Chinese). Soils and Fertilizers (1), 36–38.

Zhang, C L, Zhang, Y D, Gao, Z M Xu G H, Wang, L Y and Zhou, Q S 1988. Effect of combined application of organic manures and chemical fertilizers on the yield and quality of tomato crop. (in Chinese). Chinese J. Soil Sci. 19:276–279.

Zhang, D H and Jin, Z P 1988. Effect of vetch-wheat intercropping-animal system on fertility of coastal saline soil. Second National Symposium on Soil Organic Matter and Organic Manure, Taian, China. (in Chinese).

Zhang, F D 1984. Combined use of organic manure and inorganic fertilizer is the orientation of modern fertilizer application technique. (in Chinese). Soils and Fertilizers (1):16–19.

Zhang, M P, Zhang H Y and Leu M Z 1989b. Appropriate ratio of organic manure and inorganic fertilizer. (in Chinese). Chinese J. Soil Sci. 20:221–223.

Zhou, B Y and Wang, Y Q 1989. Ways of amelioration and utilization of Shajiang black soil. (in Chinese). Soils and Fertilizers (1), 30–35.

Zhu, G Q 1959. Investigation on the nutritive value of farm manures. 1. The effect of farm manures on rice. (in Chinese). Acta Pedol. Sin. 7:180–189.

Zhu, Z L 1978. Soil nitrogen. In: Nanjing Inst. Soil Sci. (ed.), Soils of China. (in Chinese). pp. 360–375. Science Press, Beijing.

Zhu, Z L and Xi, Z B 1990. Recycling phosphorus from crop and animal wastes in China. In: IRRI (ed.), Phosphorus Requirements for Sustainable Agriculture in Asia and Oceania. pp. 115–123. IRRI, Los Baños, Philippines.

13
Nitrogen management and environmental and crop quality

MA LI-SHAN

13.1. Introduction

Nitrogen pollution of the environment is a problem of great concern, and many international symposia on this topic have been held since 1970 (e.g. IAWPR 1975; Giorgini and Zingales 1986; Rosswall 1978). In China, a symposium on 'Nitrogen pollution and nitrogen cycling in the environment' was held in Dalian, reviewing the progress in the related research fields in China (EBESIN 1983). This chapter will discuss the problems for environmental quality caused by N loss from fertilizer applications to field crops, their counter measures, and the interrelationships between field N management, crop quality and animal health.

13.2. Effect of nitrogen loss on water quality

13.2.1. Eutrophication

Nitrogen loss from fertilizer applications may induce a series of problems for water quality, for instance, eutrophication, groundwater pollution by nitrate and nitrite and aquaculture damage caused by ammoniacal N.

Eutrophication generally stands for the enrichment of closed or semi-closed water bodies such as lakes, reservoirs, inlets or some stagnant rivers (water flow velocity <1 m min^{-1}) with nutrient elements such as N, P and C. This enrichment causes abnormal propagation of specific algae, which leads to a decrease in water transparency and dissolved oxygen, so that fish and other animals cannot survive and the odour and taste of the water are affected.

Eutrophication is a water pollution problem that perplexes developed countries, and a practical environmental problem that developing countries are confronted with. Eutrophication deprives the water body of its due functions, and changes a favourable aquatic ecosystem to an environment unfit for mankind.

Waste water discharged from factories and mines, and domestic sewage from residential areas are the main causes of eutrophication. However, recent studies have shown that discharge of nutrients from agricultural non-point sources has become an important cause of water pollution and eutrophication. Consequently, many countries and regions of the world consider that the control of agricultural pollution is an essential component of water quality management.

Zhu Zhao-liang et al. *(eds.): Nitrogen in Soils of China, 303–321.*

The criteria for eutrophication vary with the country and the generally accepted indices are listed in Table 13.1. In China the regulation for surface water quality stipulates that to prevent eutrophication, the total inorganic N in closed waters (lakes, reservoirs, etc.) should be controlled within 0.1 mg N L^{-1}, and total inorganic P 0.025 mg L^{-1} (for Grade II water quality) and 0.05 mg N L^{-1} (for Grade III water quality). In general, the upper limit for the concentrations of inorganic N and P are higher than those in developed countries. It is believed that along with the progress in the research on water pollution and quality control in China, modifications will be made to the set criteria in the near future.

Studies show that algal blooms occur when the total inorganic N and phosphate concentrations in closed waters increase beyond 0.2 mg N L^{-1} and 0.015 mg P L^{-1}, respectively. In terms of soil research, these concentrations may be small, but in aquatic ecosystems they can bring about unexpected disasters (Gao 1986).

The problem of eutrophication in China has become apparent. In 1977, red tide occurred in the sea near Dagukou, Tianjin, covering an area of 56 km^2 for more than 20 days. In Liaoning, Dahuofang Reservoir was formerly entitled as the only expanse of clear water in the province, but recent observations show the total inorganic N and P in the reservoir have increased to 1.15 and 0.06 mg L^{-1}, respectively, algal blooms occurred many times, dissolved oxygen in part of the reservoir dropped drastically, blue algae propagated rapidly, and the yields of fish and other aquatic products decreased year by year. The integrated analyses of physical, chemical and biological indices indicated that part of the water body in the reservoir had already started the process of eutrophication.

The problem of eutrophication in lakes and reservoirs has become particularly prominent in recent years in China, as shown in Table 13.2. The analytical results listed indicate that of the 20 lakes and reservoirs in various parts of the country, 13 are highly eutrophic, 4 are moderately eutrophic or becoming highly eutrophic, and only 3 are slightly eutrophic. Table 13.2 also shows that the 13 lakes and reservoirs with eutrophic water are mostly located in the plains region of East China, particularly in

Table 13.1. Water eutrophication criteria.[1]

Item	Low	Moderate	High
Inorganic N (mg L^{-1})	< 0.1	0.1–0.3	> 0.5
Inorganic P (mg L^{-1})	< 0.001	0.001–0.01	> 0.03
Biological O_2 demand (mg L^{-1})	< 1	1–5	> 10
Dissolved O_2 (mg L^{-1})	> 5	5–1	< 1
Transparency (m)	> 5	5–1	< 1
Water colour	Bluish green	Green	Yellowish green
Algal density (10^4 L^{-1})	< 30	30–100	> 100
Bacteria (per ml)	< 100	100–100,000	> 100,000
Planktonic fauna (per L)	< 100	1,000–2,000	
Benthic fauna (per m^2)	More species, smaller number 300–1,000	More species, greater number 1,000–2,000	Less species, greater number 2,000–10,000

[1] Jilin Library (1984); Zhang (1989).

Table 13.2. Water quality evaluation and eutrophication level of some major lakes in China.[1]

Lake	Area (km^2)	Total N ($mg\ L^{-1}$)	Total P ($mg\ L^{-1}$)	Biomass ($10^4\ L^{-1}$)	Dominant species	Transparency (M)	Eutrophication level
Poyang L	3,583.0	0.42	0.01	–	–	1.20	Moderate–high
Dongting L.	2,740	0.86	0.02	8.7	Blue algae	0.40	High
Tai L.	2,425.0	0.90	0.02	100.0	Blue algae	0.50	High
Hongze L.	1,960.0	0.46	0.10	11.5	Diatom & blue algae	0.30	Moderate–high
Chao L.	820.0	1.67	0.03	25.3	Blue algae	0.25	High
East L. (Wuhan)	30.75	1.76	0.14	1913.7	Blue & green algae	0.40	High
West L. (Hangzhou)	6.03	2.76	0.13	6920.0	Blue & green algae	0.35	High
Wuli L. (Wuxi)	3.7	1.56	0.06	–	Blue algae	0.48	High
Xuanwu L. (Nanjing)	3.2	2.18	0.17	–	Blue & green algae	0.45	High
Mochou L. (Nanjing)	0.37	2.98	0.14	–	Blue & green algae	0.25	High
Thin West L.	0.30	2.84	0.12	–	Blue algae	0.90	High
Hulun L.	2,315.0	0.13	0.08	11.6	Diatom	0.50	High
Qinghai L.	4,635.0	0.22	0.02	14.6	Diatom	4.5	Moderate
Dianchi L.	330.0	0.23	0.02	189.2	Green algae	0.50	Moderate–high
Erhai L.	250.0	0.10	0.01	–		4.00	Low
Fuxian L.	212.0	0.21	0.02	19.0	Diatom & blue algae	7.00	Low
Jingpo L.	95.0	0.38	0.04	40.8	Diatom & blue algae		High
Tianchi L. (Xingjiang)	2.8	0.41	0.01	–	–	9.00	Low–moderate
South L. (Changchung)	0.91	3.60	0.32	5038	Blue algae		High
Dahuofang Reservoir	53.3	1.15	0.06	–	Diatom & dinophytes	1.20	High

[1] Annual means (Peng and Chen 1988).

the intensive agricultural regions along the mid and lower reaches of the Changjiang River, and in adjacent areas between urban or residential areas and suburbs. This shows the importance of agricultural pollutants and waste water from industries and domestic sewage on eutrophication.

Ma and Qian (1987) showed that the mean annual inorganic N concentration in the water bodies of some major lakes (including Tai Lake, Dushu Lake, Jinji Lake, Ge Lake and Jiu Lake) in the Tai Lake region, South Jiangsu were all above 0.25 mg N L^{-1}, an indication of imminent eutrophication. Tai Lake, the third biggest freshwater lake in China, has 1/3 of its water body falling below the quality criteria for third grade surface water, and 1/5 of its offshore waters suffering frequent algae blooms.

The recent rapid deterioration in the water quality of the lakes mentioned above is primarily due to the booming township and village enterprises, extensive use of phosphate-containing detergents, discharge of untreated domestic sewage and industrial waste water, and uncontrolled development of aquaculture. However, the contribution of lost N from agroecosystems should not be ignored.

In the Tai Lake watershed, for example, the N which is transported from the field to the surface water when the flooded field is drained prior to transplanting of rice, and that which is lost by runoff during plant growth, amounts to 13.6–16.6% of the fertilizer N applied (Tables 13.3, 13.4 and 13.5). Omemik (see Peng and Chen 1988) showed in his study that the losses of soil N and P through erosion in agriculture-dominated regions were 7 and 10 times, respectively, as much as those in forest covered regions in the USA.

Du (1987) reported that in the United States of America, 57–64% of its rivers and lakes were affected by non-point source pollutants, and that these were mainly agricultural. In Illinois, fertilizer N consumption increased by a factor of 10.8 in the period 1949–1969, and the mean nitrate concentration in the rivers increased from 3.1 to 10.9 mg N L^{-1}. In Czechoslovakia, Prochazkova (1975) showed that the nitrate concentration in the water bodies of the reservoirs was strongly correlated with the area of farmland and the N application rate in the catchment of the two reservoirs.

13.2.2. Nitrate and nitrite pollution of ground and surface water

According to the WHO the nitrate concentration of drinking water should not exceed 10 mg N L^{-1} (Jilin Library 1984). However, the standard used varies from country to country and ranges from 4.5 to 22.5 mg N L^{-1}. While the WHO limit is used at present in China, investigations are being conducted to develop a standard more suitable to the national conditions.

Nitrate *per se* is not toxic, but it is the main source of nitrite. After entering the human body, nitrate is reduced to nitrite in the stomach by microorganisms. The reaction of nitrite with haemoglobin to produce methaemoglobinemia can be fatal in infants (Hu 1986). Nitrite can react with secondary amines or amide compounds to form carcinogenic compounds, viz. N-nitroso compounds and C-nitroso compounds inside or outside the human body. Consequently nitrate or nitrite pollution

Table 13.3. Nitrogen pollution through drainage of flood water prior to transplanting rice in the Tai Lake region, South Jiangsu (means for 1987–1988).[1]

Location	Drainage ($t\ ha^{-1}yr^{-1}$)	Concentration ($mg\ N\ L^{-1}$)	Pollutant load ($kg\ N\ ha^{-1}\ yr^{-1}$)
Xinchang, Liyang	245	31.2 ± 4.82	7.7 ± 1.2
Lujiaxiang, Wujin	291	42.0 ± 4.98	12.3 ± 1.5
Luwan, Zhangjiagang	342	37.1 ± 3.88	12.8 ± 1.4
Yifeng, Yixin	249	46.6 ± 4.28	11.7 ± 1.1
Shili, Wujiang	239	48.3 ± 3.96	11.4 ± 0.9
Mean	273	41.1 ± 4.38	11.3 ± 1.2

[1] Unpublished data of Ma Li-shan.

Table 13.4. Nitrogen pollution through surface runoff in rice growing areas of the Tai Lake watershed, South Jiangsu (1987, annual precipitation 1340–1460 mm).[1]

Location	Runoff ($t\ ha^{-1}yr^{-1}$)	Concentration ($mg\ N\ L^{-1}$)	Pollutant load ($kg\ N\ ha^{-1}\ yr^{-1}$)
Xinchang, Liyang	4,419	6.41 ± 0.88	28.4 ± 3.9
Lujiaxiang, Wujin	5,708	8.32 ± 0.96	47.6 ± 5.6
Luwan, Zhangjiagang	7,386	8.22 ± 0.94	60.8 ± 6.9
Yifeng, Yixin	7,032	7.29 ± 0.89	51.3 ± 6.3
Shili, Wujiang	5,934	6.62 ± 0.99	39.3 ± 5.9
Mean	5,922	7.37 ± 0.93	45.5 ± 5.7

[1] Unpublished data of Ma Li-shan.

Table 13.5. Nitrogen pollution through surface runoff in upland areas of the Tai Lake region, South Jiangsu.[1]

Location	Vegetation	Runoff ($t\ ha^{-1}yr^{-1}$)	Concentration ($mg\ N\ L^{-1}$)	Pollutant load ($kg\ N\ ha^{-1}\ yr^{-1}$)
Xinchang, Liyang	Soybean	833	37.3 ± 5.28	31.1 ± 4.4
Lujiaxiang, Wujin	Mulberry	750	35.0 ± 5.10	26.3 ± 3.8
Luwan, Zhangjiagang	Cotton	908	40.8 ± 6.19	36.9 ± 5.6
Means		830	37.7 ± 5.52	31.4 ± 4.5

[1] Unpublished data of Ma Li-shan.

is a potential hazard to human and animals (Ma 1979, 1983; Hu and Ma 1978, 1980).

Under certain conditions, nitrate can be leached from the soil–plant system. Leaching of N depends upon a series of factors. In the study on nitrate leaching and its prevention in the Beijing-Tianjin-Bohai Bay area, the Forest Soil Institute, Chinese Academy of Sciences found that in the paddy rice system, the growth stage of rice and water management were two important factors controlling nitrate leaching from the soil–plant system (NGPED 1983a, b, c).

In the Tai Lake region, the author and co-workers laid down 20 field trials, distributed over 16 counties, using self-made PVC field lysimeters to collect shallow groundwater samples, and at the same time took 1,220 samples from drinking water wells, rivers and lakes. The results showed that 57.9% of the samples from drinking water wells had nitrate concentrations above the limit, and 93.3%, 70.6% and 38.2% of the samples from the rivers, lakes and other wells, respectively, had nitrate concentrations above the limit. In addition the nitrate concentration in percolating water at a depth of 0.5–1 m was significantly related to the N application rate (r = 0.508–0.841; Ma and Qian 1987). A study of the percolating water in paddy fields indicated the N-pollutant load amounted to 2.45–6.13% (Table 13.6) of the mean annual N input (fertilizers and organic manures) to the field, and about 70% of the load was nitrate.

Singh and Sekhon (1979) concluded that vegetation was a very important factor controlling nitrate leaching and pollution of groundwater. In a 5-year study on movement of fertilizer N in a brown forest soil, Nikitishen (1980) showed that when the N application rate was 60 kg N ha^{-1}, movement of N occurred within the 0–30 cm soil layer with soil water moving upward and downward, alternately. The downward movement occurred mainly in fall and early spring whereas the upward movement took place in winter and summer, thus limiting the deep movement of nitrate into the groundwater. However, the threat of nitrate pollution of groundwater increased with the application rate of fertilizer N.

The Nitrogen Group, Forest Soil Institute, Chinese Academy of Sciences, conducted a column experiment using ^{15}N to study the leaching of N added as ammonium sulfate, urea and ammonium nitrate to an undisturbed calcareous light meadow soil growing flooded rice in cylinders 10 cm in diameter, and 90 cm in height. The experiment had 3 application rates, 602 mg N (high), 460 mg N (moderate) and 154 mg N (low). The results showed that during the rice growing season, 71–89% of the total N in the leachate was present as nitrate, 8–21% was nitrite, and 1–7% was ammonium (Table 13.7). When urea and ammonium sulfate were added to the cylinders 0.23–0.51% of the applied N was lost by leaching,

Table 13.6. Leaching loss of nitrogen from paddy fields in South Jiangsu in 1987.

Location	Percolation rate[1] (t $ha^{-1}yr^{-1}$)	Concentration (mg N L^{-1})	Pollutant load (kg N ha^{-1} yr^{-1})
Xinchang, Liyang	5,595 ± 1185	2.45 ± 0.61	27.5 ± 6.9
Lujiaxiang, Wujin	3,000 ± 818	1.67 ± 0.40	10.1 ± 2.4
Luwan, Zhangjiagang	9,000 ± 2242	1.89 ± 0.48	34.1 ± 8.7
Yifeng, Yixin	7,500 ± 1620	1.98 ± 0.51	29.7 ± 7.7
Shili, Wujiang	5,003 ± 833	2.09 ± 0.50	20.9 ± 5.0
Mean	6,023 ± 1335	2.02 ± 0.50	24.5 ± 6.2

[1] The percolation rate was calculated from the percolating coefficient (mm day^{-1}) and the time for which the field was over-saturated with water.
Unpublished data of Ma Li-shan.

Table 13.7. Effect of application of nitrogen to flooded rice on the form of nitrogen in the leachate.[1]

	Rate of application (mg N column^{-1})	N in leachate (% of total)		
		Nitrate	Nitrite	Ammonium
Control	0	76.10	17.95	5.95
AS	154	78.30	17.14	4.55
	460	79.22	16.41	4.37
	602	88.61	8.53	2.86
Urea	154	82.64	13.06	4.30
	460	79.43	15.28	5.29
	602	71.49	21.12	7.38
AN	154	87.45	9.20	3.35
	460	87.01	10.23	2.76
	602	79.29	19.66	1.04
AS	602 (no rice)	88.98	7.76	3.26

[1] NGPED (1983a).

whereas 1.4–9.3% of the applied N was lost when ammonium nitrate was applied (Table 13.8). The results also showed that leaching of fertilizer N was closely related to the time of application. When fertilizer N was applied to the cylinders without rice, or at the time when rice was still in the seedling stage, with undeveloped roots low in absorption capacity, the groundwater was highly polluted (NGPED 1983a).

Zhu (1983), Jiang (1984), and Zhang (1985) reported that excessive application of fertilizer N is the main cause of nitrate pollution of groundwater in many areas in the world. Studies in the former Democratic Republic of Germany, which was one of the countries with the highest application rate, showed that the nitrate concentration in the groundwater was related to the amount of N applied. In 1973, the N application rate in that country reached 360 kg N ha^{-1}, and as a result the nitrate

Table 13.8. Effect of fertilizer form on loss of nitrogen by leaching during a rice growing season.[1]

Form	Rate of application (mg N column^{-1})	Leaching loss			
		Total (mg N)	Fertilizer (mg N)	Soil (mg N)	Fertilizer (% of N applied)
Control	0	27.74		27.74	
AS	154	45.67	0.78	44.89	0.51
	460	45.28	1.38	43.90	0.30
	602	46.43	2.94	43.49	0.43
Urea	154	52.30	0.78	51.52	0.51
	460	38.16	1.08	38.08	0.23
	602	47.82	3.36	44.46	0.49
AN	154	67.18	2.91	64.27	1.88
	460	51.82	6.71	45.11	1.45
	602	131.21	63.87	67.34	9.30
AS (no rice)	602	79.64	8.30	71.34	1.20

[1] NGPED (1983c).

concentration in the water of 50% of the farming wells surpassed 60 mg N L^{-1}. In Bulgaria in 1978, investigations on 20 residential regions in the countryside showed that the nitrate concentration in the well water was greater than 50 mg N L^{-1}. In France, long-term application of fertilizer N has resulted in serious nitrate pollution of the drinking water. For instance, in the Baus Region near Paris, the nitrate concentration of the groundwater has increased to 180 mg N L^{-1}. In the period from 1972 to 1982 the nitrate concentration in the groundwater of the suburbs of Beijing was related to the fertilizer N application rate (Zhu 1983; Zhu and Tian 1986).

Take Wu County of Jiangsu Province as another example. During the period 1980–1984, the amount of fertilizer N applied each year averaged 28,000 tonnes with a mean application rate of 349 kg N ha^{-1} yr^{-1} (Table 13.9), and the nitrate concentration in the well water, 9 important rivers and Yangcheng Lake varied from a trace to 79 mg N L^{-1} (Table 13.10), 0.13–5.18 mg N L^{-1} (Table 13.11) and 0.05–0.7 mg N L^{-1} (Table 13.12), respectively.

Table 13.9. Amount of fertilizer nitrogen applied in Wu County, Jiangsu Province.[1]

Year	Fertilizer applied (t N)	Mean application rate (kg N ha^{-1} yr^{-1})
1980	27,764	347
1981	27,785	347
1982	28,010	350
1983	28,051	351
1984	28,072	351

[1] Ma and Qian (1987).

Table 13.10. Nitrate and nitrite concentrations in well water in Wu County, Jiangsu Province.[1]

Location	No of wells	Nitrate			Nitrite		
		Concentration		Wells above limit	Concentration		Wells above limit
		Range (mg N L^{-1})	Mean (mg N L^{-1})		Range (mg N L^{-1})	Mean (mg N L^{-1})	
Tongan	14	0.02–40.20	17.81	10	0.02–0.50	0.12	12
Baoan	13	0.02–79.35	24.73	8	Tr–0.58	0.19	8
Huangqiao	11	Tr–20.00	3.90	1	0.02–0.28	0.03	4
Dongzhu	10	0.09–25.80	6.54	1	Tr–0.19	0.02	1
Huangdi	8	0.02–26.80	12.22	2	0.02–0.25	0.07	8
Wangting	6	0.02–23.20	7.07	1	Tr–0.21	0.04	1
Zhenhu	5	0.02–23.20	8.00	1	Tr–0.22	0.04	1
Jinshan	3	6.80–13.20	9.28	1	0.02–0.14	0.06	3
Likou	2	5.80–5.90	5.85	0	0.02	0.02	2
Yuexi	2	38.25–40.60	39.48	2	0.18–0.23	0.21	2
Lumu	2	15.50–19.60	17.55	2	0.10–0.12	0.11	2
Total	76	Tr–79.35	14.00	29	Tr–0.58	0.08	44

[1] Ma and Qian (1987).

Table 13.11. Nitrate in the main rivers of Wu County, Jiangsu Province.[1]

River	Nitrate concentration (mg N L^{-1})			
	Normal WL	High WL	Low WL	Annual mean
Wusongjiang	0.22	0.17	0.13	0.17
South Wusongjiang	0.20	0.13	0.23	0.19
Yuanhetang	0.30	0.22	0.30	0.27
Loujiang	0.15	0.24	0.48	0.29
Grand Canal	0.20	0.15	0.62	0.32
Sudonghe	0.25	0.28	0.60	0.37
Huguang Canal	0.21	0.23	1.53	0.65
Muguanghe	0.24	0.21	4.68	1.70
Xujiang	0.22	5.18	1.46	2.29

[1] Ma and Qian (1987); WL = water level.

Table 13.12. Nitrate concentration in the water of Yangcheng Lake.[1]

Portion of lake	Nitrate (mg N L^{-1})						
	Late April		Late Sept.		Late Dec.		
	Range	Mean	Range	Mean	Range	Mean	Mean
Central	0.10–0.15	0.10	0.20–0.28	0.25	0.35–0.45	0.40	0.25
East	0.05–0.15	0.10	0.15–0.25	0.20	0.25–0.70	0.48	0.26
West	0.05–0.10	0.06	0.20–0.45	0.25	0.13–0.55	0.30	0.20

[1] Ma and Qian (1987).

According to WHO the concentration of nitrate and nitrite in drinking water should be less than 10 mg N L^{-1} and 0.02 mg N L^{-1}, respectively. However, in Wu County when the water level was low, 38.2% of the wells used for drinking water had nitrate concentrations higher than the recommended level, and 26.3% of the wells contained more than twice the recommended level (Table 13.10). Approximately 58% of the wells used for drinking water had nitrite concentrations which were greater than the recommended level. It is apparent that pollution of drinking water in Wu County by nitrate and nitrite is already serious.

Singh and Sekhon (1979) reported that 73% of the wells in Washington County of Southern Illinois, USA, had nitrate concentrations higher than the critical level. Spalding *et al.* (1978) (in Jiang 1984) investigated 1,233 km^2 of irrigated farm land in Merrick County, Illinois, took 293 groundwater samples for analysis, and found that the nitrate concentrations in 50% of the samples were 1.5–3 times higher than the permissible concentration. Investigations showed that N fertilizer was the source of the pollution. The data indicated that nitrate concentration in the groundwater was strongly related to the square root of the number of irrigation wells per unit area × soil drainage. It was believed that this regression could be used to predict nitrate pollution in a certain area.

The nitrate concentrations in percolating water at various depths, floodwater of rice fields, drainage water and rainfall, and the fertilizer N application rates for 20 tested sites in the Tai Lake region are given in Table 13.13. The nitrate concentration of the percolating water at 1m depth was markedly higher than that at 0.25–0.5 m depth, and 2–6 times higher than that of the floodwater. Correlation analysis indicated:

1. The nitrate concentrations of the percolating waters at 1 m (Y_2) and 0.5 m (Y_1) depth and the mean value of the two are significantly correlated to the annual application rate of fertilizer N (X_3).
2. The nitrate concentration of the percolating water at 0.5 m depth is significantly correlated to the fertilizer N application rate for the early rice crop (X_1). In addition the mean value of the nitrate concentrations at 0.5 m and 1 m depths is significantly correlated to the fertilizer N application rate for the early rice crop.
3. The nitrate concentration at 1 m depth is significantly related to the fertilizer N application rate for the late rice crop (X_2).

The regression equations for the above relationships are as follows:

$$Y_1 = -38 + 28X_1 + 21X_2 + 13\,[X_3 - (X_1 + X_2)]$$

$$Y_2 = -173 + 23X_1 + 9X_2 + 15\,[X_3 - (X_1 + X_2)]$$

$$Y_3 = -283 + 25X_1 + 16X_2 + 15\,[X_3 - (X_1 + X_2)]$$

where X_1, X_2, X_3, Y_1 and Y_2 are as described above, and Y_3 denotes the mean value of Y_1 and Y_2 (Ma and Qian 1987).

Having carried out the monitoring of the groundwater pollution problem, Zhu (1983) and Zhu and Tian (1986) showed that the nitrate concentration of the ground water in the suburban counties of Beijing was related to the fertilizer N application rate. In areas where groundwater tables were less than 10 m, between 10–20 m and more than 20 m in depth, the mean nitrate concentrations in the groundwater were 17.6, 23.1 and 30.0 mg N L^{-1}, respectively, showing that the nitrate concentration in the groundwater increases with depth. In the vegetable growing areas of the suburbs of Beijing, the nitrate concentration in the groundwater was as high as 61.6–124.0 mg N L^{-1} due to the heavy application of fertilizer N. It was also found that N loss in fertilized areas through runoff and percolation was 3 to 10 times greater than that in areas where no fertilizer N was applied.

13.2.3. Pollution of surface water by ammoniacal nitrogen

Ammonium does not normally move in terrestrial ecosystems, because soil colloids can adsorb and retain it. However, when it is transported to surface waters it can cause pollution problems, because as the pH of the water body increases, the ammonia concentration increases. When ammonia in surface water reaches 0.3–5.0 mg N L^{-1}, it may seriously affect aquaculture. Fortunately, in water bodies rich in dissolved oxygen, ammonia is oxidized to nitrate by nitrifying organisms, and the toxicity of

Table 13.13. Nitrate concentration in water bodies, and nitrogen application rates at 20 sites in the Tai Lake Region.

Location	Nitrate concentration (annual means mg N L^{-1})						N applied (kg ha^{-1})		
	FW	DR	RF	PW_1	PW_2	PW_3	ER	LR	AT
Changxing	0.07	0.10	0.29	0.20	0.20	0.33	157.5	120.0	480.0
Xinmao	0.06	0.12	0.29	0.29	0.39	0.44	190.5	138.0	553.5
Dongting	0.06	0.07	0.19	0.21	0.25	0.38	193.5	138.0	435.0
Wujin	0.06	0.09	0.21	0.19	0.25	0.39	178.5	121.5	480.0
Xinzhuang	0.08	0.13	0.32	0.21	0.30	0.30	177.0	120.0	465.0
Tongluo	0.09	0.19	0.48	0.19	0.29	0.35	181.5	121.5	454.5
Guangfu	0.06	0.06	0.23	0.19	0.22	0.26	157.5	121.5	456.0
Shengpu	0.10	0.12	0.25	0.21	0.31	0.44	198.0	121.5	525.0
Shangta	0.15	0.14	0.23	0.29	0.38	0.40	216.0	121.5	538.5
Guanqiao	0.09	0.11	0.32	0.28	0.30	0.36	211.5	138.0	448.5
Jiaxing	0.08	0.09	0.27	0.21	0.33	0.44	190.5	123.0	507.0
Tongxiang	0.05	0.08	0.19	0.31	0.33	0.35	177.0	138.0	478.5
Zhuze	0.06	0.07	0.13	0.20	0.29	0.38	150.0	118.5	432.0
Fengxian	0.10	0.12	0.25	0.18	0.25	0.42	157.5	138.0	475.5
Baoshan	0.06	0.07	0.27	0.25	0.30	0.41	172.5	133.5	501.0
Pinghu	0.08	0.08	0.31	0.12	0.21	0.29	190.5	120.0	451.5
Wangfang	0.08	0.09	0.26	0.13	0.16	0.22	159.0	120.0	343.5
Yuexi	0.07	0.08	0.24	0.17	0.17	0.21	150.0	132.0	417.0
Xiangxuehai	0.08	0.08	0.28	0.15	0.25	0.27	177.0	157.5	450.0
Likou	0.08	0.09	0.39	0.18	0.45	0.42	228.0	216.0	553.5

[1] Ma and Qian (1987) : FW = floodwater; DR = drainage; RF = rainfall; PW = percolating water (PW_1 = PW at depth of 0.25 m; PW_2 = PW at depth of 0.5 m; PW_3 = PW at depth of 1 m); ER = early rice; LR = late rice; AT = annual total.

ammonia to fish is mitigated. Nevertheless, in North China, with its alkaline environment, the pH of the surface water is generally >8.0 and may exceed 9.0 in some areas affected by the discharge of industrial wastes. Under these conditions, ammonia toxicity remains a problem (Gao 1986).

In a study on agricultural pollution of the Tai Lake region, Ma Li-shan *et al.* (unpublished data) found that the concentration of ammonium-N in floodwater drained from the field prior to transplanting rice and in runoff from rice growing areas varied from 2.85 to 28.6 mg N L^{-1} and 0.5 to 37.5 mg N L^{-1}, respectively. As the annual precipitation in this region is 1,100–1,400 mm, the amount of runoff is considerable, hence the contribution of runoff to ammonia pollution of surface water bodies is also significant. However, the pH values of the surface water and soil in this region range from 5.7 to 7.8, and thus the damage to aquaculture (fish, shellfish, etc.) is not as important.

13.3. Atmospheric effects

Gaseous nitrogen oxides (NO_x), mostly in the form of NO, NO_2, etc., are the products of biochemical reactions of fertilizer N in agroecosystems. Like sulfur dioxide, NO_x can create acid rain endangering aquatic and terrestrial ecosystems.

For instance, when acid rain comes in direct contact with a forest it could be seriously damaged or destroyed. When deposited into an impoverished acidic lake, changes in the amount of dissolved Al would occur, leading to decreased productivity (Zhang 1985; Gao 1986).

The problem of N_2O damaging the ozone layer has already become the focus of world attention. The ozone layer is a protective shield which prevents excess cosmic ultraviolet radiation reaching the earth's surface. Once this shield is damaged, the increased ultraviolet radiation leads to an increase in the incidence of skin cancer in animals, interferes with the normal growth of animals and plants, and affects climate adversely. Nitrous oxide and nitric oxide are produced from fertilizer N in soil by bacterial nitrification and denitrification. It is estimated that by early next century, the N generated in industrial and agricultural production and released into the atmosphere could be as high as 1,500 Tg, sufficient to reduce stratospheric ozone by several percent. By the end of the next century, the ozone level may be reduced by 8%. As a consequence, the ultraviolet radiation reaching the land surface will increase by 15% (Gao 1986).

Moreover, under irradiation by sunlight, gaseous N oxides generate photochemical smog, which irritates the aspiration system and corrodes the lungs of humans (Hu 1986).

13.4. Significance of farm management on crop and environmental quality

13.4.1. Effect of nitrogen fertilizer application on crop quality

Excess uptake of nitrate increases the susceptibility of crops to pests and insects and affects the quality of the harvested product, e.g. by reducing the resistance of potato to mechanical damage during harvest and transportation. It also degrades vegetables and fruits by affecting flavour and storage endurance in winter, and lowers the nutritional value of the produce by decreasing the concentration of methionine, and decreases the quality of protein in the produce. Moreover, excessive nitrate increases the amount of protease in crops which may induce cancer in animals (Zhu, 1983; Zhang 1989; Hu 1986).

Zhang *et al.* (1984) indicated in their study that the combined application of organic manure and fertilizer N would not only contribute to soil fertility and crop yield, but would also improve the quality of grains and vegetables. Biological quality can be defined as the sum of mineral elements (macro- and microelements), organic compounds (carbohydrates, proteins, fat, etc.), vitamins, hormones, enzymes, etc. in a given biological product used as food to sustain normal physiological metabolism. For example, in modern irrigated pastures, even though heavy applications of fertilizer N may increase the yield of forage grasses, in the long run it may cause severe deficiency of Cu in the forage grasses. Animals feeding on these grasses will suffer physiological disorders. The solution to the problem is to apply the correct rate of Cu to the pasture soil. The problem of degradation in environmental and crop quality, resulting from long-term abuse of fertilizer N, is worthy of close attention. However, because of the complexity of the subject, research in this field lags behind the development of the problem.

13.4.2. Accumulation of nitrate in vegetables

Considerable research has been conducted on factors controlling nitrate concentration in vegetables and measures to overcome the problem Shen *et al.* 1982; Barker 1974; Cantliffe 1972; Maynard 1974; Olday 1976). Results indicate that the nitrate taken up by vegetables transforms into amino acids and proteins through a series of enzymatically catalysed reactions. When the soil lacks available molybdenum, nitrate accumulates in the plant, and this is particularly obvious in vegetables. It was also found that the nitrate concentration in vegetables varies with species, variety, growth period, plant part and cultivation practice. For instance, in celery, lettuce, turnip and spinach, the concentration varies from 2,000 to 3,000 mg NO_3^- kg^{-1}, and in shallot, carrot and sweet pepper from 100 to 500 mg NO_3^- kg^{-1}. The nitrate concentration in the seeds of these vegetables is very low. The nitrate concentration in vegetables and fruit decreases in the following order: leaf vegetables > roots and tubers > cucurbits > fruit.

Shen *et al.* (1982) investigated 34 of the main vegetables of China and found that nitrate accumulation varied greatly between species, varieties and plant parts. For example, the concentrations in ginger, spinach and turnip were 2,650, 2,350 and 2,127 mg kg^{-1}, respectively, while the concentration in tomato was very low, being only 15 mg kg^{-1}. In vegetables, the nitrate concentration differs with part sampled and decreases in the order root > stem > leaf > seed (fruit). For the 34 vegetables studied, nitrate accumulation varied greatly with the greatest variation occurring in lettuce, spinach and cabbage.

Vegetable crops have relatively high nitrate concentrations in the early growth stages; for instance, the nitrate concentration in young buds is relatively high and it decreases when the crop matures. The nitrate concentration in vegetables is related to the cultural conditions, such as the rate of application of fertilizer N and the amount of sunlight (Table 13.14). Excess fertilizer N, increased plant density (overlapping and shading) and application of herbicides are factors which can adversely affect the N metabolism of the plant and lead to accumulation of nitrate (Xu and Mao, 1982).

Table 13.14. Nitrate concentration in spinach and sugarbeet in relation to fertilization and illumination.[1]

N applied (mg N kg soil^{-1})	Nitrate concentration in dry matter (mg N kg^{-1})			
	Spinach		Sugar beet	
	Illuminated	Shaded	Illuminated	Shaded
0	911	1,417	607	1,518
50	3,542	11,031	1,822	5,060
100	7,286	16,293	11,334	12,751

[1] Xu and Mao (1982).

The toxic level of nitrate for humans is 3,099 mg. In 1973 WHO and FAO stipulated that the ADI of nitrate was 3.6 mg per day per kg body weight (Jilin Library 1984). Vegetables are the main source (about 81.2%) of nitrate for humans (Cantliffe 1972), and the type of food crop most likely to be enriched in nitrate. Modern science has shown that nitrate enrichment in vegetables does not cause any adverse effect on the plant, but can harm the people and animals that eat them.

13.4.3. Effect of nitrogen pollution on animal health

The perniciousness of the threat of environmental N pollution to the health of animals has already aroused attention the world over. Epidemiological investigations on malignant tumours revealed that gastric carcinoma was possibly related to the high nitrate concentration of vegetables and drinking water (Cuollo 1976). In Qidong County of Jiangsu, the hepatoma fatality rate is correlated with the nitrate and nitrite concentrations of drinking water. In Nantong Prefecture of Jiangsu, the hepatoma fatality rate is correlated with the nitrate concentration of the soil ($r = 0.886$, $P = 0.05$; Ma 1979; Hu *et al.* 1983; Table 13. 15). In Lin County of Henan Province, the incidence of hyperplasia of the oesophageal epithelium and oesophageal carcinoma is correlated with the nitrate and nitrite concentrations of drinking water (Wang 1979).

The most common product formed as a result of N pollution is nitrite. Nitrate in drinking water and the food chain is readily reduced to nitrite by microorganisms, and the conditions in man's oral cavities are favourable for this transformation. In addition, the pH of man's gastric juice is 1–3, which is favourable for the formation of carcinogenic N-nitroso compounds. N-nitroso compounds can be ingested with drinking water or food, and can be formed inside the viscera when secondary or tertiary amines or amide-type compounds are ingested with nitrate or nitrite. Consequently, the nitrate and nitrite content of food and drinking water which result from N in the water and soil environment are critical concerns for human health.

Nitrogen pollution of the environment also threatens livestock. A number of forage crops can absorb and accumulate a large amount of nitrate from soil and water. A nitrate concentration greater than 1g N kg^{-1} in the fodder, or 1,500 mg N L^{-1} in the

Table 13.15. Relationship between hepatoma fatality rate and soil nitrate in Nantong, Jiangsu Province.[1]

Location	Fatality rate (per 100,000)	Nitrate in 0–25 cm layer (mg N kg^{-1})
Huiping & Hehe, Qidong County	61	34.3 ± 8.56 (14)
Bingfang, Rudong County	47	17.4 ± 3.16 (14)
Pingshan & Chengdong, Haiman County	34	18.0 ± 4.07 (15)
Jiali, Rugao County	22	13.0 ± 0.28 (14)
Tienfen & Xining, Qidong County	16	17.0 ± 3.60 (14)
Xingnan, Haian County	8	6.8 ± 2.70 (15)

[1] Date sampled, Aug. 1975; values in parentheses denote number of samples (Ma 1983; Hu *et al.* 1983).

drinking water, is toxic for animals. Microorganisms in the stomach of cattle and sheep, and in the colon of horse, donkey and mules, reduce nitrate to nitrite. This is absorbed into the blood where it interferes with the circulation of oxygen and causes various diseases and death (Zhu 1983; Zhang 1989).

13.5. Techniques for reducing nitrogen loss

13.5.1. Improved fertilizer application techniques

Loss of fertilizer N from farmland and the techniques for reducing it have already been discussed in Chapter 11, but will be discussed again here in terms of environmental protection. In the design of fertilizer management practices, it should be emphasized that maximum economic profit is the important aim rather than maximum yield.

Many management practices have been suggested to improve the efficiency of use of fertilizer N, including split application, deep placement, slow release fertilizers, nitrification inhibitors and addition of lime. Split application is effective in some situations, as was demonstrated in a field experiment with ^{15}N-labelled ammonium sulfate (Table 13.16). The results showed that N recovery in the plant was improved and N loss was reduced, thus mitigating the environmental pollution by N.

Deep placement of fertilizer is an effective method of reducing N loss (Zhao *et al.* 1983). Chen *et al.* (1989) suggested that flooded rice be fertilized by broadcasting the fertilizer onto the drained paddy field before flooding. This technique is rather practical and applicable. Application of slow-release fertilizers is another possible approach to reducing N loss.

13.5.2. Balanced fertilization

The economic and environmental benefits of a soil testing and fertilizer recommendation system are significant, because it would help to overcome the following defects of the current fertilization practice in China (Mao 1987; Zhou 1987), viz. (i) the bias towards N, which results in an extreme imbalance of N, P and K, (ii) the overdosing with N in areas with highly intensive agriculture, and (iii) the ignorance in current fertilization practices which are based on experience only. Soil testing will improve and maintain an optimum N, P, K ratio, and control the over-application of

Table 13.16. Effect of split application on nitrogen balance (% of N applied).

Treatment	Retained in soil + roots	Recovered in plant	Loss
Bs[1]	36.2	38.7	25.1
1/2 Bs + 1/2 TD–Ti	32.7	42.5	24.8
1/3 Bs + 1/3 TD–Ti + 1/3 TD–PI	29.6	49.6	20.8

[1] Bs = applied as basal fertilizer; TD–Ti = topdressed at tillering; TD–PI = topdressed at panicle initiation (Zhu *et al.* 1984).

N in high yield regions. Consequently, fertilization costs will be lowered, lodging controlled, damage by diseases and insects will be reduced, biological quality of agricultural products will be improved, and environmental problems will be greatly reduced.

Compound fertilizers help ensure a balanced supply of N, P, and K to crops. Moreover, the introduction and application of B, Mo, Zn, Cu and other micronutrient fertilizers not only increases crop yields, but also improves the quality of agricultural products. For example, research in Qidong County of Jiangsu Province indicated that molybdenum greatly increased the yield of leguminous crops, corn and cotton, and significantly reduced the nitrate concentration in soybean and corn (unpublished data of Hu R. M. *et al.*)

13.5.3. Application of nitrification inhibitors

Nitrification inhibitors have the potential to reduce nitrate leaching from the soil-plant system. Li (1986) and Li *et al.* (1983) showed that application of inhibitors sharply reduced the population of nitrifying bacteria in wheat and rice fields to 13–47% of the control, thus inhibiting formation of nitrate in the soil.

The Nitrogen Group, Institute of Forest Soil, Chinese Academy of Sciences, found that 3 nitrification inhibitors, nitrapyrin, quanylthiourea and dicyandiamide were effective for reducing the leaching of nitrate from the soil–plant system (NGPED 1983b; Table 13.17).

13.5.4. Rational irrigation

The higher the concentration of mineral N in the soil solution, the greater the N loss. In order to lower the N pollutant load in paddy fields, through surface runoff, measures should be taken to reduce the amount of N in the surface water of the field after fertilization and to optimize irrigation. In general, the amount of irrigation water is calculated in accordance with the water budget principle, to reduce drainage prior to transplanting rice and surface runoff caused by rainfall. Unpublished data of Ma Li-shan indicated that irrigation with a small amount of water, rather than flooding the paddy field, would make drainage unnecessary, except after storms, save costs and energy, reduce environmental pollution, and increase yield by 7.2–8.6 %.

In short, whether in upland areas or paddy regions, water saving practices should be advocated for three reasons, (i) to save valuable water resources, (ii) to save energy, and (iii) to reduce environmental pollution.

13.5.5. Control of land erosion

Nitrogen pollution from non-point sources in agriculture comes from soil loss, i.e. soil erosion, and hence is closely related to land use patterns. A reasonable land use pattern is the key to controlling this type of pollution.

Table 13.17. Effect of nitrification inhibitors on nitrate leaching from a calcareous light meadow soil.[1]

Treatment	Loss (mg N cylinder^{-1})	
	Nitrate	Nitrite
Control	44.0	7.8
Urea	68.5	29.8
Urea + DCD	45.3	5.1
Urea + GTU	21.4	2.5
Urea + Nitrapyrin	29.9	4.0
Ammonium sulfate	82.8	33.7
Ammonium sulfate + DCD	41.7	8.7
Ammonium sulfate + GTU	40.2	13.9
Ammonium sulfate + Nitrapyrin	35.3	15.9
Ammonium bicarbonate	62.9	26.1
Ammonium bicarbonate + DCD	41.8	10.1
Ammonium bicarbonate + GTU	44.3	11.7
Ammonium bicarbonate + Nitrapyrin	42.7	7.0

[1] 6.75 kg soil was filled into a PVC cylinder, 10 cm in diameter and 90 cm deep. Rice was planted and 700 mg N and 21 mg nitrification inhibitor were added. Each treatment had 3 replicates, and the cylinders were irrigated with distilled water throughout the growing period (NGPED 1983b).

The amount of N and P lost as a result of soil erosion depends on various factors such as fertilization rate, crop variety, soil texture, soil permeability, land slope and rainfall. Therefore, minimum or zero tillage, the building of terraced fields, and improved cultivation techniques to maintain proper vegetation are effective measures to mitigate environmental pollution.

In conclusion, increasing the plant recovery of fertilizer N and reducing N loss from agriculture will help solve the problems of food, energy and environmental pollution. From a strategic and long-term point of view, in order to attain a thorough solution to the problem of environmental pollution by non-point sources of N, important breakthroughs need to be made in the field of nitrogen-fixation, gene engineering and plant breeding, with a view to changing the traditional management of N.

13.6. References

Barker, A V 1974. Variations in nitrate accumulation among spinach cultivars. J. Am. Soc. Hort. Sci. 99:132–134.

Cantliffe, D J 1972. Nitrate accumulation in vegetable crops affected by photoperiod and light duration. J. Am. Soc. Hort. Sci. 97:414–418.

Chen, R Y, Chen, W and Zhang, J C 1989. Improvement and efficiency evaluation of deep placement techniques in paddy field. (in Chinese). Soils 21:125–129.

Cuollo, C 1976. Gastric cancer in Colombia. I. Cancer risk and suspect environmental agents. J. Natl. Cancer Institute 57:1015–1020.

Du, F R 1987. Study and control of non-point source pollution in foreign countries. (in Chinese). Overseas Agricultural Environment Protection (1):1–3.

EBESIN (Editorial Board of Environmental Science Information Network), Chinese Academy of Sciences. 1983. Proceedings of Nitrogen Pollution and Nitrogen Cycling in Environment. (in Chinese). Institute of Environmental Chemistry, Chinese Academy of Sciences. 284 p.

Gao, Z M 1986. Nitrogen loss from soil-plant system and environmental protection. In: Soil Agricultural Chemistry and Soil Biology and Biochemistry Committees, Soil Science Society of China (eds.), Advances and Prospects for Soil Nitrogen Research in China. (in Chinese). pp. 82–91. Science Press, Beijing.

Giorgini, A and Zingales, F 1986. Developments in Environmental Modeling, 10. Agricultural Nonpoint Source Pollution: Model selection and application. pp. 283–322. Elsevier Science Publishing Co. Inc., New York.

Hu, R M and Ma, L S 1978. Generalization of the study on chemical carcinogenic substance: N-nitrosamine compounds. (in Chinese). Jiangsu Medicine (6):21–26.

Hu, R M and Ma, L S 1980. Analytical chemistry of N-nitroso-compounds. (in Chinese). pp. 134–173. Science Press, Beijing.

Hu, R M, Ma, L S and Xu, Z Y 1983. Soil N, nitrosamine and hepatoma, In: Editorial Board of Environmental Science Information Network, Chinese Academy of Sciences (ed.), Proceedings of Nitrogen Pollution and Nitrogen Cycling in Environment. (in Chinese). pp. 275–282. Institute of Environmental Chemistry, Chinese Academy of Sciences.

Hu, H S 1986. Environmental Medicine. (in Chinese). pp. 73–76. China Environmental Science Publishing House

IAWPR 1975. (International Association on Water Pollution Research). Conference on Nitrogen as Water Pollutant, Vol. 1–3. Specialized Conference, Copenhagen, Denmark.

Jiang, D A 1984. Nitrogen pollution of the water-soil system. (in Chinese). Environmental Science Collection (5):59–63.

Jilin Library 1984. Selection of Environment Standards of Other Countries. (in Chinese). pp. 11–14. China Standards Publishing House.

Li, L M, Zhuang, S, Zhou, X R and Pan Y H 1983. Nitrification and denitrification in soil. In: Editorial Board of Environmental Science Information Network, Chinese Academy of Sciences (ed.), Proceedings of Nitrogen Pollution and Nitrogen Cycling in Environment. (in Chinese). pp. 160–168. Institute of Environmental Chemistry, Chinese Academy of Sciences.

Li, L M 1986. Status and outlook of the study on nitrification and denitrification in soil in China. In: Soil Agricultural Chemistry and Soil Biology and Biochemistry Subcommittees, Soil Science Society of China (ed.), Advances and Prospect for Soil Nitrogen Research in China. (in Chinese). pp. 68–79. Science Press, Beijing.

Ma, L S 1979. Nitrosamine and cancer. (in Chinese). Env. Prot. (5):40–43

Ma, L S 1983. Study on the soil environment of regions with a high incidence of hepatoma and carcinogenic N-nitrosamine. In: Editorial Board of Environmental Science Information Network, Chinese Academy of Sciences (ed.), Proceedings of Nitrogen Pollution and Nitrogen Cycling in Environment. (in Chinese). pp. 44–45. Institute of Environmental Chemistry, Chinese Academy of Sciences.

Ma, L S and Qian, M R 1987. Nitrate and nitrite pollution of water environment in the Tai Lake region. Environment Science. (in Chinese). 8:60–65.

Mao, D R 1987. Modern Principles and Techniques for Fertilization. (in Chinese). pp. 64–76. Science Press, Beijing..

Maynard, D N 1974. Nitrate accumulation in spinach as influenced by leaf type. J. Am. Soc. Hort. Sci. 99:135–138.

Nikitishen, V I 1980. Downward migration of nitrogen in soil. Proceedings of USSR Conference on Pollutant Migration in Soil from Neighbouring Mines. (in Russian). pp. 131–135. Hydrometeorology Publishing House. Leningrad.

NGPED 1983a. (Nitrogen Group, Pollution Ecology Department, Institute of Forest Soil, Chinese Academy of Sciences). Nitrogen in the soil-plant system. (I) Study on dynamics of N loss in the soil plant system by ^{15}N technique. In: Editorial Board of Environmental Science Information Network, Chinese Academy of Sciences (ed.), Proceedings of Nitrogen Pollution and Nitrogen Cycling in Environment. (in Chinese). pp. 68–75. Institute of Environmental Chemistry, Chinese Academy of Sciences.

NGPED 1983b. (Nitrogen Group, Pollution Ecology Department, Institute of Forest Soil, Chinese Academy of Sciences). Nitrogen in the soil-plant system. (II) Effect of nitrification inhibitors in reducing N loss from the soil-plant system. In: Editorial Board of Environmental Science Information Network, Chinese Academy of Sciences (ed.), Proceedings of Nitrogen Pollution and Nitrogen Cycling in Environment. (in Chinese). pp. 75–84. Institute of Environmental Chemistry, Chinese Academy of Sciences.

NGPED 1983c. (Nitrogen Group, Pollution Ecology Department, Institute of Forest Soil, Chinese Academy of Sciences). Dynamic simulation on nitrate leaching in the soil-plant system in relation to sewage irrigation in the Jing-Jin-Bo Region. In: Editorial Board of Environmental Science Information Network, Chinese Academy of Sciences (ed.), Proceedings of Nitrogen Pollution and Nitrogen Cycling in Environment. (in Chinese). pp. 84–91. Institute of Environmental Chemistry, Chinese Academy of Sciences.

Olday, F C 1976. A physiological basis for different patterns of nitrate accumulation in two spinach cultivars. J. Am. Soc. Hort. Sci. 10:217–219.

Peng, J X and Chen H J 1988. Water Eutrophication and Its Control (in Chinese). pp. 50–65, 180–181. China Environmental Science Publishing House.

Prochazkova, L 1975. Long term studies on nitrogen in two reservoirs related to field fertilization. Proc. of Conf. on Nitrogen as a Water Pollutant. Vol. 1. Analysis-Sources-Public Health.

Rosswall, T 1978. Nitrogen Cycling in West African Ecosystems. Reklam and Katalogtryk, Uppsala, Sweden. 450 p.

Shen, M Z, Zhai, B J, Che, H R and Li, J G 1982. Studies of nitrate accumulation in vegetables. I. Evaluation of nitrate and nitrite concentrations in various vegetables. (in Chinese). Horticulture (4):41–48.

Singh, B and Sekhon, G S 1979. Nitrate pollution of groundwater from farm use of nitrogen fertilizers – A Review. Agriculture and Environment. 4:207–225.

Wang, Y L 1979. Determination of nitrate and nitrite concentrations in the drinking water from wells of Yaocun Commune, Lin County. (in Chinese). Oncology Fascicle, Chinese Medical Science 5:201–204.

Xu, H X and Mao, W Y 1982. Nitrate, nitrite and human environment. (in Chinese). Environmental Science 3:59–63.

Zhang, F D, Jin, W X, Yu, Y N, Wang, X P, Zeng, M X, Zhao, X Y and Yao Y X 1984. Effect of combined application of organic manure and nitrogen fertilizer on yield and quality of wheat and corn. (in Chinese). Soil and Fertilizer (3):11–16.

Zhang, F D 1985. Trend and countermeasures of chemical fertilizers pollution. (in Chinese). Environmental Science 6:54–59.

Zhang, Z J 1989. Environmental Pollution Ecology. (in Chinese). China Environmental Science Press.

Zhao, Z D, Zheng, J S and Ren, S R 1983. Improving nitrogen recovery in plant. In: Editorial Board of Environmental Science Information Network, Chinese Academy of Sciences (ed.), Proceedings of Nitrogen Pollution and Nitrogen Cycling in Environment. (in Chinese). pp. 124–141. Institute of Environmental Chemistry, Chinese Academy of Sciences..

Zhou, M Z 1987. Soil testing and fertilizer recommendation in China. (in Chinese). Chinese Journal of Soil Science 18:7–13.

Zhu, J C 1983. Causes for nitrate pollution of ground water. In: Editorial Board of Environmental Science Information Network, Chinese Academy of Sciences (ed.), Proceedings of Nitrogen Pollution and Nitrogen Cycling in Environment. (in Chinese). pp. 210–217. Institute of Environmental Chemistry, Chinese Academy of Sciences.

Zhu, J C and Tian, Y L 1986. Chemical nitrogen fertilizer and groundwater pollution. (in Chinese). Hydrologic Geology and Engineering Geology (5):38–41

Zhu, Z M, Wu, T B, Zheng, S G and Dai, Z Q 1984. Crop response to split application of nitrogen fertilizers. (in Chinese). Soil and Fertilizer (3):29–32.

14
Nitrogen balance and cycling in agroecosystems of China

ZHU ZHAO-LIANG

14.1. Introduction

Nitrogen cycles in agro-ecosystems by entering into the system through different routes, undergoing a series of interlinked transformations and transfer processes and exiting from the system by different routes. The total input may be higher, equal to or lower than the total output, thus resulting in different balances.

Nitrogen cycling in agro-ecosystems is open in nature. The N in the system undergoes continuous exchange with the N in the atmosphere and hydrosphere. Therefore, in developing strategies for exploiting the potential productivity of an agro-ecosystem, not only the economic efficiency but also the environmental impacts have to be taken into consideration.

On the basis of the discussions in previous chapters on the main transformations and transfer processes in soils and agro-ecosystems, the input and output of N in Chinese agriculture will be evaluated quantitatively in this chapter. In addition, strategies for managing N in the development of China's agriculture will be discussed.

14.2. Input of nitrogen through biological fixation

The parameters for estimating the N fixed symbiotically and non-symbiotically and the amount of N fixed in China will be addressed in this section.

14.2.1. Symbiotically fixed nitrogen

Fried (1978) summarized the data relevant to the N symbiotically fixed by leguminous crops. The amounts fixed in soybeans, peanuts, and broad beans ranged from 3–102, 82–106 and 20–34 kg N ha^{-1}, and averaged 51, 96 and 28 kg N ha^{-1}, respectively. LaRue and Patterson (1981) indicated that the N fixed symbiotically in leguminous pastures ranged mainly between 110–190 kg N ha^{-1}, which is higher than that reported by Fried (1978). The great range in the estimates of N fixation by different plants and authors can be attributed to differences in plant biomass and symbiotic activity and methods used for estimation.

In assessing N fixation by legumes it is generally assumed that N in the tops accounts for two-thirds of the N in the whole plant, and that two-thirds of the total

Zhu Zhao-liang et al. *(eds.): Nitrogen in Soils of China, 323–338.*

plant N has been fixed symbiotically. Consequently, the amount of N fixed (Ndfa) is approximately equal to that accumulated in the aerial parts of the plants.

However, it is found that the ratio of the N in tops to that in whole plant varies greatly for different legumes. For annual legumes it was 90–96% (Allison 1973), which is much higher than the assumed ratio (2/3). Even in pot experiments, where the root density is much higher than that in the field, the ratio for annual legumes was as high as 77–90% (Wen 1989). It was only in the case of perennial legumes that the ratio was close to the figure of 2/3 (Allison 1973). As the dominant leguminous crops grown in China are annual it is suggested that a more suitable proportion is 90%.

The Ndfa for legumes varies greatly under different experimental conditions, as shown in Tables 14.1–14.3. Excluding data relevant to high rates of N application, Ndfa for grain legumes (mostly soybeans) varied from 23 to 91% (usually >60% in pot experiments), and from 21 to 77% (weighted mean 60%) in field experiments.

Table 14.1. Proportion of nitrogen in soybean derived from symbiotic fixation (Ndfa%).[1]

Treatment	Ndfa (%)	Reference
0.083 g N 10 kg soil^{-1}	75,78	Ma *et al.* (1983)
0.05 g N 11 kg soil^{-1}	91	Ma *et al.* (1987)
0.10 g N	85	
0.50 g N	75	
1.00 g N	62	
2.00 g N	53	
4.00 g N	0	
0.25 g N 15 kg soil^{-1} at seeding	74	Zhang *et al.* (1985)
0.80 g N at seeding	69	
0.80 g N + P at seeding	65	
0.80 g N + K at seeding	65	
0.80 g N at seeding + 0.80 g N at flowering	51	
0.25 g N at seeding + 0.25g N at flowering	48	
0.25 g N 15 kg soil^{-1} at seeding + 0.25g N at pod-setting	62	
0 N, 0 P 3 kg soil^{-1}	79	Ho *et al.* (1985)
0 N, 0.45 g P	83	
0.225 g N + 0.45 g P	67	
0.45 g N + 0.45 g P	61	
0.065 g N (as AS) 4 kg soil^{-1}	42	Witty and Ritz (1984)
0.065 g N (as KNO_3)	36	
0.065 g N (as KNO_3 + gypsum pellet)	38	
0.065 g N + glucose	31	
0.065 g N (as oxamide)	63	
0.494 g N (as soybean tops)	64	
0.1 g N 0.86 kg Andosol^{-1}	50	Chiu & Yoshida (1986)
0.1 g N 1.58 kg Yellow soil^{-1}	23	

[1] Pot experiment; AS = ammonium sulfate.

Table 14.2. Proportion of nitrogen in legumes other than soybean derived from symbiotic fixation.[1]

Legume	Treatment	Ndfa %	Reference
Peanut	75 kg N as AS ha^{-1}	65	Zhang *et al.* (1988)
	75 kg N as urea ha^{-1}	61	
	75 kg N as ABC ha^{-1}	60	
	75 kg N as NH_4Cl ha^{-1}	40	
Milk vetch	15N-labelled soil, zero-N	84	Wen (1989)
Sickle alfalfa	"	87	
Vetch	"	88	
Broad bean	"	89	
	"		
Trifolium subterraneum	low fertility soil	95	Ledgard *et al.* (1985)
	high fertility soil	40	

[1] Pot experiment; AS = ammonium sulfate; ABC = ammonium bicarbonate.

Table 14.3. Proportion of nitrogen in legumes derived from symbiotic fixation.[1]

Legumes	Ndfa (%)	Reference
Soybean	21	Ma *et al.* (1983)
	24	Zhang *et al.* (1983)
	66, 77	Wada *et al.* (1986)
	26, 28, 29, 37	Armarger *et al.* (1979)
	48	Johnson *et al.* (1975)
	67 (n = 36)	Kucey *et al.* (1988)
	34, 50	Fried & Broeshart (1981)
Phaseolus bean	38, 61, 63, 65, 65	Rushel *et al.* (1982)
	54	Witty (1983)
Peas	57	Witty (1983)
Field bean	62	Witty (1983)
Lucerne	65,70	Zhang & Luan (1987)
Red clover	59	Witty (1983)
Weighed mean	60	

[1] Field experiments.

In leguminous green manure crops Ndfa ranged mainly from 84 to 95% in pot experiments. However, Ndfa was greatly reduced by high N application rates (Armarger *et al.* 1979; Duque *et al.* 1985; Fried and Broeshart 1981; Johnson *et al.* 1975; Norhayati *et al.* 1988). As the rate of application of N to legumes is generally low, it is considered that the value of 60% be used for assessing N fixation. Thus, the contribution of legumes in China to the N economy may be assessed by multiplying the N in the aerial parts of the legumes by 0.90×0.60, i.e. 0.67. No relevant Ndfa data are available for azolla.

Based on the relevant data from the Almanac of China's agriculture (ECACA 1980, 1984, 1988), an assessment of the N fixed in China by leguminous green manure crops and grain legumes can be made (Table 14.4).

Table 14.4. N fixed symbiotically in China.[1]

Year	Leguminous green manure			Soybean and peanut		Other grain legumes[3]		Total
	Area (10^3 ha)	Yield[2] (t ha^{-1})	N fixed (10^3 t)	Yield (10^3 t)	N fixed (10^3 t)	Yield (10^3 t)	N fixed (10^3 t)	(10^3 t)
1952	2,481	22.5	131	11,666	466	8,435	282	879
1979	8,493	18	358	10,282	410	11,568	387	1,155
1983	5,150	15	181	14,510	546	11,132	373	1,100
1987	4,145	12	117	18,355	719	9,883	331	1,167

[1] N in green manure is assumed to be 0.35%; N accumulated for each ton of soybean, peanut or other grain legume is taken as 0.06 ton, 0.058 ton, and 0.05 ton, respectively; N fixed = N accumulated × 0.67.
[2] Fresh weight.
[3] Broad bean, pea, mung bean, etc.

Symbiotic fixation by legumes and azolla has provided a considerable amount of N and played an important role in the development of China's agriculture (Chen 1988). However, since 1980 along with the increase in cultivation index, the area growing green manure has declined gradually, and was reduced to 4,145 thousand hectares in 1987, the majority being in rice-based cropping systems. It is well-known that in the rice-rice-green manure cropping system a large proportion of the fields growing green manure is used as seedling beds for the early rice crop. As the area has to be plowed in early spring, the yield of green manure is very low. Therefore, the area in which green manure can grow for a satisfactory length of time is much less than the area given above. This has already been taken into consideration in assessing N fixation in Table 14.4. The results in Table 14.4 show that N fixed by green manure crops in 1979 was 358 thousand tonnes, but it decreased dramatically to 117 thousand tonnes in 1987, a level which was slightly less than that fixed in 1952. On the other hand the amount of N fixed by grain legumes has increased from 797 thousand tonnes in 1979 to 1050 thousand tonnes in 1987. This is mainly due to the increased area sown to grain legumes and their yield. The increase in N fixed by grain legumes compensated for the decrease in N fixed by green manure crops. As a result the amount of N fixed symbiotically by these crops remained almost unchanged since 1979 (Table 14.4). This indicates that increasing the area sown to grain legumes and improving the efficiency of fixation are important approaches to exploiting the contribution of symbiotic N fixation to the agriculture of China, especially as the area sown to green manure crops is likely to be reduced even further.

14.2.2. Non-symbiotically fixed nitrogen

Nitrogen input through nonsymbiotic fixation differs greatly with different agro-ecosystems. In general, it is higher in flooded rice than in upland cropping systems (Xi 1986), and this is partly attributed to the higher phototrophic fixation of N in flooded rice (Moore 1966; Roger *et al.* 1986).

Wetselaar (1981) indicated that the net gain of N in balance sheets for the zero-N plots of long-term field experiments with rice conducted in Japan, Thailand and the

Philippines ranged from 38 to 59 kg N ha^{-1} yr^{-1}, and averaged 49 kg N ha^{-1} yr^{-1}. The majority of the N was derived from non-symbiotic N fixation. According to Watanabe *et al.* (1981a), the net non-symbiotic N fixed in the absence of fertilizer N in field experiments conducted in Japan and the tropics ranged from 6 to 26 kg N ha^{-1} yr^{-1} and 64 to 100 kg N ha^{-1} yr^{-1}, respectively. In China the relevant data are not yet available. In a pot experiment with rice and ^{15}N-labelled soils Zhu *et al.* (1986) found that the contribution of non-symbiotic N fixation to total N uptake in rice at maturity was 21.4% (mean of 3 soils), and that Ndfa correlated positively with the dry weight of the rice plants. A similar relationship was found by App *et al.* (1986) in a study of the non-symbiotic fixation by different rice varieties. Consequently, it seems reasonable to use this value (21.4%) for assessing non-symbiotic N fixation in flooded rice. Based on this parameter, Zhu *et al.* (1986) estimated that non-symbiotic fixation in the Taihu region to be in the range of 57 to 62 kg N ha^{-1} yr^{-1}.

Non-symbiotic fixation is affected markedly by the presence of ammonium in soil. However, the results obtained in pot experiments with rice showed that the non-symbiotic fixing activity was not reduced, and may even be slightly improved by applying 5.6 mg ammonium-N kg $soil^{-1}$. It was reduced only by high concentrations of ammonium (Qian *et al.* 1985; Charyuler and Rao 1981; Ventura *et al.* 1986). Furthermore, observations in field plot experiments conducted at the International Rice Research Institute showed that the application of 60 kg N ha^{-1} did not reduce the activity of heterotrophic N fixation (ARA) and only phototrophic N fixation was slightly inhibited (Watanabe *et al.* 1981b). Therefore, it seems that the inhibition of non-symbiotic N fixation by ammonium may not be large if the rate of N application is not high, or the N is correctly deep-placed. However, when high rates of N are broadcast into the floodwater, with or without incorporation, the non-symbiotic fixing capacity will be reduced considerably.

Based on the above discussion, it may be reasonable to use a value of 45 kg N ha^{-1} yr^{-1} to obtain a rough estimate of the non-symbiotic N input to flooded rice. Then, from the total area of paddy rice in China (25 million hectares) it can be estimated that the total input of N through non-symbiotic fixation would be 1,140 thousand tonnes.

The net gain of N in zero-N plots growing wheat was found to be in the range of 15–27 kg N ha^{-1} yr^{-1} (Moore 1966). The non-symbiotic fixation of N in the zero-N plots in the Broadbalk wheat experiment averaged 20 kg N ha^{-1} yr^{-1} (range 18–23 kg N ha^{-1} yr^{-1}; Jenkinson 1977), which is close to the value given by Moore (1966). However, if the loss of fixed N through denitrification is taken into account, non-symbiotic fixation would be ~35 kg N ha^{-1} yr^{-1} (Jenkinson 1977). From these data it seems appropriate to use 15 kg N ha^{-1} yr^{-1} for assessing non-symbiotic fixation in wheat. If this value is used to assess non-symbiotic N fixation in other upland cropping systems, then the total amount of N fixed non-symbiotically in upland cropping systems would be 1,110 thousand tonnes (the total area of cropped upland soils is 74 million hectares).

Therefore, we conclude that 2.25 million tonnes of N are fixed non-symbiotically in China. Thus, the input of N through symbiotic and non-symbiotic fixation was estimated to have been 3,417 thousand tonnes in 1987. Of this, 3.4% resulted from symbiotic fixation by green manure, 30.7% came from grain legumes, and the remaining 65.8% was from non-symbiotic N fixation.

14.3. Reuse of nitrogen in harvested crops

Except for N in leguminous green manure, the N in organic manure is reused N from harvested crops. The bulk of the organic manure is the excreta of humans and animals, and the remainder consists of part of the harvested straw and stalk that is returned to the field directly or after being used as bedding material, and part of the seed cake.

The amount of organic manure-N applied in China has been estimated by different authors (ISFCAAS 1986; Zhang 1984; Wen 1989; Zhu and Xi 1990). The present estimate is given in Table 14.5. Nitrogen in poultry manure is not included due to lack of information for poultry and thus, organic manure-N may be underestimated. However, the error is likely to be small because the amount of poultry manure-N is much less than that in human and livestock excreta.

The recycling ratio refers to the proportion of N in the harvested crop that is reused as manure in agriculture. The total amount of N in organic manure used in China and the recycling ratio was estimated by ISFCAAS (1986) for different years. In 1983, 38% of the harvested material and 4.23 million tonnes of N was cycled through animals. These values are close to those given in Table 14.5.

Table 14.5. Amount (10^3t) of nitrogen in harvested crop reused as organic manures in China's agriculture.[1]

Manures	1952	1979	1983	1987
Human excreta	458	538	562	565
Cattle excreta	642	809	886	1,074
Horse and other draft animals	180	98	88	96
Pig excreta	289	1,028	960	1,054
Sheep excreta	33	201	223	242
Subtotal	1,602	2,674	2,719	3,051
Green manure	195	535	270	174
Straw and stalk	332	526	623	679
Seed cake	142	293	533	623
Total	2,271	4,028	4,145	4,527
N in harvests	4,165	9,915	11,569	12,298
Recycling ratio, %	54.5	40.6	35.8	36.0

[1] The following assumptions were used in calculating the proportion available for reuse, after correcting for losses during collection and storage: (i) Population × 0.85 = number of adults; N in excreta for cities and towns = 0.44 kg N $adult^{-1}$ yr^{-1} ; for countryside = 0.694 kg N $adult^{-1}$ yr^{-1}; (ii) N in cattle, horse, pig and sheep excreta = 11.35, 9.79, 3.217 and 0.7 kg N $head^{-1}$ yr^{-1}, respectively; (iii) 80% of leguminous crop was used directly as manure; the remainder was used as fodder, thus included in animal excreta; (iv) rice and wheat straw, or maize stalk = grain yield × 1.0, 1.2 or 1.5, respectively; assuming 30% of the straw and stalk was returned to field directly or used as bedding material or for composting; the remainder was used as fodder and included in excreta; (v) only 90% of the cotton and rape seed cake were included, with an N concentration of 5%.

The agro-ecosystem is an open one, and nutrients are unavoidably lost during recycling. The loss is particularly severe for N and thus, the recycling ratio for N is markedly lower than that for P and K (Zhu and Xi 1990). Consequently, if fertilizer N is not added, and the input of N is solely from N reuse, then the input of N to agriculture will reduce, and the productivity of agriculture will decline. Therefore, the input of fertilizer N is indispensable for sustained productivity. On the other hand, along with the increase in fertilizer N, the reuse of N in harvested material will become more and more important, because of the increased N in crop harvests (see Table 14.8). Otherwise it would be a great waste of a N resource, which is not only of economic significance, but is also of environmental concern. However, as shown in Table 14.5, the recycling ratio declines with the increase in input of fertilizer N. This is a problem which deserves immediate attention.

Zhang (1984) indicated that the contribution of leguminous green manure-N to the total amount of applied N in China varied from 3 to 5%. From Tables 14.5 and 14.6 it can be seen that the contribution of green manure-N to organic manure-N and total applied N was 3.8–13.3% and 1.0–8.4%, respectively, in the years under estimation. This percentage has been on the decrease since 1952. Evidently, it resulted from the sharp decrease in the area sown to green manure. This implies that the contribution of green manure-N to the total has been low, and became very small in recent years. It is reasonable to expect that this contribution will not increase in the foreseeable future.

Along with the increase in input of fertilizer N, the amount of organic manure other than green manure has increased considerably (Tables 14.5 and 14.6). However, the percentage contribution of organic manure N to total applied N has decreased. Table 14.5 further shows that N in human and livestock excreta is the major form of organic N, accounting for 66 to 71%. Nitrogen in cattle and pig excreta has been the most important form accounting for 58–70% of the N in human and animal excreta, and 41–47% of the total manure N. The high contribution of cattle and pig excreta is primarily due to the large population, the high concentration of N in the excreta, and to the fact that they are mostly raised in feedlots so that more is reused.

Table 14.6. Contribution (%) of different sources of nitrogen to agriculture of China.

Applied N	1952	1979	1983	1987
Chemical fertilizer	2.5	67.6	75.5	78.5
Organic manure	97.5	32.4	24.5	21.5
Human excreta	19.7	4.3	3.6	3.3
Cattle excreta	27.6	6.5	5.6	6.0
Horse and other draft animals	7.7	0.8	0.6	0.5
Pig excreta	12.4	8.3	6.1	5.9
Sheep excreta	1.4	1.6	1.4	1.4
Green manure	8.4	4.3	1.7	1.0
Straw and stalk	14.2	4.2	3.9	3.8
Seed cake	6.1	2.4	3.3	3.5
Total (million t)	2.33	12.4	15.8	17.8

The large amount of human excreta available in China provides a big N resource. However, reuse is low, particularly in big cities. Jin (1986) indicated that only about one-fifth of the N transported into a city as food is reused in the countryside as manure. Therefore, reuse of the N in human excreta will decrease further as the cities in China increase in population. Then, human excreta in the city will be considered mainly as an environmental problem.

The N in straw and stalks returned to the field directly, or used as bedding for livestock and as material for composting accounted for 13–15% of the total organic manure-N. It is foreseeable that along with the development of industry in the countryside, the collection and reuse of human and animal excreta will decrease because it is labor-consuming. Under these conditions the direct return of straw and stalks to the field may become more common to save labour costs. However, from the point of view of N recycling, the significance of straw and stalk reuse is limited, because most of N assimilated by crop plants accumulates in the grain. Therefore, efforts have to be made to improve the reuse of human and animal excreta, even if the amount of straw and stalk returned directly to field is increased.

Table 14.6 shows that the contribution of fertilizer N to total applied N was increased to 78.5% in 1987. Thus, improvement in the recovery of N by the plant and efficiency of use of fertilizer N are obviously of primary importance. In addition it was estimated (IFSCAAS 1986) that the input of fertilizer N to China's agriculture will increase to 18 million tonnes in the year 2000. Evidently, this is essential for the further development of agriculture in China, but it will result in a large increase in the amount of N reusable as organic manure. Thus, the importance of efficient use of the N in organic manures will be increased rather than decreased. This indicates that the prevailing system of combined application of nitrogen fertilizer and organic manure will continue to be the basic strategy for nutrient management in China.

14.4. Loss of fertilizer and organic manure nitrogen

Loss of organic manure and fertilizer N is an important output from agroecosystems. In China, organic manure is usually applied in combination with N fertilizer. Therefore, any interaction between the organic manure and fertilizer N in the combined application has to be considered in assessing the loss of N. However, few investigations on the interaction of farmyard manure, compost or straw with fertilizer N have been made (Mo and Qian 1981). The results of a pot experiment with rice showed that N lost from *Crotalaria* and ammonium sulfate in combined application was almost equivalent to the sum of the N lost from each of them when applied separately (Huang *et al.* 1981), implying that there was no interaction on N loss. Therefore, the loss of N from fertilizers and organic manure applied in combination can be assessed from the data obtained in field microplot experiments where the sources were applied separately.

In Chapter 11 the loss of fertilizer N from cropland was estimated to be ~60% for flooded rice and 45% for upland crops. Considering the fact that the area devoted to upland crops is much larger than that of flooded rice, it is suggested that 50% of the fertilizer N used in China is lost from the plant–soil system.

Field investigations on the fate of N in organic manures in crop–soil systems are rare. The data available are given in Table 14.7. The degradability and fate of N in organic manure is closely related to the chemical composition and N content of the material (Shi *et al.* 1980). Leguminous green manure (e.g. milk vetch), due to its high N and low lignin content, decomposes faster than straw and azolla, but more N is lost during its decomposition (Shi *et al.* 1980). However, the loss of N from cattle manure (Rempe *et al.* 1981) seems to be comparable with that from milk vetch. Moorhead *et al.* (1988) found that little N was lost from the dried plant materials used in their experiments. From these results it is estimated that N lost from leguminous green manure and human and animal excreta applied to cropland is of the order of 15%, and that little is lost from straw.

14.5. Input and output of nitrogen in water

14.5.1. Input in rain and irrigation

Peng (1937) showed that the mean input of N in rain in Guangzhou for the years 1932–1936 was 7.64 kg N ha^{-1} yr^{-1}, which is close to the worldwide mean value of 8.8 kg N ha^{-1} yr^{-1} estimated by Allison (1965). Recent investigations conducted at a number of sites in southern China showed that the mean concentration of N in rainwater ranged from 1 to 2 mg L^{-1}, and that the N input in rainfall ranged from 8 to 20 kg N ha^{-1} yr^{-1} (Lu and Shi 1979; Liu *et al.* 1984; Zhang and Gong 1987; Ma and Non 1988; and unpublished data of Liu, Y. C.). These values are considerably higher than those reported by Peng (1937). The increased N content in rain-water may be attributed to the dramatic increase in application of fertilizer N in China, the majority being in the form of ammonium bicarbonate. This is indicated by the observation that the concentration of ammonia in the air at the time of fertilizing rice at Danyang, Jiangsu Province, was in the range of 10–30 μg N m^{-3} (Freney *et al.* 1987).

Table 14.7. Fate of organic manure nitrogen applied to crops (% of applied N).[1]

Manure	Crop	Recovery in plant	Residual in soil	Loss	Reference
Milk vetch	Rice	30	51	19	Mo & Qian (1983)
		42	46	12	Wen *et al.* (1988)
Azolla	Rice	20	74	6	Wen *et al.* (1988)
Blue-green algae	Rice				Tirol & Roger (1982)
Surface broadcast		14	59	27	
Incorporated		28	64	8	
Rice straw	Rice	10	96	-6	Reddy & Patrick (1980)
Cattle manure	Winter wheat	14	70	16	Rempe *et al.* (1981)

[1] Field microplot experiments with ^{15}N.

The N brought in through irrigation depends on the N content of the water and irrigation scheduling. The total N concentration in some Chinese Rivers ranged from 0.43 to 1.52 mg N L^{-1}. However, the concentration in the Huaihe River was found to be 5.96 mg N L^{-1}. Ammonium and nitrate generally ranged from 0.04 to 1.30 mg N L^{-1} and 0.20 to 0.87 mg N L^{-1}, respectively, while the nitrate concentration in the Huaihe River in Huainan, and the Bahe River in Xi'an was greater than 1.1 mg N L^{-1} (Zhang and Gong 1987). In northern China, well-water is usually used for irrigation but no data are available for the N concentration of the well water in that region. Zhang, S. L. and Zhu, Z. L. (unpublished data) showed that N concentration of the well-water in Fengqiu, Henan Province, in October just after the wet season was only 0.22 mg N L^{-1} (with a trace of ammonium). However, in spring, during the dry season the concentration increased to 2.5 mg N L^{-1} (0.66 mg ammonium N L^{-1} and 1.79 mg nitrate N L^{-1}). Based on the data for well water and assuming 1,800 m^3 of irrigation water ha^{-1} was applied to winter wheat (as is usual for the region), then the N input in irrigation water would be only 4.5 kg N ha^{-1}. This value can be taken as the annual input of N through irrigation, because summer crops (such as maize) are not irrigated in this region. Similar amounts of N (4.35–5.7 kg N ha^{-1} yr^{-1}) were added in irrigation water at 5 sites in the Taihu region (average 5.1 kg N ha^{-1} yr^{-1}; unpublished data of Liu, Y. C.). Therefore, it can be concluded that N input to cropland in irrigation water is not high, except when water with high N concentration is used for irrigation.

In arid or semi-arid areas there is a possibility that nitrate from deep layers is moved upwards to the rooting zone, thereby contributing N to the system (Peng *et al.* 1981). No data are available to allow estimation of the importance of this mechanism.

14.5.2. Loss through leaching and runoff

In China leaching of fertilizer N has generally been evaluated from the distribution of labelled fertilizer N in the soil profile at the end of a microplot experiment in the field. As indicated in Chapter 11 leaching of fertilizer ^{15}N was usually negligible during the growth of the current crop. However, this did not mean that leaching of soil N was also low. In order to quantify the total leaching loss of N from soil, organic manure and fertilizer sources, in situ measurements of percolation rate and the N concentration of leachate are required.

In situ measurements conducted at 5 sites in the Taihu region showed that the percolation rate in different rice-based cropping systems ranged from 3,615 to 5,055 m^3 $ha^{-1}yr^{-1}$. The ammonium, nitrate and total N concentrations in the leachates ranged from 0.37 to 0.46, 0.29 to 0.40, and 0.73 to 0.78 mg N L^{-1}, respectively. The loss of N by leaching estimated from these data ranged from 2.85 to 3.75 kg N ha^{-1} yr^{-1} (mean 3.3 kg N ha^{-1} yr^{-1}; unpublished data of Liu Y. C.). In the same investigation, it was found that runoff ranged from 4,140 to 5,295 m^3 $ha^{-1}yr^{-1}$, and N loss through runoff ranged from 1.95 to 4.5 kg N ha^{-1} yr^{-1} (mean 3.45 kg N ha^{-1} yr^{-1}). From these figures it was concluded that total N loss by leaching and runoff was 6.6 kg N ha^{-1} yr^{-1} (range 4.8 to 8.25 kg N ha^{-1} yr^{-1}). Integrating N input through rainfall and irrigation and N loss through leaching and runoff resulted in a net input of 12 kg N ha^{-1} yr^{-1} (range 7.2 to 16.5 kg N ha^{-1} yr^{-1}) through water cycling in the croplands of the Taihu region. Although the amount was not high, it still contributed a considerable proportion to the net gain of 38–59 kg N ha^{-1} yr^{-1} in the region. Consequently, it is necessary to take this into account in estimating the amount of non-symbiotic N fixation.

Little data have been collected on leaching and runoff losses of N from upland cropping systems. Investigations conducted on an old manured loessial soil in the Loessial Plateau (Peng *et al.* 1981) showed that, the nitrate moved downwards 2 m in the wet season of August–October under fallow conditions. Each 2–3 mm rainfall leached nitrate to a depth of 1 cm. While nitrate leaching may be important under fallow conditions, it was found that there was little nitrate in the soil profile of the plot planted to summer maize. Therefore, loss of N by leaching from this type of soil when cropped may be small, especially when the rate of application of N is not excessive.

Erosion may result in a large loss of N from agroecosystems in certain regions. For example, Peng *et al.* (1987) estimated that N loss through erosion in the agricultural regions of the Loessial Plateau was 16.5 kg N ha^{-1} yr^{-1}, which is 5 times as high as that found in the Taihu region. The difference was mainly due to the large amount of N eroded in the solid phase (13.5 kg N ha^{-1} yr^{-1}); the N lost in runoff water (3 kg N ha^{-1} yr^{-1}) was close to that found in the Taihu region. It should be emphasized that N loss through erosion in agricultural regions of the Loessial Plateau was severe, and equivalent to 22% of the N in harvested crop; the corresponding figure for the Taihu region was only 1%. Control of erosion is obviously one of the most important strategies for improving the N economy of the Loessial Plateau.

From the preceding discussion it seems reasonable to take the mean value of 12 kg N ha^{-1} yr^{-1} as the input of N through rainfall. For N input through irrigation, it is necessary to take into account the fact that the area used for flooded rice is much larger and the irrigation water used for it is much greater than that used for upland crops (mostly confined to winter wheat). In 1987 the effective irrigation area was 21.2 million hectares (ECACA 1988), accounting for 21.3% of the total cropland. Then, even if the N input through irrigation in the Taihu region is used for assessing the input for the whole cropland of China, it would amount to only 1 kg N ha^{-1} yr^{-1} on average. Based on the estimated leaching and runoff losses from the croplands in the Taihu region and the Loessial Plateau, N loss from China's croplands would, on average, amount to 9 kg N ha^{-1} yr^{-1}.

14.6. Nitrogen balance in China's agriculture

Based on the preceding discussions, N balance sheets for different years have been calculated and are given in Table 14.8. The percentage contributions of the different inputs and outputs are presented in Table 14.9. The results indicate that in 1952 the total input of N was low (about 65 kg N ha^{-1} yr^{-1}). The majority of this came

Table 14.8. Nitrogen balance in China's agriculture (million t).

Category	1952	1979	1983	1987
Input				
Chemical fertilizer	0.059	8.4	11.925	13.964
Organic manure, excluding green manure	2.14	3.67	3.964	4.41
Symbiotically fixed	0.879	1.155	1.1	1.167
Non-symbiotically fixed	2.42	2.25	2.25	2.25
Rainfall and irrigation	1.3	1.3	1.3	1.3
Seed	0.375	0.375	0.375	0.375
Sub-total	7.173	17.150	20.914	23.466
Output				
Crop harvests	4.165	9.915	11.569	12.298
Loss, chemical fertilizer	0.03	4.2	5.963	6.982
Loss, organic manure, excluding straw and stalk	0.279	0.511	0.513	0.562
Loss, soil	0.32	0.3	0.3	0.3
Leaching and runoff	0.9	0.9	0.9	0.9
Sub-total	5.694	15.826	19.245	21.042
Balance	+ 1.479	+ 1.324	+ 1.669	+ 2.424
N in harvested crop/N input (%)	58.1	57.8	55.3	52.4

Table 14.9. Contribution of the different inputs and outputs of nitrogen in China's agriculture (%).

Category	1952	1979	1983	1987
Input				
Chemical fertilizer	0.8	49.0	57.0	59.5
Organic manure, excluding green manure	29.8	21.4	19.0	18.8
Symbiotically fixed	12.3	6.7	5.3	4.9
Non-symbiotically fixed	33.7	13.1	10.8	9.6
Rainfall and irrigation	18.1	7.6	6.2	5.5
Seed	5.2	2.2	1.8	1.6
Sub-total	99.9	100.0	100.1	99.9
Output				
Crop harvests	73.1	62.7	60.1	58.4
Loss, chemical fertilizer	0.5	26.5	31.0	33.2
Loss, organic manure, excluding straw and stalk	4.9	3.2	2.7	2.7
Loss, soil	5.6	1.9	1.6	1.4
Leaching and runoff	15.8	5.7	4.7	4.3
Sub-total	99.9	100.0	100.1	100.0

from non-symbiotic N fixation and the reuse of N in harvested crop. The N in harvested crops accounted for 58.1% of the total input of N and 73.1% of the total output, implying that the efficiency of the N input for crop production was high, although the productivity was low.

The amount of N reused in agriculture increased along with the consumption of fertilizer N, but the proportion reused decreased rapidly. In 1979 the proportion reused was only 21.4% of the total input, while fertilizer N made up 49.0% of the total. Nitrogen reuse decreased even further to 18.8% in 1987, while the contribution of fertilizer N increased to 59.5%.

In addition, along with the increase in consumption of fertilizer N, the efficiency of N in crop production decreased rapidly. This is shown by the increased proportion of fertilizer N in the total output of N and the decreased proportion of added N in the harvested crop. In 1979 the former increased to 26.5% while recovery in the crop decreased to 57.8%. However, even in 1987 when the consumption of fertilizer N accounted for 59.5% of the total input of N, the N in harvest material was still equivalent to 52.4% of the total input of N in the system, which is slightly higher than the corresponding figure of a system with a high rate of N application (Frissel 1977). This is obviously due to the fact that the rate of N application in China was only about 140 kg N ha^{-1} yr^{-1} or 95 kg N ha^{-1} $crop^{-1}$ in 1987, which was less than that in Japan and western Europe.

The positive N balances from 1952 to 1987 demonstrate that the N reserves in China's soils increased. However, the situation is different for different regions. For example, the input of N in the suburbs of Shanghai was much higher than the output and the soil N content in the region has increased significantly (Xi 1986). On the other hand, Peng *et al.* (1987) found a negative balance of 0.2 million tonnes of N in the agricultural region of the Loessial Plateau due to the large loss of N through erosion. In addition, a positive N balance may induce eutrophication of waters through leaching and runoff, particularly in those regions where the rate of N application is very high (see Chapter 13).

14.7. Conclusions

Several conclusions may be drawn from the previous discussion. Nitrogen input has been greater than N output in China's agriculture, and thus there is a tendency for the soil N reserves to be increased. This is favourable for the development of agriculture in China.

The rapid increase in the rate of application of nitrogen has resulted in a dramatic increase in N loss, and a considerable decrease in the efficiency of N for production. This is not only a problem of economic importance, but is also of environmental concern. Therefore, improvement in efficiency and reduction of losses from cropping systems are of primary importance. The rational distribution of N fertilizers and improved techniques for application need to be considered further.

It is necessary to increase the consumption of fertilizer N in order to develop the agriculture of China. The reuse of N in harvested crop, will then become more important not only to conserve the N resource but also to protect the environment. The protection of the environment will become essential along with the development of agriculture. Therefore, the combined application of fertilizer and organic material will still be the essential nutrient management system in the future.

In order to exploit the potential efficiency of the increased fertilizer N in crop production it is important to develop and use new techniques in agriculture, improve the facilities for agricultural production, such as irrigation and drainage, introduce high yielding varieties, and use balanced fertilization. In regions already using a high rate of N application, the cultivation of high N response varieties, and increasing the cropping index are approaches for improving N efficiency. Protection from erosion is of primary importance in erodible regions.

Although addition of fertilizers and organic manure is the main input of N into agriculture in China, the use of symbiotic N fixation should not be overlooked. Recently, the area sown to leguminous green manure crops has been reduced dramatically and will be further reduced in the future. Consequently, the problem of exploiting the potential of symbiotic N fixation has gained more attention than before. It appears that it will be essential to extend the area under cultivation with grain legumes.

14.8. References

Allison, F E 1965. Evaluation of incoming and outgoing processes that affect soil nitrogen. In: Bartholomew, W V and Clark, F E (eds.), Soil Nitrogen, pp. 573–606. Agronomy Series No. 10, Soil Sci. Soc. Am. Madison, Wisconsin.

Allison, F E 1973. Soil organic matter and its role in crop production. p. 637. Elsevier Scientific Publ. Co., New York.

App, A A, Watanabe, I, Ventura, T S, Bravo, M and Jurey, C D 1986. The effect of cultivated and wild varieties on the nitrogen balance of flooded soil. Soil Sci. 141:448–452.

Armarger, N, Mariotti, A, Mariotti, F, Durr, J C, Bourguignon, C and Lagacherie, B 1979. Estimate of symbiotically fixed nitrogen in field grown soybean using variations in 15N natural abundance. Plant Soil 52:269–280.

Charyuler, B N and Rao, V R 1981. Influence of ammonium nitrogen on nitrogen fixation in paddy soils. Soil Sci. 131:140–144.

Chen, L Z 1988. Green manure cultivation and use for rice in China. In: Green Manure in Rice Farming, pp. 63–70. IRRI, Manila.

Chiu, C Y and Yoshida, T 1986. Fate of nitrogen in soils planted to soybean crops. Soil Sci. Plant Nutr. 32:273–284.

Duque, F E, Noves, M C P, Franco, A A, Victoria, R L and Boddey, R M 1985. The response of field grown Phaseolus vulgaris to Rhizobium inoculation and the quantification of N_2 fixation using ^{15}N. Plant Soil 88:333–343.

ECACA 1980. (Editorial Committee of Almanac of China's Agriculture). Almanac of China's Agriculture (in Chinese). Beijing.

ECACA 1984. (Editorial Committee of Almanac of China's Agriculture). Almanac of China's Agriculture (in Chinese). Beijing.

ECACA 1988. (Editorial Committee of Almanac of China's Agriculture). Almanac of China's Agriculture (in Chinese). Beijing.

Freney, J R, Trevitt, A C F, Zhu, Z L, Cai, G X and Simpson, J R 1987. Methods for estimating volatilization of ammonia from flooded rice fields. (in Chinese). Acta Pedol. Sin. 24:142–151.

Fried, M 1978. Direct quantitative assessment in the field of fertilizer management practices. 11th Congr. Inter. Soc. Soil Sci. 3:103–129.

Fried, M and Broeshart, H 1981. A further extension of the method for independently measuring the amount of nitrogen fixed by legume crop. Plant Soil 62:331–336.

Frissel, M J 1977. The nitrogen cycle in agricultural systems. Landbouwkundig Tijdschrift 89:380–388.

Ho, L B, Li, Q Z and Sun, K Y 1985. Studies on the absorption and utilization of different nitrogen sources by soybean by means of 15N isotope tracer. (in Chinese). J. Crop Sci. 11:181–189.

Huang D M, Gao, J H and Zhu, P L 1981. The transformation and distribution of organic and inorganic fertilizer nitrogen in rice-soil system. (in Chinese). Acta Pedol. Sin. 18:107–121.

ISFCAAS 1986. (Institute of Soils and Fertilizers, Chinese Academy of Agricultural Sciences). Regionalism of chemical fertilizers in China. (in Chinese). China Agricultural Science and Technology Publishing House, Beijing.

Jenkinson, D L 1977. The nitrogen economy of the Broadbalk experiments. 1. Nitrogen balance in the experiments. Report Rothamsted Experimental Station, 1976, part 2, 103–109.

Jin, W X 1986. Organic manure- and inorganic fertilizer-N in China and the efficiency in combined application. In: Soil Agricultural Chemistry and Soil Biology and Biochemistry Committees, Soil Science Society of China (eds.), Advances and Prospects for Soil Nitrogen Research in China. (in Chinese). pp. 116–125. Science Press, Beijing.

Johnson, J W, Welch, L F and Kurtz, L T 1975. Environmental implications of N-fixation by soybeans. J. Environ. Qual. 4:303–306.

Kucey, R M N, Snitwongse, P, Chaiwanakupt, P, Wadisirisuk, P, Sirpaibool, C, Arayangkool, T, Boonkerd, N and Rennie, R J 1988. Nitrogen fixation (^{15}N dilution) with soybeans under Thai field conditions. 1. Developing protocols for screening Bradyrhizobium japonicum strains. Plant Soil 108:33–41.

LaRue, T A and Patterson, T G 1981. How much nitrogen do legumes fix? Adv. Agron. 34:15–38.

Ledgard, S F, Simpson, J R, Freney, J R, Bergersen, F J and Morton, R 1985. Assessment of the relative uptake of added and indigenous soil nitrogen by nodulated legumes and reference plants in the ^{15}N dilution measurement of N_2 fixation: Glasshouse application of method. Soil Biol. Biochem. 17:323–328.

Liu, C Q, Cao, S Q and Chen, G A 1984. Contents of nutrient elements in precipitation of Fujian and Yunnan provinces. (in Chinese). Acta Pedol. Sin. 21:438–442.

Lu, R K and Shi, T J 1979. The content of plant nutrients of precipitation in Jinhua district of Zhejiang province. (in Chinese). Acta Pedol. Sin. 16:81–84.

Ma, C L, Wen, X F, Yao, Y Y and Zhou, Z X 1983. Measuring the symbiotic nitrogen fixation by legumes with 15N-AN value. (in Chinese). Application of Atomic Energy in Agriculture No. 2, 18–25.

Ma, C L, Yao, Y Y, Zhou, Z X, Liu, X L and Chen, M 1987. Effect of the rate of N application on the estimation of nitrogen fixation with the method of AN. (in Chinese). Application of Atomic Energy in Agriculture No. 1, 48–53.

Ma, M T and Non, Z Y 1988. Nutrient element content of the rain in Liuzhou area, Guangxi Zhuang Autonomous Region. (in Chinese). Soils 20:35–38.

Mo, S X and Qian, J F 1981. Uptake of nitrogen by rice plant from straw manure, urea and soil. Institute of Soil Science, Academia Sinica (ed.), Proc. Symp. on Paddy soil, pp. 800–804. Science Press, Beijing.

Mo, S X and Qian, J F 1983. Studies on the transformation of nitrogen of milk vetch in red earth and its availability to rice plant. Acta Pedol. Sin. 20:12–22.

Moore, A W 1966. Non-symbiotic nitrogen fixation in soil and soil-plant systems. Soils Fertil. 29:113–128.

Moorhead, K K, Graetz, D A and Reddy, K R 1988. Mineralization of carbon and nitrogen from freeze- and oven-dried plant material added to soil. Soil Sci. Soc. Am. J., 52:1343–1346.

Norhayati, M, Mohd Nor, S, Chong, K, Faizah, A W, Herridge, D F, Peoples, M B and Bergersen, F J 1988. Adaptation of methods for evaluating N_2 fixation in food legumes and legume cover crops. Plant Soil 108:143–150.

Peng, J Y 1937. Five years results on the composition of rain-water in the vicinity of Aanton. (in Chinese). Soils Fertil. 1:1–14.

Peng, L, Peng, X L and Lu, Z F 1981. The seasonal variation of soil NO_3-N and the effect of summer fallow on the fertility of manured loessial soil. (in Chinese). Acta Pedol. Sin. 18:212–222.

Peng, L, Yu, C Z and Zhang, W 1987. Reserve of soil nitrogen and its change in terrestrial ecosystems in Loessial Plateau. Proc. 6th Congr. Soil Sci. Soc. China (Abstracts) (in Chinese). 133–134. June, 1987. Nanchang.

Qian, Z S, Min, H and Mo W Y 1985. Influences of different NH_4^+-N levels on the nitrogenase activity in rhizosphere of rice. (in Chinese). Acta Pedol. Sin. 22:144–149.

Reddy, K R and Patrick, W H Jr 1980. Losses of applied ammonium 15N, urea 15N, and organic 15N in flooded soils. Soil Sci. 130:326–330.

Rempe, E Kh, Vayushkina, N M, Kirpaneva, L I and Nikiforova, M V 1981. Nitrogen transformations of mineral and organic fertilizers and microbiological processes in a sandy derno-podzolic soil. (in Russian). Agrokhimiya No. 7, 3–9.

Roger, P A, Tirol, A, Ardales, S and Watanabe, I 1986. Chemical composition of cultures and natural samples of N_2-fixing blue-green algae from rice fields. Biol. Fertil. Soils, 2:131–146.

Rushel, A P, Vose, P B, Matsui, E, Victoria, R L and Tsai Saito, S M 1982. Field evaluation of N_2-fixation and N-utilization by Phaseolus bean varieties determined by 15N isotope dilution. Plant Soil 65:397–407.

Shi, S L, Wen, Q X and Liao, H Q 1980. The availability of nitrogen of green manures in relation to their chemical composition. (in Chinese). Acta Pedol. Sin. 17:240–246.

Tirol, A C and Roger, P A 1982. Fate of nitrogen from blue-green algae in a flooded rice soil. Soil Sci. Plant Nutr. 28:559–569.

Ventura, T S, Bravo, M, Daez, C, Ventura, V, App, A A and Watanabe, I 1986. Effects of N-fertilizers, straw, and dry fallow on the nitrogen balance of a flooded soil planted with rice. Plant Soil 93:405–411.

Wada, E, Imaizumi, R, Kabaya, Y, Yasuda, T, Kanameri, T, Saito, G and Mishimune, A 1986. Estimation of symbiotically fixed nitrogen in field grown soybeans: An application of natural 15N/14N abundance and a low level 15N-tracer technique. Plant Soil 93:269–286.

Watanabe, I, Craswell, E T and App, A A 1981a. Nitrogen cycling in wetland rice fields in south-east and east Asia. In: Wetselaar, R, Simpson, J R and Rosswell, T (eds.), Nitrogen Cycling in South-East Asian Wet Monsoonal Ecosystems, pp.4–17. Australian Academy of Science, Canberra.

Watanabe, I, De Guzman, M R and Cabrera, D A 1981b. The effect of nitrogen fertilizer on N_2 fixation in the paddy field measured by in situ acetylene reduction assay. Plant Soil 59:135–139.

Wen, Q X, Cheng, L L and Shi, S L 1988. Decomposition of Azolla in the field and availability of Azolla nitrogen to plants. In: IRRI (ed.), Azolla Utilization, pp. 241–254.

Wen, Q X 1989. Management of farm-grown nutrient sources for rice. In: IRRI (ed.), Progress in Irrigated Rice Research, pp. 165–172. IRRI, Manila.

Wetselaar, R 1981. Nitrogen inputs and outputs of an unfertilized paddy field. In: Clark, F E and Rosswell, T (eds.), Terrestrial Nitrogen Cycles, Processes, Ecosystem Strategies and Management Impacts, Ecol. Bull. (Stockholm) 33:573–583.

Witty, J F 1983. Estimating N_2-fixation in the field using 15N-labelled fertilizers: some problems and solutions. Soil Biol. Biochem. 15:631–639.

Witty, J F and Ritz, K 1984. Slow-release 15N fertilizer formulations to measure N_2-fixation by isotope dilution. Soil Biol. Biochem. 16:657–661.

Xi, Z B 1986. A brief discussion on nitrogen cycling in agro-ecosystem. In: Soil Agricultural Chemistry and Soil Biology and Biochemistry Committees, Soil Science Society of China (eds.), Advances and Prospects for Soil Nitrogen Research in China. (in Chinese). pp. 217–227. Science Press, Beijing.

Zhang, E S, Liu, G Z, Wan, Z X and Yu, M Y 1988. Investigations on the uptake of different fertilizer nitrogen by peanut with 15N tracer technique. (in Chinese). Shandong Agric. Sci. 9–11.

Zhang, F D 1984. The combined application of organic manure and inorganic fertilizer is the orientation of the development of modern techniques of fertilizer application. (in Chinese). Soils Fertil. 16–19.

Zhang, H, Xu, B and Gao, J F 1983. The activity of soybean-rhizobia symbiosis and the assessment of nitrogen fixation. In: Soil Science Society of China (ed.), Rational Utilization and Improvement of Fertility of the Soils in China. (in Chinese). Vol. 2, 225–228.

Zhang, H, Zhang, G L, and Wan, X M 1985. Effect of application of ammonium sulfate and ammonium nitrate on the nitrogen fixation of soybean-rhizobia symbiosis and the yield of soybean. (in Chinese). Jilin Agric. Sci. 43–48.

Zhang, W X and Luan, S 1987. Improvement of soil nitrogen fertility and the determination of annual nitrogen fixation by lucerne. (in Chinese). J. Northeast Agricultural College 18:85–88.

Zhang, X P and Gong, Z T 1987. Geochemical characteristics of rain-water, ground water and surface water in south China. (in Chinese). Soil Bull. No. 41, 169–187.

Zhu, Z L, Chen, D L, Zhang, S L and Xu, Y H 1986. Contributions of nonsymbiotic nitrogen fixation to the nitrogen uptake by growing rice under flooded conditions. (in Chinese). Soils 18:225–229.

Zhu, Z L, and Xi, Z B 1990. Recycling phosphorus from crop and animal wastes in China. In: IRRI (ed.), Phosphorus Requirements for Sustainable Agriculture in Asia and Oceania, pp. 115–123. IRRI, Los Baños.

Zeitfracht Medien GmbH
Ferdinand-Jühlke-Straße 7
99095 Erfurt, Deutschland
produktsicherheit@kolibri360.de